Pesticide and Nutrient Management

Books by Frederick Whitford

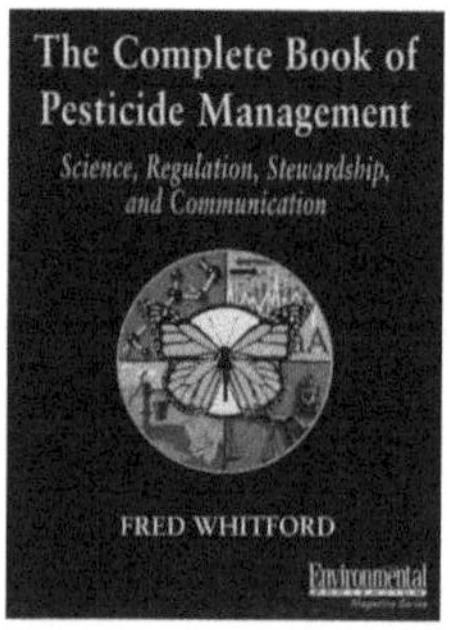

The Complete Book of Pesticide Management

Pesticide and Nutrient Management: Application-Based Problem Solving in Agriculture, Volume 2

The Grand Old Man of Purdue University and Indiana Agriculture: A Biography of William Carroll Latta

The Queen of American Agriculture: A Biography of Virginia Claypool Meredith

Planting the Seeds of Hope: Indiana County Extension Agents During the Great Depression and World War II

Scattering the Seeds of Knowledge: The Words and Works of Indiana's Pioneer County Extension Agents

Enriching the Hoosier Farm Family: A Photo History of Indiana's Early County Extension Agents

Memories of Life on the Farm Through the Lens of Pioneer Photographer J.C. Allen

Pesticide and Nutrient Management

Application-Based Problem Solving in Agriculture

VOLUME 1

Frederick Whitford

Purdue University Press
West Lafayette, Indiana
www.press.purdue.edu

Cataloging-in-Publication Data is on file at the Library of Congress.

978-1-62671-277-5 (hardcover)
978-1-62671-278-2 (epdf)

Designer: Dawn Minns

Editor: Kim Miller

Indexer: Marilyn Augst

The writing of *Pesticide and Nutrient Management: Application-Based Problem Solving in Agriculture* was in fact a 34-year process: The content first took form as a series of Extension publications. The two-volume book depended on the sharing of expertise from a wide array of professionals. Each chapter has a list of contributing authors who provided important insights into the facts and experiences. These people at the top of their professions taught me the ins and outs of their science, which in turn allowed me to share those facts with a wide range of audiences. As I start to wind down my Extension career at Purdue University, it is my hope that the sharing of this information has provided worthwhile guidance that has helped move farmers', agricultural retailers', and others' operations forward in a safe and effective manner. It has been my pleasure to have served all of you for such a long time.

CONTENTS

INTRODUCTION

Reliance on technology is not new to production agriculture. In the early twentieth century, steel plows replaced wooden plows, allowing for more efficient plowing; hybrid corn supplanted open-pollinated corn, significantly increasing crop yields; herbicides usurped plowing, allowing the expansion of farm acres; and tractors displaced horses, changing how farms were powered. Agriculture has long depended on technological innovations.

The inclusion of newer technology and the translation of agricultural research into production practices continue into the twenty-first century as global positioning systems allow for more precise applications, the incorporation of drones for applying inputs has become mainstream, and spray equipment is outfitted with high-tech computers and sensors. Agricultural producers have made significant strides in adopting and modifying production practices based on research, which has helped farming operations remain viable and sustainable enterprises as they continue to face stiff competition from domestic and international markets. The advances include but are not limited to:

- Planting cover crops to protect against soil erosion and to build soil organic matter
- Tiling fields to remove excess water
- Performing precision soil sampling to determine lime and nutrient needs
- Sampling plant tissues to determine nutrient deficiencies
- Purchasing improved hybrid lines of crops
- Planting earlier to capture the full growing season
- Seeding at higher rates to maximize the capture of sunlight and space
- Splitting fertilizer applications to feed plants when needs are greatest
- Rotating pesticide classes to combat weed, insect, and disease resistance
- Building grain storage bins to take advantage of market prices
- Farming with the aim of reducing the carbon footprint on the environment
- Following labels that specify how to protect endangered plant and animal species

From a competitive viewpoint, farmers and agricultural retailers in the United States continue to demand that manufacturers bring more practical innovations into the marketplace. However, adopting unproven practices can be expensive if the cost of the new practice exceeds the benefits provided. This is why there is such a reliance on scientific investigations from universities, the US Department of Agriculture, and industries to answer questions asked by those who farm or who supply services to farmers.

Pesticide and Nutrient Management: Application-Based Problem Solving in Agriculture is a two-volume set designed to bring together science and theory with practice in the field. Pesticides, fertilizers, and fuel are discussed in numerous books and journals, but the details of practical use are often lacking. The goal is to get those who use pesticides to better appreciate the science through the explanation of the practical. By blending and weaving both science and experience together, the user of pesticides would be better prepared to handle them in a responsible manner. This book aims to be technically sound and easy to understand.

Inputs such as pesticides, diesel, manures, organic fertilizers, and lime play important roles in the production process, but they also carry risks that must be managed when they are transported on roads, stored on the farm and at the agricultural dealer, and applied in the field. The present volume was written to blend the use of technologies with protection of human health, soil, water, and wildlife. This approach is never technology over safety, or vice versa. Both must be considered during all phases of farming.

This approach is best exemplified with the eight chapters in section 1, dealing with tank construction, design, placement, and inspection. Whether on the farm or at the retailer, storing inputs to last the season is critical to many operations. Having them on hand can save significant money by locking in seed, fertilizer, and chemical costs in the offseason when prices are normally at their lowest. When stored on-site, these inputs are available for immediate use at the time of the application; this saves considerable time when farmers do not have to wait for them to be transported.

While storing these inputs provides the tangible and measurable benefits of having them on hand at a lower cost, a ruptured tank that releases its contents around the storage site or along a highway can lead to serious public safety, environmental, and regulatory repercussions. The costs associated with environmental remediation and regulatory fines could at times exceed the savings from buying large quantities in the offseason. This is why selecting an appropriate tank to store specific products is critically important—to preserve the benefits and reduce potential risks.

Determining the best tank(s) to purchase is more complicated than just calculating the number of gallons needed for the season. Questions asked could include:

- What are the strengths and weaknesses of the materials used to construct the tanks?
- Is the tank needed immediately, or is there time to have it manufactured to your specifications?

- Will the tanks be placed on trucks or trailers to transport the inputs, or will the tanks be used for stationary storage?
- Are the materials to be stored in a tank corrosive?
- Is the cost of renting a crane to offload a larger tank, if the seller does not provide one at the time of delivery, included in your price quote?
- Does the addition of a new tank place your operation under federal and state containment regulations?
- Do you know how to size a tank so that it will fit indoors in a building or in the concrete containment outdoors?
- Can the tanks be moved away from environmentally sensitive areas?
- How can I protect the tanks against vehicular collisions?
- What inspection steps can be taken to reduce the likelihood of a tank rupturing?
- What do I do and who do I call in the event a tank ruptures?

There are literally dozens of questions that arise when purchasing the right tank to store a specific product to fit an exact location. Questions like those just presented can be developed for any technology, tool, or practice, whether mixing products, storing products, using application equipment, securing the farm against trespassers, driving across railroad tracks, or keeping application booms away from power lines.

Section 2 of this volume takes the same no-nonsense approach, offering suggestions for those transporting, storing, and handling pesticides. Each chapter aims to help farmers achieve the full value of these expensive products by targeting the application, taking steps to keep products from expiring in storage, avoiding mixing errors, using adjuvants to maximize a product's efficacy, and understanding how the water used in mixing can influence a pesticide. The mixing and application process is more complicated than most assume, but understanding the individual components helps to increase efficacy, while simultaneously protecting the environment.

Section 3 focus on spraying technology. The first two chapters provide a detailed discussion of advances in spraying technology: Nozzle technology is often overlooked but deserves attention for how it can improve spray performance, while spraying with drones is an exciting possibility that is rapidly becoming more functional. Maintenance is less exciting than technological changes, but as every farmer and agricultural dealer knows, keeping equipment maintained lets you avoid costly repairs and reduce the hours of downtime. The final two chapters describe what items to check on and how to properly rinse a spray tank to avoid crop contamination, as well as how to winterize equipment for an easy spring start-up. Once again, being efficient and reducing risks to people and the environment play a key role in the recommendations that we provide.

Volume 2 covers topics related to general maintenance and security, important foundations for those who use pesticides. It begins with often-overlooked aspects of maintenance: oils, lubricating grease, nuts, hoses, and tires. It then moves on to transportation safety, such as securing a load, driving a spray rig, and extracting a stuck vehicle. The third section discusses security planning and emergency response.

The contributors who participated in the development of this book represent all facets of the agricultural industry, including practicing farmers, university and industry scientists, regulatory officials, agricultural retailers, risk coordinators, and manufacturing personnel. With their years of practical field experience, they provided invaluable insights, helping focus the information, and recommendations on taking full advantage of a technology, while using it in a manner protective of people and the environment.

The intent of *Pesticide and Nutrient Management: Application-Based Problem Solving in Agriculture* is to provide a framework for making decisions based on commonsense recommendations. We have succeeded if the reader says, "I never knew that or even thought to ask those questions!" or "The recommendations were right on the mark and were exactly what I was looking for."

SECTION 1

Selecting and Maintaining Tanks

1

Managing Storage and Transport Tanks

Many farms and businesses need to store or haul various products over the course of the year. But tanks bring both costs (purchasing the tank but also installing and maintaining it) and risks (spills, damage to expensive tanks). This chapter discusses how to evaluate your options, so that you get the best tank for your location and product. It then discusses how to prepare your site and install your tanks.

Figure 1.1. Poly, fiberglass, and steel tanks all have advantages and disadvantages. Many businesses and farms will have a variety of tanks to meet different needs.

CHOOSING TANKS

1. What Do I Plan to Store?

The product you intend to store will limit your choice of tanks, since not all tanks are compatible with all products. And if you plan to store several different products in the same tank over the course of a season, it needs to be compatible with anything you might put in it, and it must be able to be drained completely and cleaned.

2. What Is the Specific Gravity of the Material I Plan to Store?

The weight of the product you plan to store will affect your choice of tanks. Specific gravity is one way of talking about the weight of liquid products.

Understanding the significance of specific gravity for tanks begins with the weight of water: 8.3 pounds per gallon. The specific gravity of water is 1.0, and the specific gravity of all other substances is relative to the weight of water. In general, the weight of liquid fertilizers ranges from 10 to 12 pounds per gallon. Specifically, 10-34-0 liquid fertilizer weighs 11.7 pounds per gallon, while 28-0-0 liquid fertilizer weighs 10.7 pounds per gallon. Most pesticides weigh slightly below or just above the weight of water.

Poly tanks have a specific gravity rating that indicates what weight of materials they can handle. Fiberglass and steel tanks do not have such ratings, but a steel tank that holds a heavier product will need to have thicker sidewalls. It is always advisable to talk to the tank manufacturer to make sure its construction can hold the weight of the material.

NARROWING THE FIELD

Facts to consider to help you narrow the choices for selecting a storage tank

3. How Much Storage Should I Plan For?

Often, the amount of storage capacity you should have depends on whether your fertilizer or product vendors can offer you price breaks for bulk purchases. If vendors do not have to accept delivery of loads at their retail facilities and then reload them back onto smaller trucks to bring to your farm, they can often reduce the costs for fertilizers or other products by delivering them directly to you.

Another factor is how much inventory you want on hand at the start of the season and how much you will need over the growing season. You have to consider the trade-offs between, on one hand, having product readily available on the farm and getting a price break for a semi load and, on the other, having the bulk plant store the material and deliver it as you need it.

Think about having enough capacity to take that 5,000-gallon semi load of liquid fertilizer at one time. For example, one semi load can fill two 2,500-gallon tanks if they are empty. It might be best to have even more capacity so you can enjoy the price breaks of full semi loads of fertilizer even if the tanks are not empty at the time of delivery.

4. Is Bulk Containment Required?

State and federal regulations require containment structures if you store certain quantities of pesticides or fertilizers. You must include the cost of containment when you analyze the costs of storing large volumes of product. When building new containment structures, doubling the size of the area under containment will allow for future expansion to include more tanks.

Having larger storage areas also allows you to inspect and maintain both the tanks and the containment structures provided that the tanks are set away from containment walls and from each other.

Figure 1.2. These tanks were set in a dike, and the building is being constructed around them. Storing tanks indoors lengthens their life but adds to the cost if you have to build a new structure or modify an existing one.

5. What Tank Material Do I Want (Poly, Steel, or Fiberglass)?

When buying a tank, you will need to choose among poly, fiberglass, carbon steel, and stainless steel tanks. The following chapters go into detail on the benefits and drawbacks of each material. Here is an overview to get you started:

	PROS	CONS
Poly	• Is the cheapest option • Does not corrode (pH of stored products does not matter) • Does not rust • Is constructed with HDLPE or XLPE based on weight stored • Can be transparent in most cases, allowing level of liquid to be viewed	• Has a shorter lifespan than other tanks • Shatters when it breaks, potentially causing a large spill • Should not be used for gasoline or diesel fuel • Degrades from UV exposure • Holds only up to around 11,000 gallons • Cannot be repaired
Fiberglass	• Is more durable than poly • Is a reinforced material made of glass fibers • Is compatible with a wide range of materials depending on the resin used internally to line the tank • Offers a choice of sizes • Can be repaired • Leaks slowly or weeps, giving you time to detect problems before it's a major spill	• Generally needs to be custom ordered • Needs gel coating to handle UV exposure • Can break down if the stored product is not compatible with the resin used in a particular tank
Carbon (mild) steel	• Has a high carbon content • Is less expensive than stainless steel • Has a very long life if protected from corrosion and rust • Can be used for very large tanks • Can have multiple compartments • Can be repaired	• Can rust or corrode from exposure to acids • Needs maintenance (e.g., painting) to protect against rust
Stainless steel	• Comes in low- and high-carbon versions • 316 (low carbon) is more refined and better resists corrosion • Potentially has the longest life if protected from corrosion • Many configurations possible • Can have multiple compartments • Can be repaired	• Is generally the most expensive of the tank materials

6. What Size Tanks Do I Need?

The area of your flat space and the size of the tanks will determine how many tanks you can place on the site. One large tank may take up less surface area than two smaller tanks. For indoor tanks, the amount of storage may be limited by the size of your building. Size constraints are tighter for indoor tanks than for outdoor tanks. However, a single large tank (e.g., with a capacity of 18,000 gallons) will take nearly twice the containment area of two 10,000-gallon tanks.

It also is important to consider the height of the tanks. One larger tank will be taller than two smaller tanks. However, many older barns and buildings are not tall enough to hold some of the taller tanks. To use these tanks indoors, you can excavate and drop below the existing floor level. Building a pit to accommodate larger tanks in a smaller barn seems easy enough. However, consider the following when contemplating digging down in a preexisting building:

Figure 1.3. Digging down inside a building can accommodate taller tanks and also provide containment.

- What is the seasonal water table?
- How am I going to pump water out of the pit?
- Where is the best place to dig the pit?
- If I have a spill in the pit, how will I clean it up?

Another important factor is the door dimensions. Doors have to be wide and tall enough to allow the tank to be brought into the building standing up vertically on a forklift.

Check with the manufacturer, who will have information about the dimensions for each tank that will help you calculate volumes, diameters, heights, and weights.

Figures 1.4–1.5. These poly tanks barely fit inside the building. A few inches taller, and they would not have fit. Once a tank is sold, most manufacturers will not accept returns. This is especially true when the tank has been custom designed and ordered.

7. What Tank Shape Do I Want?

The shape of a tank affects the cost, the ease of emptying and cleaning, and the volume that can be stored in a given amount of space.

Flat-bottom tanks are less expensive and can hold more product than cone-bottom tanks of the same material, height, and diameter. But it can be difficult to remove all the liquid from them. This can be a problem if the tank will hold different materials, or if sludge builds up in the bottom of the tank. So flat-bottom tanks are best used when you plan to refill the tank with the same product.

Cone-bottom tanks are ideal if you need to hold different products during a growing season—such as 28 percent fertilizer, starter fertilizer, water, or pesticide. Cone-bottom tanks have walls that taper toward the center bottom, which allows for full drainage and lets you more easily clean tanks between fillings. Remember, because a cone-bottom tank has tapered walls, it will hold less material than a flat-bottom tank with the same height and diameter. Be sure to account for this reduced volume when making your decision.

Dish-bottom tanks are a hybrid of flat- and cone-bottom tanks. Dish-bottom tanks have a 3- to 6-inch slope toward the center, which allows their contents to be fully drained. They hold more than cone-bottom tanks with similar dimensions but slightly less than flat-bottom tanks.

Figure 1.6. A cone-bottom tank.

Figure 1.7. A dish-bottom tank: The underside has a 3- to 6-inch slope to the center.

8. How Much Will the Fully Loaded Tank Weigh?

When tanks are full of product, that places a significant load in pounds per square foot on the concrete pad. That load will not be an issue if the concrete pad is designed properly and its bedding has the proper gravel base. In addition, the area under the concrete must be well drained to prevent water from freezing underneath it, which can cause it to crack and fail.

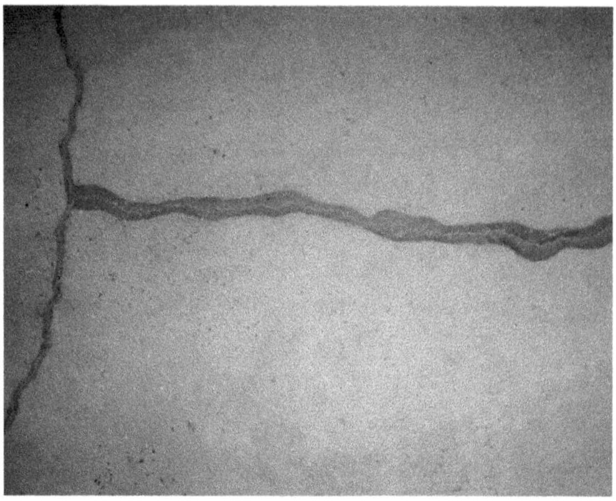

Figure 1.8. Make sure the concrete can hold the weight of filled tanks. If not, you will always be repairing cracks.

If placing tanks on existing concrete, make sure the concrete is strong enough to handle the load before ordering tanks. The thickness or quality of the existing concrete may limit the size of the storage tanks, requiring several small tanks rather than just a few large tanks. In some cases, you can remove existing concrete and pour a new slab that will support the increased weight of larger tanks.

When considering the concrete pad, ask yourself these questions:

- How much weight will the completely filled tank exert on the concrete?
- How thick does the concrete need to be?
- Will the added cost of concrete be worth the benefit of larger tanks?
- What reinforcements will be needed?
- How much slope will be needed to allow for complete drainage into a sump?
- What surface finish is best?
- Are concrete sealers helpful?

9. How Will the Tanks Be Delivered and Unloaded?

When considering tank expenses, don't forget to include the cost of hiring a semi to transport the tanks from the company to your business or farm, because the driver will require a commercial driver's license (CDL) and the truck will need an oversize load permit.

Delivery trucks may come equipped with knuckle booms to unload the tanks, but don't assume this. Some companies do not have this capability. If you must hire someone else to unload the tanks, that will add significant cost. Cranes cost hundreds of dollars per hour. When you price a tank, be sure to ask whether the total includes transportation charges and unloading (taking the tanks off the semitrailer) and whether the contractor is insured.

Figure 1.9. This tank delivery requires an oversize load permit, shown on the back of the tank.

Figure 1.10. There are significant transportation costs associated with moving tanks on the highway. When comparing quotes from manufacturers, be sure to see they include transportation costs.

Figure 1.11. Cranes must be on-site on the day that tanks are delivered.

10. What Accessories Are Important to Me?

Manufacturer catalogs list many different options. One of the more important considerations is the manway, since it will allow access to the inside of the tank for cleaning. The higher the manway, the better, because there is less pressure at higher points. You even need to consider the manway's shape. Curved shapes are stronger than flat shapes.

Don't forget to get the proper-size vent for your tank. Tanks without vents (or with undersized vents) collapse when adding or removing fluids, or even during heat expansion in the spring.

11. Where Do I Want the Plumbing Fittings Placed?

Consider the location of plumbing fittings and ensure that they will work for your setup. For example, if the tank will be in containment, be sure the fittings will not be too close to the wall. A tank with two fittings will allow the tank to be filled and unloaded at the same time.

You should also consider how the fittings will allow you to fill, drain, and clean a tank. Installing a fitting about 6 inches from the bottom of the tank that has threads on both ends, to allow you to attach a hose on the inside, can help pull more product out of the tank. Even so, there can be areas on the bottom of the tank where the fluid or sludge can't be removed.

Figure 1.12. A fitting on the bottom of this stainless tank allows it to be emptied completely.

You can also install drain fittings that allow more fluid to be drawn out. By installing couplings higher on the tank, you can pipe them on the inside with suction tubes to drain the fluid off the bottom. In other cases, you can use sump pumps or wet-dry vacuums to remove material on the bottoms of tanks through the manway.

12. Does the Product You Plan to Store Require Agitation?

Some products require agitation to remain effective. For example, fertilizer ingredients can separate (or salt out) when it's cold: 28-0-0 will salt out at 0°F, 32-0-0 at 28°F, and 10-34-0 at 10°F. When a material salts out, it deposits residue on the tank floor, which can speed up corrosion.

Residue tends to accumulate around the tank's perimeter in particular because pumping fertilizer into or out of a tank agitates the center, and the solid product gets resuspended by the flowing liquid. Adding an agitation system or recirculating vents helps keep material from salting out along the edges and reduces corrosion.

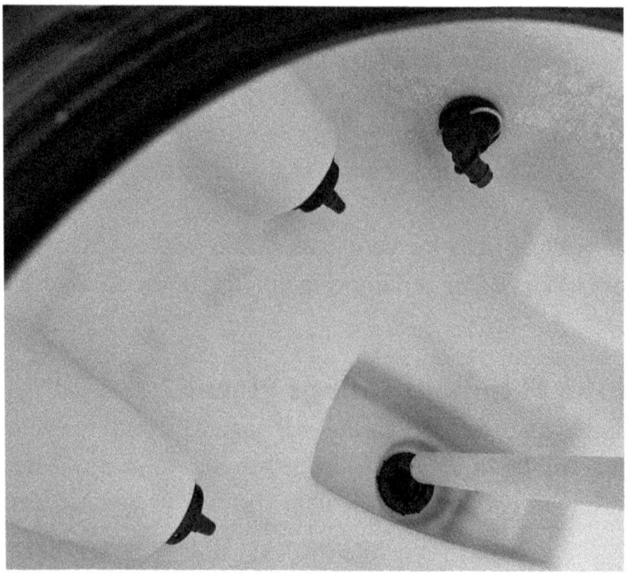

Figure 1.13. A sparger tube that provides agitation.

13. What Type of Inventory Monitoring Gauges Do I Want?

The choices for inventory monitoring gauges may include a sight gauge, a float, or even an ultrasonic system. State or federal law may regulate the gauge you select, so be sure to ask plenty of questions.

A sight tube is simply a 1-inch clear vinyl hose hooked onto two 90-degree threaded couplings at the top and bottom of the tank. For safety, you should replace these inexpensive tubes every two years. The standard external fittings are PVC elbows. Consider upgrading to stainless steel fittings. Use a locking stainless steel valve at the bottom of the sight gauge in case the sight gauge hose gets a hole in it. Keep the valve locked when not using the tank to avoid vandalism.

For outdoor tanks, consider ordering embedded studs or hooks so you can secure external sight gauges to the tanks at intervals. Storm winds can whip the tubes, which can cause them to fail—these studs can help hold the tubes in place.

14. What Tie-Downs Do I Need?

Today there is more emphasis on securing outdoor storage tanks. Strong winds can knock tanks over, which can crack tanks or cause them to fail.

Anchoring is also important to prevent tanks from floating in water. Just a few inches of rain can float an empty tank within containment, which could then strike other tanks or equipment and cause serious damage. More information about flood safety is available in volume 2 of this book.

Transport tanks must be securely anchored in ways that prevent them from moving but also don't damage the tank. Transport tanks bring a greater risk of leaks. Whether you use straps, hoops, angle iron, or other tie-downs, keeping tanks locked in place in transportation is very critical. These tanks are built with more points of connection for tie-downs.

INSTALLING VERTICAL STATIONARY (STORAGE) TANKS

Location

The decision on where to locate a tank usually centers on accessibility: You have to be able to get to it easily. But if you're installing a replacement tank, take time to reconsider the location; placing it where the old one stood might not be the best decision.

Consider for a moment what would happen if a tank of diesel or fertilizer were to split open, releasing its contents. Where would the product go? Could it flow into a well or creek? While convenience and accessibility are valid reasons for selecting a location, the potential environmental impact of a spill or leak is equally important. The ideal location is clear of drainage tiles, wells, septic tanks, ditches, and ponds.

In a real-life example, a poly tank broke open and released 2,500 gallons of liquid starter fertilizer onto the ground. It was unfortunate, yes, but the tank had been placed in a perfect location, and the fertilizer flowed down a dirt driveway, accumulating at the edge of an agricultural field. Cleanup was easy, and the spilled material did not contaminate drinking or surface water nor kill any fish.

Vertical stationary tanks need to be in a well-drained area not subject to flooding. Tanks have been known to float off-site or shift as heavy rains wash away the fill, creating an unstable base.

Figures 1.14–1.15. These photos show the aftermath of a release. Due to the placement of the poly tank, its contents flowed into a nearby ditch. If your tank were to fail, where would the contents go? Think through the possibilities before choosing its placement.

Surface Considerations

Before installing a tank, make sure the ground where it will be placed is level so that the pressure is evenly distributed within the tank. The area beneath the tank has to be free of sharp rocks and other objects that could cut into the tank over time and cause a leak.

Figures 1.16–1.18. Every tank needs to sit on a solid foundation, unlike those shown here. This places undue stress on the tanks, which can lead to spills.

Figure 1.19. An unstable base will shorten the life of any tank. Any spill can have serious repercussions for the environment and lead to expensive cleanups and regulatory fines from state and federal agencies.

Figure 1.20. This tank sits on a pad of pea gravel, which helps to move water away from the tank.

Figure 1.21. When tanks sit in water, the bottom of the tank can rust. This tank was put out of service. When it was moved, the bottom of the tank came off.

Small stationary tanks (less than 2,000 gallons) and medium-size tanks (2,000 to 6,000 gallons) should be installed on compacted soil, sand, pea gravel, or concrete. The base must be solid, level, not subject to erosion, and at least 4 inches deep (6–10 inches is better); if loose fill is used, the base must be framed by a material that will not decay. Tanks that hold 6,000 gallons or more should always be placed on a properly designed concrete pad.

The pad should keep the bottom of the tank out of standing water. To give the tank more stability, make the pad's diameter slightly larger than the diameter of the tank's base.

All outdoor vertical tanks should be anchored to the ground using the tie-down connectors on the tank, to prevent the wind from tipping them over and potentially cracking them or damaging valves.

Figure 1.22. Note how each tank here has its own pad that raises it slightly above the floor.

Figure 1.23. Tie-downs are important for all tanks. Even a tank that is full can be blown over by high winds. The tank is more prone to blowing off its foundation when it is partially filled.

Also, just a couple of inches of rain accumulating in an outdoor containment area can lift tanks. If the tank moves only a few inches, it can break valves, fittings, or pipes, which can cause leaks and environmental problems.

However, be careful not to overtighten the tie-downs, as doing so can damage tanks. Tighten the cables just enough to keep the tanks in place during a storm. Check the cables annually for tension, and retighten as needed.

Containment

If you store large amounts, there may be legal requirements for containment. Call your state department of agriculture for specific containment regulations. For contact information, see the "Control Officials" page of the Association of American Pesticide Control Officials: https://aapco.org/2015/07/28/resources-2/.

Some states specify how far tanks must be from containment walls (a typical rule may call for a 4-foot setback from walls). These specifications can allow enough distance between the tank and wall so that any leak with head pressure cannot spurt over the dike wall. It also allows enough room to inspect the outside of the tank.

Figure 1.24. Vertical storage tanks properly installed on a level surface and secured to the base. Note the concrete containment walls. The details of the containment system are often left to the state's agricultural agency.

Delivery and Unloading of New Tanks

Before you have a tank delivered, sketch out the area where it will be set. If the new tank will be placed next to other tanks, leave enough space to allow for inspections, repairs, and painting. Before placing the tank, mark the ground in the containment pad with spray paint to indicate where the tank should go. Cut and lay spacer boards to indicate the spaces you want to leave relative to other tanks, walls, and other features.

Your sketch should also indicate where you want the manhole and fittings to be located. Remember that you must place a tank on a smooth, level, well-compacted surface. Placing a tank on soft or unlevel ground can create instability by placing more weight on one side of the tank, which may cause the tank to sink into the soil on the soft side. When that happens, the tank can topple unless you straighten it. It is best to put a steel tank on a pad designed to carry the tank's weight when loaded.

When your tank is delivered, inspect it before it's unloaded. Look for buckles, flat spots, or distortions on the sides of the tank that may indicate more than cosmetic damage.

You need a plan for how your tanks will be delivered, unloaded, and moved into place. Be sure to check whether the delivery driver will be able to unload the tank or whether you need to hire a crane prior to the tanks showing up on-site.

Figures 1.25–1.28. A knuckle boom helps with unloading the tank outdoors. Then a forklift or front-end loader is used to position the tank. Know how much each tank weighs, and make sure you have a forklift that can handle the weight. Strap tanks to the forklift, and use spotters to bring them indoors.

Figure 1.29. Tall tanks may require a crane, not just a knuckle boom. Cranes can add significant cost to the overall purchase and installation price.

Figure 1.30. This brand-new, 40-foot-tall tank with a 14-foot diameter was put up on a Friday. Unfortunately, the crane that was to put the tank in the dike did not arrive until Tuesday. Over the weekend the wind pushed the tank over. The damage to the 46,700-gallon tank was beyond repair. In all, the cost of not having the crane the day the tank was delivered came to $34,000 for a new tank, plus the cost of the crane and operator.

Knuckle booms that unload tanks from semitrucks make your job much easier. Each tank will have lift points near its top and along the sides; it can be physically lifted straight up and then placed on the ground. The boom operator then switches the hooking to the top and sets the tank up vertically. You also will need a forklift to move the tanks to their final location.

Valves and other plumbing fixtures are installed after most tanks are delivered. While there are many choices for valves and fittings, be sure to choose materials that are compatible with the materials that will be stored in the tank. In addition, ask the manufacturer how much compression force (torque) you can apply as you tighten the fittings and valves.

Once the system is plumbed, fill a new tank with water for a day or two, and check for leaks around the plumbing or through the tank. If you need to test several tanks, once the first tank has passed the test, you can pump the water from that tank into the next tank and repeat the process until you have evaluated all your tanks. Contact the retailer who sold you the tank immediately if you notice a leak or other problem with the recently purchased tank.

CALCULATE TIPPING HEIGHT

When installing a new tank inside, it isn't enough for the structure to be as tall as or taller than the tank you wish to install. You also need to account for the tank's tipping height.

To calculate the tipping height, think of a right triangle. The tank's height (A) squared, plus the tank's width (B) squared will be equal to the tipping height (C) squared. For a tank that is 22.375 feet tall (A=22.375) and 12 feet wide (B=12), the tipping height (C, the height the building's ceiling and inside space must be in order to position the tank upright) will be at least 25.4 feet.

A = 22.375 feet
B = 12 feet
C = Tipping Height

$A^2 + B^2 = C^2$
$22.375^2 + 12^2 = C^2$
$500.641 + 144 = C^2$
$644.641 = C^2$
C = 25.4 feet

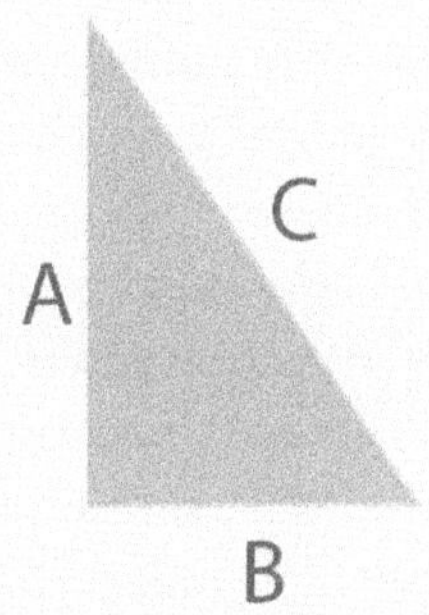

INSTALLING HORIZONTAL (TRANSPORT) TANKS FOR STABILITY AND LOAD WEIGHT

When mounting a tank to a truck or trailer, it's important to answer these questions:

- How will I keep the tank from moving around?
- How will I keep the tank from rubbing against metal and sharp objects?
- How will I keep the tank from vibrating?

Failure to properly address these three issues will cause your new tanks to fail prematurely.

Once a tank is installed on a vehicle, completely inspect the tank, including the nuts and bolts, to make sure all of the attachments are tight. Fill the tank with water, and check it for leaks. Turn the pump on at a low pressure for a few minutes. Then apply a higher pressure and let it run for 30 to 60 minutes. Look for signs of leaks around any fittings and seepage through any crack that may have occurred during transportation from the manufacturer to your operation.

If you will occasionally remove a tank during the season, secure the tank with web tie-downs. Be aware the tank may still move, which may allow it to come in contact with sharp objects. It is very important that you mount tanks on surfaces that are not going to flex during operation. Transport tanks under flexing stress can crack.

Fittings on the tank can often be overlooked. The fittings need to be in safe places where they are not going to be broken off. Be sure to provide plenty of clearance between the fittings and the vehicle bed to avoid damaging the fittings.

You need to inspect the tank each day, making sure that there are no leaks around the truck or residues in the bed. Remove and thoroughly inspect every tank (and the pumps and hoses) each offseason. The offseason is also a good time to check the bottoms and sides of tanks to see if any portions have rubbed.

Tie-Downs

Tanks used for transportation must be tied down securely to keep them from shifting when transported. Shifting could damage the tank, release the contents, or cause an accident. Tie-downs must be installed carefully to avoid damaging the tank but must be strong enough to hold the tank in place. The difficulty is that the shifting can be forward and backward as well as side to side. The appropriate type of tie-down varies by tank material; see chapters 2–4 for what tie-downs to use for what tank type.

Many horizontal storage tanks are placed in a concrete or steel saddle or in a supporting frame. Make sure to place a piece of rubber (or a similar product) between the tank and the saddle to reduce corrosion and abrasion.

Load Capacity Considerations

Hauling full tanks on trucks and trailers raises important legal and safety issues. In most states, a special permit is required if the combined weight of the vehicle(s) and the load exceeds 80,000 pounds. Most drivers intend to stay below the limit; since the average empty tractor/trailer combination weighs 24,000 to 28,000 pounds, only about 52,000 pounds of cargo can be hauled legally. If your overloaded truck or trailer were declared out of service by your state's motor vehicle enforcement division, you would have to offload some cargo.

Always consider the number of loaded tanks plus all support equipment (mini-bulks, hoses, pumps, etc.) when estimating the weight of your load. To get a rough estimate of the loaded weight of a tank, multiply the total gallons the tank can hold by 10 pounds per gallon (a precise number for 10-34-0 is 11.4 pounds per gallon, in which case a 2,500-gallon poly tank could weigh up to 28,500 pounds). Add up the weight of the tank and its contents to get the total weight that the truck will have to carry in the bed or tow on a trailer.

From a safety perspective, trucks and trailers need to be equipped to handle the maximum allowable weight they can carry; that is, their steering, tires, wheels, bearings, suspensions, and brakes must be designed to withstand the stress imposed by the load. Manufacturers assign gross vehicle weight ratings (GVWRs) for trucks and trailers; the GVWR represents the maximum load the truck or trailer can safely carry, including the weight of the vehicle itself. If the combined weight exceeds the GVWR, it means that you cannot safely transport the load—or, perhaps, even stop it efficiently. If you are involved in an accident under these circumstances, your insurance company may not honor your claim (see volume 2 for more on transportation security).

Transportation Security

Attention has shifted over the years to safeguarding loads transported on our highways. Concern about theft of product, acts of vandalism, and even threats of terrorism has increased our diligence in protecting the chemicals we haul.

Figures 1.31–1.32. Exposed tanks are subject to vandalism and damage during transport. Enclosed tanks can be secured and locked.

- Vehicles used to transport pesticides should be kept locked when not in use.
- Pesticide and fertilizer application equipment should be locked, if possible, and attended at all times if it contains any product whatsoever; all unattended equipment should be empty.
- Tank valves should be closed and secured with chains, padlocks, or other devices when unattended.
- Lids on unattended tanks must be closed and secured.
- Tanks and equipment should be inspected daily before each use; look for signs of tampering, vandalism, or leaks.
- Procedures should be in place for notifying appropriate law enforcement officials when tank tampering or vandalism is discovered.
- Tanks should not be left in unsecured areas, along roadsides, or in fields.

PREVENTING AND RESPONDING TO SPILLS AND OTHER PROBLEMS

Most warranties cover the tank if it is used according to the manufacturer's specifications and if failure occurs during the warranty period. Coverage is for workmanship and materials only. Under these terms, a defective tank would be repaired (if possible) or replaced at no cost to the original owner. Warranties do not cover the cost of lost product or environmental remediation due to a release or spill, nor do they cover tank failure if the material stored had a specific gravity higher than the tank was built to store. Read the manufacturer's warranty to find out what is and is not covered.

Know What Your Insurance Policy Covers

Many farm and commercial business insurance policies do not cover pollution from fuel, fertilizer, or pesticide spills. It is critical for business owners and farmers to consult with their insurance representatives to confirm what is and is not included in their coverage. Some insurance companies now offer specialized pollution coverage, but policyholders must read all endorsements to make sure their coverage is not compromised or negated by exclusions.

Your insurance agent can tell you whether cleanup and product replacement costs resulting from a ruptured tank are covered under your property or vehicle policy. Ask whether policy discounts are available for things like conducting and documenting annual and biannual inspections and diking your storage tanks.

Notification That You Store Hazardous Materials

What is *hazardous* depends on the definitions from OSHA, the Department of Transportation, and state departments of agriculture and environmental agencies. Some individuals may consider pesticides as hazardous. Regardless, you may need to let local authorities like your fire department know what products you store on your farm so that if they respond to a spill or fire, they already know what they may face. For more on security, see volume 2, which provides detailed instructions on how to develop a map that will be very useful to those responding to emergencies at your farm or retail business.

Notification in the Event of a Spill

You've had that old poly tank for 10 years. You bought it new, kept it maintained as best as you could, and used it only in the spring. It's never given you any problems until now: The old tank has failed. Liquid fertilizer is spilled all over the ground, trickling down to the drainage tile at the edge of the property—and the water in a nearby ditch is discolored. Whom do you need to call? Your insurance agent? Your neighbor? The Environmental Protection Agency?

It is important to know that all spills require a spill response and cleanup regardless of reportability. Most states have a very broad definition of a spill. For instance, one state reports that a *spill* is any unexpected, unintended, abnormal, or unapproved dumping, leakage, drainage, seepage, discharge, or other loss of petroleum, hazardous substances, extremely hazardous substances, or objectionable substances. The term does not apply to releases on impermeable surfaces when the substance does not migrate off the surface nor penetrate the surface and enter the soil.

Generally, spills that damage the waters of the state, impact or may impact drinking water, are beyond a facility boundary, exceed the "reportable quantity" (as determined by EPA regulations), or are within a wellhead protection area require immediate reporting to state, local, and sometimes national officials. You can contact the EPA's National Response Center at 800-424-8802. For contact information for your own state, see the resources on the website of the Association of American Pesticide Control Officials: https://aapco.org/2015/07/28/resources-2/. It is the responsibility of owners and/or operators of facilities and modes of transportation to report spills. Failure to report spills in a timely manner can result in fines and penalties.

Immediate cleanup is required as well. Contact your Local Emergency Planning Committee for information to help you preplan notifications in case of a release.

Security of the Area

Pesticides, fertilizers, fuel, and other materials stored in tanks must be secured from theft and vandalism. Follow these advisories:

- Keep all tank valves closed and locked when not in use or when no one is on-site.
- Place tanks as far away from roads as possible to reduce the threat of vandalism.
- Place tanks in areas where there is adequate dusk-to-dawn or motion-sensitive lighting so they are visible from the house or shop.
- If possible, fence the tank area to discourage intrusion, and lock the entrance gate.

- Limit key access to authorized persons.
- Keep forklifts and other equipment away from tank sites.
- Post NO TRESPASSING or KEEP OUT signs and other regulatory signs, as required.
- Inspect facilities, tanks, gaskets, fittings, hoses, pumps, and valves regularly; watch for leaks and signs of tampering.
- Maintain an accurate inventory of pesticides or fertilizers stored in the tanks, and label the tanks.
- Ask law enforcement to conduct periodic inspections if no one is on-site on a regular basis.
- If tanks are located away from the main farmstead or facility, post contact phone numbers at the storage location.
- Keep a list of emergency numbers handy: fire department, law enforcement, medical assistance, county emergency management, insurance, state department of agriculture, and environmental agencies.
- Establish procedures for notifying appropriate law enforcement when evidence of tank tampering or vandalism is discovered.

Figures 1.33–1.34. Fencing helps to secure tanks. Some business owners fence their entire property. Lights can also deter thieves.

MAINTAINING TANKS

It is easy to take tanks for granted once they have been installed since they last for years. But tank ownership involves regular maintenance to detect problems early. Visually inspect your tanks each time you fill them; watch for physical damage such as holes, dents, and abrasions. Make sure that the base on which vertical tanks rest remains solid. Animals can burrow underneath, causing the base to become uneven. Keep a record of each inspection in case liability or warranty issues are raised.

The following are some things to check regularly:

- Does the tank show any early signs of problems? (See the following chapters for specific things to check for each tank type.)
- Does the tank need painting (steel) or gel-coating (fiberglass)? These will lengthen tank life.
- Are weeds and garbage clear of the tank?
- If the tank has a sight gauge, does it need replacing?
- Are the signs and labels on the tank still accurate and easy to read?

A TRAGIC STORY!

The nephew of a farmer entered a stainless steel tank on a sprayer to clean it out and was overcome by fumes. A farm employee went in to rescue him and also was overcome. The nephew died, and the employee had to be rushed to the hospital. This tragedy is a serious reminder to make sure you monitor the air quality in any tank or confined space before entering it and to use supplemental breathing equipment as required.

2

Poly Tanks

With contributions from Steve Salomon, Darrell Oltman, John Obermeyer, Steve Hawkins, Matt Pearson, Bryan Overstreet, Mike Titus, Shawn Lambert, Mike Hawkins, and Arlene Blessing

Plastic tanks are commonly used on farms and by commercial application businesses for efficient storage and transportation of water and other liquids such as pesticides and fertilizer. These polyethylene tanks, called *poly tanks*, are popular for several reasons:

- They are available from many suppliers and retailers.
- They cost significantly less than tanks made from other materials.
- They are compatible with many liquid products used by farmers, commercial applicators, and commercial businesses.
- Manufacturers offer a diversified product line for transportation, storage, and application uses.
- They offer design flexibility.
- They are relatively lightweight and easy to handle.
- They are corrosion resistant; that is, they won't rust.
- They are relatively impact resistant.
- They can be designed so that the amount of liquid in the tank is clearly visible.

While poly tanks have many benefits, they have a shorter lifespan on average than other tank types. A poly tank's useful life depends on a number of factors, including the quality and amount of poly material used to manufacture it, the materials stored in it, and its specific use (storage or transport).

If a poly tank is left outdoors, ultraviolet (UV) radiation from the sun degrades the polyethylene over time, changing it from a tough, resilient material to a hard, brittle one and making the tank more prone to breakage. Poly tanks degrade quicker when misused, neglected, or used beyond their design specifications.

The goal is to replace a tank before it ruptures; the hard part is assessing the structural integrity of the tank to determine when replacement is necessary. Continued use of an old tank can be a serious economic and environmental mistake when its structural soundness falls below its original design capability.

Knowing when to replace a poly tank is very, very important. No one likes spending money unnecessarily, but poly tanks can—and do—rupture "for no apparent reason." When they do, the loss is not limited to the value of the spilled material; the cost

Figure 2.1. Catastrophic failures can endanger people, property, and the environment. You can lower your probability of experiencing incidents such as these by choosing your tanks carefully, caring for them properly, and conducting timely inspections.

of cleanup can far exceed the value of the product itself. In some cases, regulatory fees for restoring the environment are added to the bill as well. Poly tanks have a limited operational life; you should view their replacement as the equivalent of changing tires, oil, filters, and hoses on your vehicles and equipment. Failing to do routine inspections and maintenance can result in a catastrophic tank failure.

This chapter explains how to evaluate tank designs, provides management strategies that can extend the tank's longevity, offers inspection techniques to identify tanks that need to be replaced, and offers disposal options for out-of-service tanks.

TANK MATERIAL AND CONSTRUCTION

Poly tanks are built with highly durable, chemical-resistant resins formulated for today's pesticides and fertilizers. They are built to the internal specifications of the manufacturer, not according to a national standard. The resins used in the manufacture of poly tanks are either high-density linear polyethylene (HDLPE) or high-density cross-linked polyethylene (XLPE). In general, both materials work well for the storage and transport of most pesticides and fertilizers, but XLPE is more chemical resistant and durable (although generally more expensive) than HDLPE. Contact the chemical manufacturer if you are uncertain whether an HDLPE tank will work for the products you store and/or haul.

Poly tanks generally are manufactured by using a process known as *rotational molding*. A powdered polymer compound with an ultraviolet (UV) protection package is poured into a two-piece mold. The amount of UV protection varies by tank type and manufacturer. The mold is clamped shut and heated in a hot oven. During the heating process, the mold is tumbled in two directions; as the powered polymer melts, the tumbling causes it to coat the inside of the mold. After that, the mold is moved into a cooling chamber where the temperature is slowly decreased. The completed tank is then removed from the mold.

This process molds poly tanks as a single piece. A visible external seam—known as the *parting line*—gives the impression that two pieces are joined together. But the parting line is merely an external cosmetic artifact from the manufacturing process; it represents the juncture of the two pieces of the mold, not a seam in the tank itself.

Figure 2.2. Notice the parting line on this tank.

Specific Gravity

A poly tank's specific gravity rating is a measure of its ability to hold materials. Specific gravity is particularly crucial for poly tanks because they have a shorter lifespan and tend to shatter when they fail, releasing all of their contents.

The specific gravity of a substance is a comparison of its weight per unit volume to that of water. Manufacturers have designed tanks with specific gravity ratings of 1.0 to 1.9 or more. Higher specific gravity ratings indicate a greater ability to withstand hydrostatic stresses caused by a stored liquid. Tank wall thickness and overall design are two of the most important factors determining a tank's specific gravity rating.

A poly tank with a specific gravity rating of 1.0 is designed to hold the weight of water or any other liquid that weighs 8.3 pounds (or less) per gallon. Most pesticides weigh slightly below or just above the weight of water. A poly tank rated at 1.5 specific gravity will withstand the internal forces of liquids weighing up to 12.5 pounds per gallon. A tank rated at 1.9 specific gravity can store products weighing up to 15.8 pounds per gallon. In general, liquid fertilizers weigh 10 to 12 pounds per gallon and so should be stored and handled in poly tanks with at least a 1.5 specific gravity rating. A good rule of thumb is to purchase a tank with a rating at least one increment higher than the specific gravity of the product you intend to put into it.

Figure 2.3. Manufacturers normally attach a tag to the poly tank. This tank tag indicates a tank with a 1.5 specific gravity rating.

If the specific gravity rating tag is no longer attached to the tank, it is impossible to guess the rating based on the tank's appearance. Ask your dealer for the specific gravity rating of the tank you're considering, review the manufacturer's catalog, or examine the tank specification sheet. In some cases, the specific gravity rating is part of the product code. Some manufacturers stamp information on the side or top of the tank. These variations make purchasing used tanks a real challenge.

Extreme Temperatures

Extreme temperatures usually do not impact good-quality polyethylene tanks. Quality tanks are designed to withstand expansion and contraction caused by extreme hot and cold. Results of tanks tested to –40°F demonstrate that freezing does not damage poly tanks. Constant, sustained temperatures above 100°F may weaken tanks.

Water Deterioration

Poly tanks do not decay or rust in standing water or in contact with the ground. Any mold or algae that might develop on the inside is removable with an algicide.

Product Compatibility

HDLPE tanks are compatible with most pesticides and fertilizers. However, poly tanks manufactured to hold fertilizers and pesticides should not be used for gasoline or diesel fuel, which can break down and soften the plastic. Pure biofuels such as B-100 can be stored in poly tanks built to American Society for Testing Materials (ASTM) D1988 standards with a 1.9 specific gravity rating. But some manufacturers' warranties do not cover tank failure due to storing 100 percent biodiesel or blended modifications.

Petroleum-based oils also permeate and soften plastic, but the process is much slower than with gasoline. Oil should be stored in poly tanks with a 1.9 (or higher) specific gravity rating.

Valves, gaskets, and hoses must also be compatible with the products to be stored in the poly tank. In certain cases, replacing a gasket with one that will stand up to biodiesel makes using an HDLPE tank acceptable. Using compatible equipment prevents component failure and product release.

Figure 2.4. Read and heed all attached warnings on the use of poly tanks.

TANK DESIGN: VERTICAL STORAGE AND HORIZONTAL TRANSPORT

Poly tanks have two distinct designs: vertical tanks (also called *stationary*, *upright*, or *hockey puck tanks*) and horizontal tanks (*side-to-side*, *transport*, *application*, or *leg tanks*). The significant design differences between vertical and horizontal tanks dictate how they should or should not be used. Using a tank outside its design specifications voids the manufacturer's warranty and increases the odds of tank deterioration, tank failure, costly cleanup, and lost inventory.

Vertical tanks are flat- or cone-bottom cylindrical tanks designed and manufactured specifically for stationary placement on a reasonably smooth, level surface. As the tank is filled, the pressure of the liquid forces the wall to flex outward. As the tank is emptied, the walls revert to their original shape. This movement of the tank wall is determined by the depth and specific gravity of the liquid. Manufacturers strengthen tanks by making the plastic thicker where the sidewalls meet the bottom; this is where the pressure is greatest.

Even though vertical tanks are for stationary use only, many come with tie-down connectors for anchoring them to the ground; otherwise, empty tanks can be set askew—or blown over—during high winds. The connectors on vertical tanks are, however, not strong enough to secure loaded vertical tanks to the bed of a truck or trailer for transport.

Figure 2.5. The storage tanks in this photo are being misused as transport tanks. While it is not illegal to do so, it is not a good practice as storage tanks are not designed to withstand the rigors of on- and off-road travel.

Figures 2.6–2.7. A storage tank used for transport can fail at the points where its tie-downs are attached. See breakage from this type of failure in the second photo. Storage tanks are not designed for the rigors of transport.

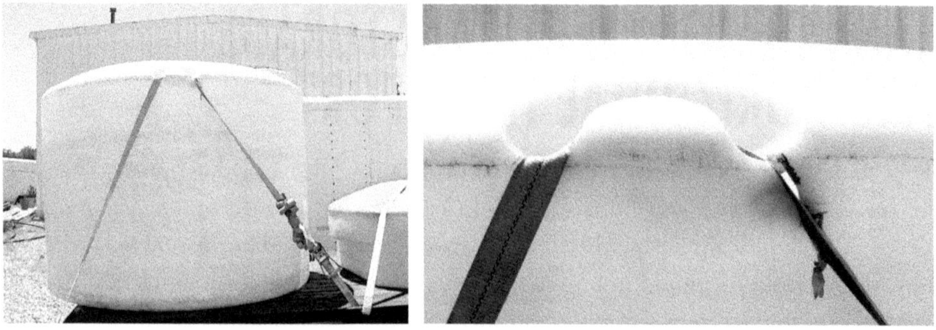

Figures 2.8–2.9. Attempts to secure a tank actually can deform the plastic, as demonstrated by the dents in the green tank. The strap on the white tank actually cut into the plastic.

Horizontal tanks are designed for placement on trucks, trailers, and field sprayers, but they also may be used as stationary tanks. Pressure points on horizontal poly tanks are much different than those on vertical tanks. Liquids move from front to back when a vehicle stops or accelerates, creating a surging effect that exerts pressure on the front and back walls. Large-capacity horizontal tanks (greater than 1,000 gallons) generally are equipped with internal plastic baffles that reduce the force of liquid as it surges from one end of the tank to the other.

The tie-down features on horizontal tanks are more substantial than those on vertical tanks. Horizontal tank tie-downs commonly are pipe hoops or metal bands capable of holding a loaded tank in place if the truck or trailer stops suddenly. The hoops and bands also support the sides of the tank.

Both vertical and horizontal tanks should be properly secured in place without bulging or looking deformed. Although tanks are designed to change shape slightly when filled, they must not be squeezed out of shape by straps or tie-downs. A deformed tank indicates too much stress on the polymer, which in combination with UV exposure and the weight of the material contained may cause the tank to fail.

WARRANTY

Manufacturers usually guarantee their poly tanks to last three to five years from the date of manufacture, not the date of purchase. Most tanks have three-year warranties, but longer warranties are provided on more durable tanks, such as those with a 1.9 specific gravity rating. The date of manufacture often is imprinted on the tank as part of the serial number but not always. If you cannot clearly determine the manufacture date of a failed tank, have the manufacturer look up the serial number.

Establish a file marked “Poly Tanks” for all shipping papers, warranties, delivery tickets, and the like; while your tank is new and clean, record serial numbers and parts numbers that are stamped on or into the plastic or written on a decal, and file them for future reference. Such information is commonly required to file a warranty claim with the manufacturer.

SIX QUESTIONS TO ASK WHEN PURCHASING A POLY TANK

Price can be a factor in deciding which poly tank to purchase, but when comparing tanks, be sure they have the same or very similar specifications. Ask about the specific gravity ratings and tank warranties; consider your intended use of the tank (storage or transport) and the products you will store in it.

Consider the following questions when purchasing a poly tank:

Question 1. *What size poly tank do I need to purchase?* Many variables affect the decision about tank size. Large-capacity tanks don't have to be filled as often, which can save time, labor, and transportation costs, and they offer the following benefits that can contribute to efficiency:

- Better product pricing because you can store bulk materials
- Smaller space requirement than for multiple small tanks totaling the same volume
- Single tank and pump versus setups for multiple smaller tanks

One disadvantage of large horizontal transport tanks is the weight they impose on trucks and trailers. Before purchasing a large tank, make sure your vehicle and/or trailer axles, tires, and brakes are capable of handling its filled weight. Never install a large tank on a truck or trailer not designed to carry the load.

Small tanks offer different advantages:

- They cost less per tank (although not necessarily per gallon stored).
- Their small capacity might exempt you from storage facility requirements such as dikes.
- Less material is spilled and lost if a release occurs.
- They can be used on a wider array of trucks and trailers.

Question 2. *How will I use the poly tank?* The choices are straightforward. Vertical stationary tanks are designed for storage at ground level only. Horizontal transport tanks are designed for mounting onto a truck, trailer, or sprayer for mobile operations, but they also can be used for storage on the ground.

Question 3. *What specific gravity rating should I select?* Determine the specific gravity of the products you will store and transport before selecting a poly tank. Store or transport fertilizers and pesticides in tanks rated at least 1.5 specific gravity. A rule of thumb for vertical stationary tanks: 1.5 specific gravity is sufficient if the tank remains indoors throughout the year, if it is placed outdoors under a roof, if it is placed in a properly constructed dike of adequate size, and/or if it is used infrequently.

A 1.9 specific gravity tank used and maintained in the same manner as its 1.5 specific gravity counterpart lasts longer and offers more security and peace of mind to the owner. A more expensive tank rated at a 1.9 specific gravity might be considered if the following scenarios are likely:

Figure 2.10. Vertical storage tanks should not be used for transport.

- Tanks will be mounted on a truck or trailer.
- Tanks will remain outdoors year-round.
- Tanks are extremely large.
- Tanks will be refilled many times during the year.
- Tanks will be placed near a water source (e.g., pond, creek, well).
- Tanks will be neither diked nor contained.
- Tanks will be used to store heavy materials such as 10-34-0 fertilizer.

Question 4. *What level of service do I expect?* Poly tanks can be purchased from internet suppliers or farm supply stores. Always consider these service questions:

- How important is the warranty?
- Will the vendor service the tank if service is required under warranty?
- Does the vendor carry compatible parts such as hoses, gaskets, and tank fittings?
- How does installation of fittings (by the user) affect the warranty?

Question 5. *How much does the poly tank really cost?* From your answers to the previous questions, you can use price to decide where to buy. First, compare all features and determine the tank best suited to your intended uses; then compare the services available through various vendors. The lowest-priced tank may or may not be the best deal.

Question 6. *How does the poly tank compare with tanks made of other materials?* Other materials used to manufacture tanks include fiberglass, mild steel, and stainless steel. Each has its own strengths and drawbacks. Tanks made from other materials may last longer than poly tanks but are generally more expensive.

POLY TANK INSTALLATION

Choose the best location, and place the tank properly. The few minutes this takes can lessen the likelihood of a spill and reduce the impact in the event of tank failure.

Vertical Stationary Tanks

The decision on where to locate a stationary tank usually centers on accessibility: You have to be able to get to it easily. But if you're installing a replacement tank, take time to reconsider the location; placing it where the old one stood might not be the best decision. Good tank placement is important for reducing the likelihood of spills, minimizing impacts of any spills, avoiding accidental tank damage, keeping tanks and their contents secure, and reducing regulatory fines or liability. See chapter 1 for more on tank placement and security.

Horizontal Transport Tanks

Horizontal poly tanks are mounted to steel skids or fastened to trailers. If they have legs, each entire leg must rest on the trailer. Those without legs must rest in a cradle that supports them from end to end and surrounds one-third of their circumference.

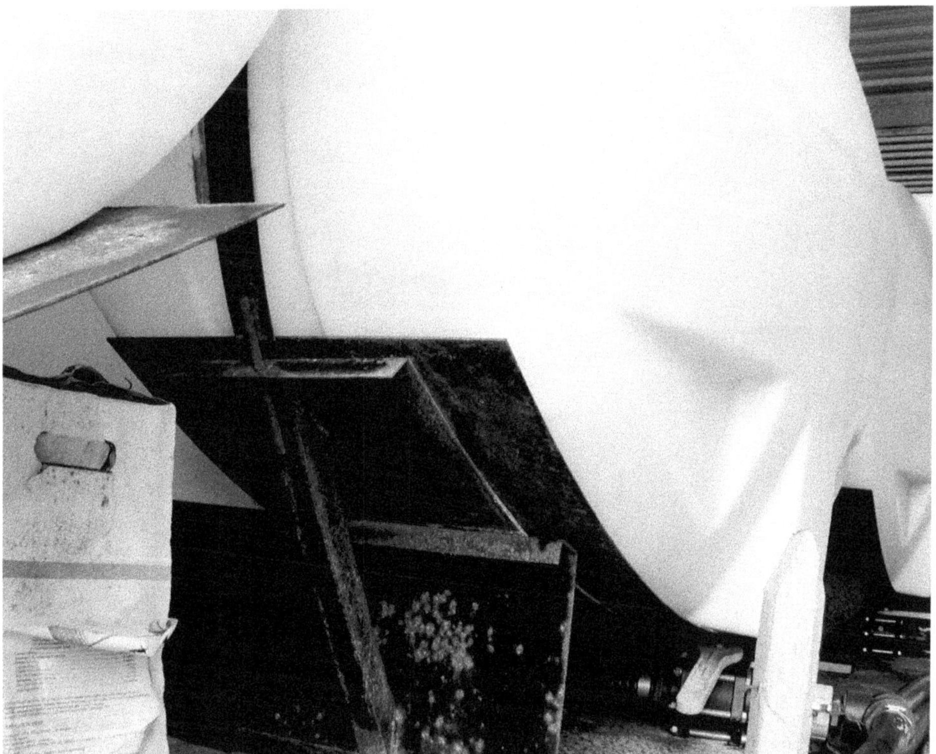

Figure 2.11. Placing a poly tank in a cradle helps to keep the tank from moving. However, make sure the tank is protected against rubbing: If the metal cradle rubs on it, that can weaken the tank.

Baffles

Poly tank manufacturers recommend that horizontal transportation tanks that hold more than 1,000 gallons be mounted lengthwise on the truck or trailer. If tanks are mounted crosswise, the baffles would not be in position to control the forward and backward surges of contents during transit.

Baffles dramatically reduce liquid surge within a horizontal tank. The tank size (length, width, height), tank volume in gallons, tank shape (round, oval), and lid opening diameter affect the size and number of baffles needed. Portable baffles are small structures placed in the tank to take up space but not much volume (less than 1 percent).

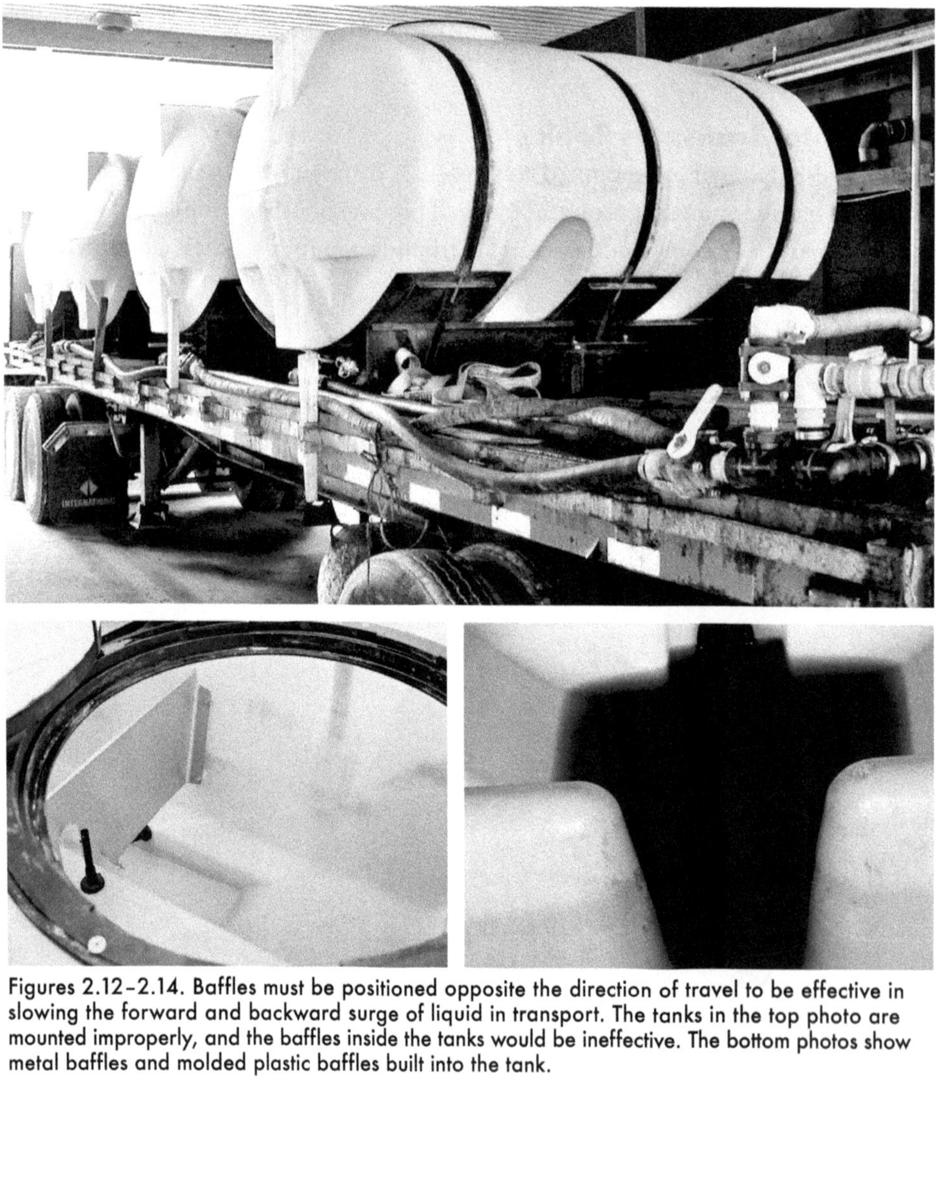

Figures 2.12–2.14. Baffles must be positioned opposite the direction of travel to be effective in slowing the forward and backward surge of liquid in transport. The tanks in the top photo are mounted improperly, and the baffles inside the tanks would be ineffective. The bottom photos show metal baffles and molded plastic baffles built into the tank.

Metal Bands and Hoops

Horizontal tanks and horizontal leg tanks sometimes require metal bands, straps, or hoops for structural strength and stabilization during transit. There is little difference in performance between bands and hoops on tanks under 750 gallons because of their smaller volumes. Bands often are made of sheet metal, which has excellent strength; however, the bolts that hold the metal to the trailer or cradle can become weakened by pressure from large tanks. Large tanks may require stronger fastening systems. Unlike bands, hoops are made of structural steel tubing or pipe and can withstand the powerful forces exerted by larger volumes of liquid when a transport vehicle stops quickly.

Metal straps or tubular hoops that run across the top and down the sides of the tank to anchor it and support the sidewalls require tightening, but always be careful not to overtighten them. Overtightening prevents poly tank walls from expanding adequately when filled. That is, when the tank fills, the sides push outward, and if the straps or hoops are too tight, the tank becomes deformed—and weakened—at the point of contact.

Figure 2.15. This transport tank has legs and is secured onto the bed of the truck with metal tie-down hoops.

Figure 2.16. This photo demonstrates the use of specific brackets and webbing to secure the tank to its saddle, which is bolted to the truck.

Figure 2.17. Instructions and warnings from a manufacturer about the use of tubular hoops.

Pull the strap or hoop over the tank so that no gap remains. Fill the tank with water and look again for gaps; tighten a little at a time until no gaps exist, being very careful not to overtighten. Watch for evidence of stress, such as deforming, warping, or bowing, and recheck straps and bolts at regular intervals.

POLY TANK ATTACHMENTS AND VENTING

Attachments placed on poly tanks can decrease tank life. A system completely plumbed with stainless steel or solid plastic pipe is too rigid, causing excessive stress on the tank valve and bung area. Flexible connections, like hoses, are preferable. Keep everything as short, lightweight, and flexible as possible.

The most common valves used on poly tank systems are stainless steel or poly ball valves, and there are two options for their placement. First, you can use flexible hose between the fittings and the ball valve. This is especially important if the ball valve is heavy: Placing the valve some distance from the poly tank keeps its weight from pulling down on the plastic. The one problem with this approach is that you cannot stop a leak if the flexible hose fails.

Figure 2.18. One of the weak points is around the bung. A hose hanging on the end places stress around the bung, which can weaken the poly material around it. This can cause the poly to break down, crack, and spill its contents.

Second, you can attach a lightweight ball valve as close to the tank as possible, followed by a flexible hose. This allows the flexible hose to pulsate as the fluid is pumped from the tank, and the valve allows the tank to be shut off in the event of a hose or pump failure. Valves can be supported with a wood block, chain, or C-clamp to alleviate tank stress from the weight of the valve and hose.

Make sure all connections allow the tank to flex. Also, make sure hoses don't hang over the trailer or truck bed, since their hanging weight can weaken the plastic around the fittings. Rigid supports and excessive weight around fittings can decrease a tank's service life; blocks placed under ball valves and hoses—after filling the tank—can reduce this type of stress.

Figure 2.19. A flexible hose between the tank and the valve can reduce the stress placed on the tank.

Figure 2.20. A lightweight valve puts no stress on the tank. Attaching a hose to this valve is what will create stress on the tank.

Figure 2.21. This fitting is supported by a metal U clamp.

Figure 2.22. (Left) Instructions associated with a poly tank. (Upper right) Putting a block under the valve to relieve the stress on the tank. (Lower right) A hose and valve pulling down on a tank, which can lead to spills.

Figure 2.23. It is easily seen that the valve is pulling down on the tank. A block is needed to support the valve on this tank.

Figure 2.24. This storage tank collapse was the result of a clogged vent.

Tanks come equipped with a vent, usually built into the lid or the tank itself, that allows the tank to breathe during loading and unloading. An improper or nonfunctional vent can cause tanks to swell excessively during loading or to collapse during unloading. A larger pump with a higher flow rate may require the installation of additional vents by either the tank manufacturer or the owner. With large-capacity pumps, additional tank venting can be accomplished by opening or removing the lid, but then you must be sure to replace the lid before moving the tank.

WATER-TESTING A NEW POLY TANK

Fittings can loosen or a tank may be damaged during shipment from the factory, and sometimes brand-new tanks from the manufacturer have an undetected defect. Therefore, many new tank warranties require a water test before the tank is used. First, install all attachments (e.g., ball valve, tank fittings, hoses) and set the tank in its designated location. Then fill it with water and let it sit for a few days to see if it leaks. If the new tank is found to be defective, contact the manufacturer for a settlement under warranty.

FREQUENT INSPECTIONS DURING USE

A few simple maintenance procedures can increase the use-life of a tank and prevent a spill or reduce its impact.

Pretrip Inspection of Horizontal Transport Tanks

It doesn't matter if you last inspected your horizontal tanks two hours, two days, or two weeks ago. Always repeat the process before sitting in the driver's seat. Look for the following:

- Missing, broken, or bent bolts on the hoops or bands
- Leaks around valves, tank bungs or lids, or plumbing
- Cracks or splits in hoses
- Improperly secured hoses
- Improperly secured covers
- Vents in the lid or tank that are obstructed or clogged

Check continually for movement of horizontal tanks on transport vehicles. Tanks held in place with metal straps should be inspected numerous times during the season since the straps can cut into the tank if it shifts forward or backward. Place thin rubber or foam underneath the metal straps to prevent this. Metal hoops usually do not need this type of protection.

Figures 2.25–2.26. Even if maintained and used correctly, a tank can degrade. Therefore, it is important to choose the proper location in the event of a spill and to conduct routine inspections to prevent a spill.

Vertical Storage Tank Site Inspection

Threaded poly fittings sometimes develop leaks when left dry over the winter, so be sure to check them in the spring. Some of the older poly tanks are plumbed with black iron fittings rather than poly fittings, and after years of use, the iron starts to decay from the inside out. Take a close look at these fittings as well; you might want to consider replacing them with the newer poly fittings.

TANK LONGEVITY: HOW LONG CAN I EXPECT THE TANK TO LAST?

Tank owners expect their storage, transport, and application tanks to last well beyond the three- to five-year manufacturer's warranty. But how long will a tank last? How old is too old? There are no definitive answers.

Keeping a tank under cover as much as possible helps protect the polyethylene from deterioration caused by UV radiation. Most UV damage occurs to the outside of the tank, beginning on the outside wall and spreading inward over a number of years, making the plastic brittle and rigid. An uncovered outdoor poly tank generally can be expected to show signs of UV damage within five to eight years.

Color plays only a very small part in reducing UV breakdown of polyethylene. The color (red, green, black, yellow, blue, or white) is added as a dry blended pigment during the molding process, but it does not add UV protection. Some poly tank manufacturers use color to separate product lines; for example, a 1.5 specific gravity tank may be white, and a 1.9 specific gravity tank may be blue. Some use color as a marketing tool. But you should not rely on tank color as an indication of specific gravity, especially when comparing tanks made by different manufacturers.

Common questions asked about slowing down the rate of UV degradation include:

- *Can a poly tank be painted to reduce UV penetration?* Polyethylene tanks are very similar to Teflon chemically, so paint does not stick well to the tank surface. However, some paints contain solvents that allow the paint to "bite" or absorb into the plastic; these solvent paints usually will not weaken the tank. In some cases, paint manufacturers recommend abrading the area to be painted; while this will help the paint stick to the tank, it will also promote increased deterioration of the plastic.

  Also, most paints will crack as the tank flexes during expansion and contraction. There are flexible paints on the market that can build up a membrane over the tank; however, these paints scratch and chip easily and require continual repainting to keep the membrane intact.

  Another disadvantage is that painting a tank prevents visual inspection of the tank using a black marker (see figures 2.32–2.33); it also prevents viewing the contents through the tank walls. There seems to be little value in painting as a means of reducing UV damage to poly tanks.

- *Does covering the tank with a tarp or black plastic reduce UV damage?* Tarps do offer limited UV protection, but even good-quality tarps need to be replaced often due to breakdown from sunlight and rain. Getting tarps to fit over tanks and getting them properly secured is difficult.
- *Does the tank have to be totally enclosed to prevent UV degradation?* The best management option is to keep the tank away from direct sunlight when not in use, preferably inside a building with the doors closed. A partially enclosed structure that does not protect the entire tank does not provide sufficient UV protection.

Factors That Influence Longevity

The only way to truly assess tank deterioration and damage is to conduct routine inspections each fall and spring. Base your decision to replace a tank on the findings—or on the warranty expiration date, if feasible. The following factors contribute favorably to tank longevity:

- High specific gravity rating
- Infrequent refilling
- Protection from UV radiation
- Stationary placement

Fall and Spring Tank Inspections

Whether a tank is a few years old or 20 years old, the only way to be sure it is structurally sound is to inspect it before use in the spring and again at the end of the application season. The spring inspection, prior to filling, provides reassurance that the tank can safely store or transport the fertilizer or pesticide you intend to place in it. Fall inspections are particularly recommended to provide forewarning of the need to purchase a new tank before spring; that is, if your tank is found to be defective or deemed unserviceable, you have ample time to consider a replacement.

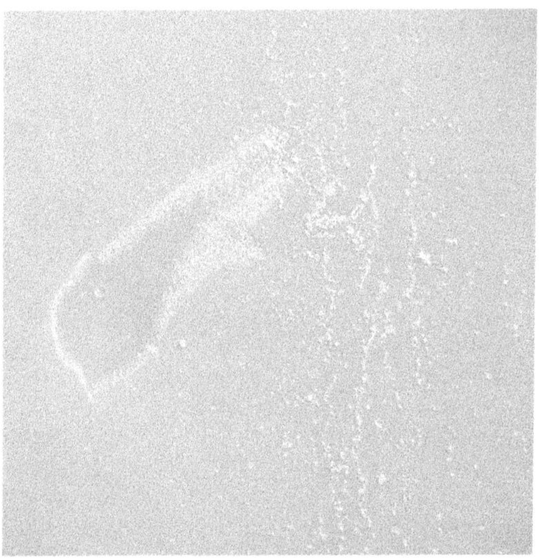

Figures 2.27–2.28. From a distance, the tank looks fine. But on closer inspection you can see evidence of UV radiation damage. The plastic material has broken down from exposure to sunlight.

It is difficult to visually distinguish a good tank from a bad tank. Three simple inspection techniques—writing with water-soluble ink, candling with light, and hitting with a baseball bat—can pinpoint weakened walls and stressed areas around the fittings.

It is important to know the difference between surface scratches, crazing within the tank wall, and cracks that extend through the tank wall. *Crazing* is the development of very fine cracks within the tank wall, usually appearing as a network of fine lines that cannot be felt with a fingernail. The tank will still hold liquids, but its structural integrity is significantly reduced.

Crazing occurs in both high-density and cross-linked poly tanks; it can be a sign of serious deterioration within the plastic, which leads to cracks and fractures. *Cracks* can be felt with a fingernail. It is common for the poly material around the crack to appear whiter than the surrounding polymer. Most *scratches* displace minute amounts of polymer but remain superficial.

- Scratches are open to the surface. Displaced plastic material is evident on the tank's surface. A fingernail catches.
- Crazing is displayed as a patchwork of fine lines.
- Cracking causes no displaced material; very abrupt lines may run parallel or cross at right angles. UV cracking has a dry-rot or alligator-skin look in advanced stages. A fingernail may catch.

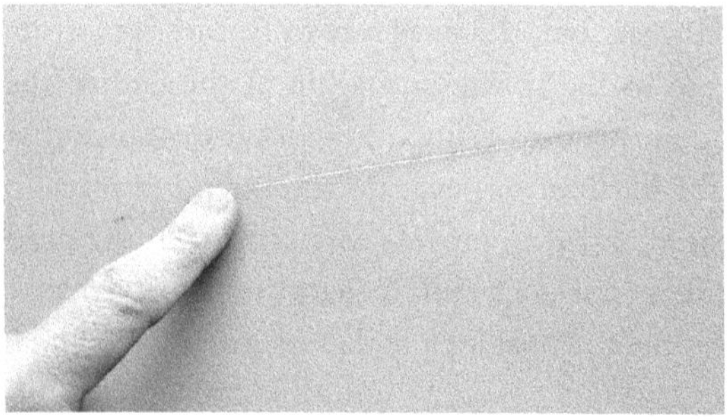

Figure 2.29. Scratch.

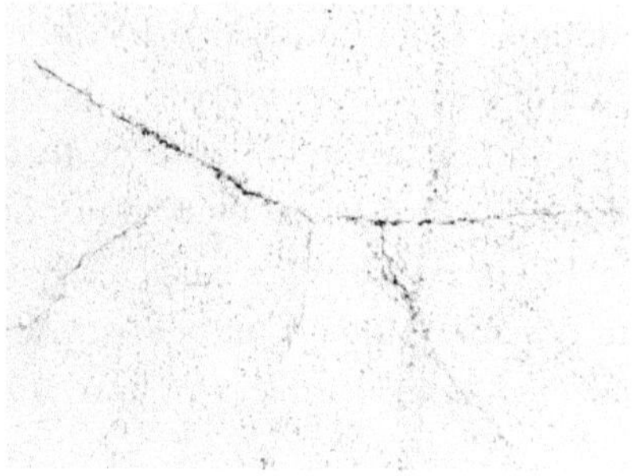

Figure 2.30. Crack.

Figure 2.31. Crazing.

Marking the Tank with a Water-Soluble Marker

Crazing may signal UV damage. UV crazing, which is very difficult to see, forms in areas where the tank gets maximum sunlight exposure. The lines become more visible if you "color" the tank with a water-soluble marker: Rub the marker over several 6-inch by 6-inch sections on the sides of the tank that are exposed to sun, on its top, and around fittings. Quickly rub off the ink with a dry cloth or paper towel. The ink left behind has penetrated the surface of the tank.

Crazing is one of the first signs of deterioration, so tanks with crazing should be checked often. If rubbing the ink off reveals no obvious signs of crazing or cracking, the tank probably is good for another season of use. If the ink reveals cracking or spiderwebbing where the lines go in all directions, classic UV radiation damage is indicated. Advanced deterioration to the plastic presents a checkered or "dry rot" appearance, indicating loss of elasticity. A tank displaying such symptoms should be replaced—or at least not used for fertilizers or chemicals. Consider using crazed tanks for water only.

The appearance of parallel lines signals early UV damage and the need for continual inspections. Tanks with parallel lines in the plastic around fittings should be replaced immediately or used for water only.

Figures 2.32–2.33. Using a marker to identify cracks and crazing from UV exposure.

Candling: Visual Inspection with a Light

Candling consists of placing a bright, cool light source inside a poly tank while conducting a visual inspection from the outside (do not use a hot lamp, as it could melt the tank). Defects and cracks usually show up as areas or lines of different light intensity.

Repeat this procedure with the light on the outside of the tank and someone looking through the fill neck or manway. Do not enter the tank. A camera, camcorder, or other optical device may be helpful in recording the inspection from the top of the tank.

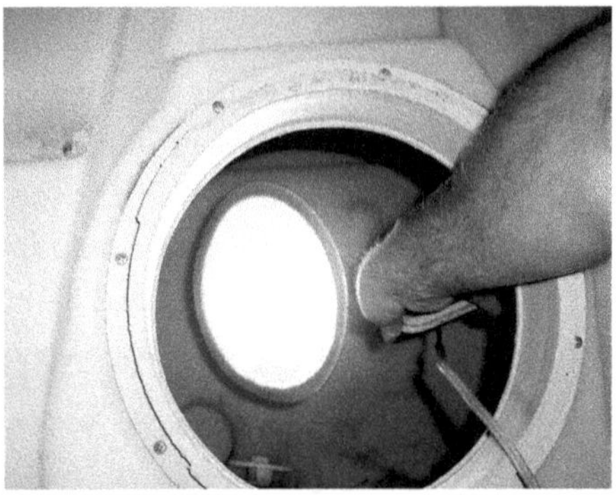

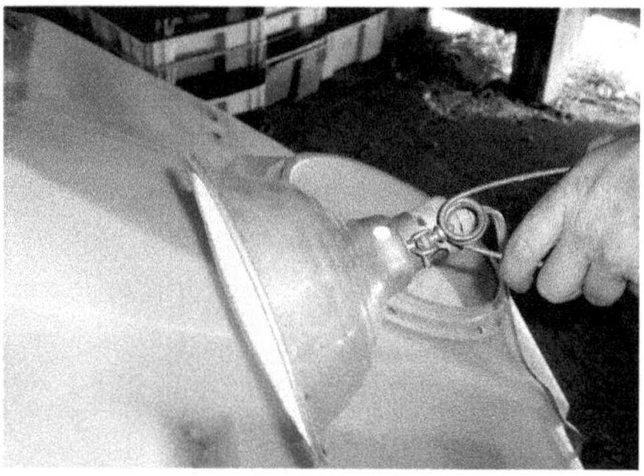

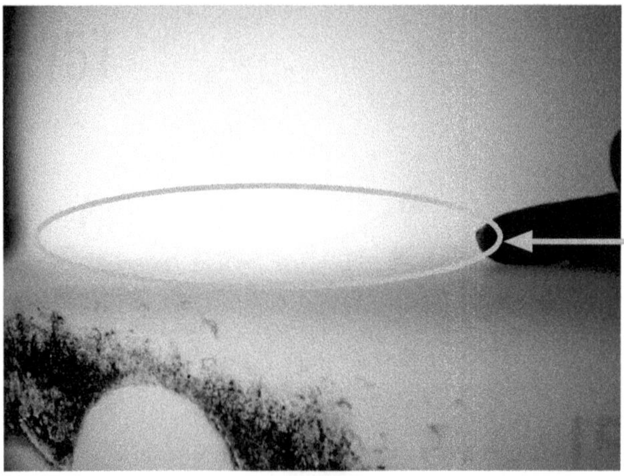

Figures 2.34–2.37. Use a light source to help locate defects when inspecting poly tanks. The technique is called *candling*.

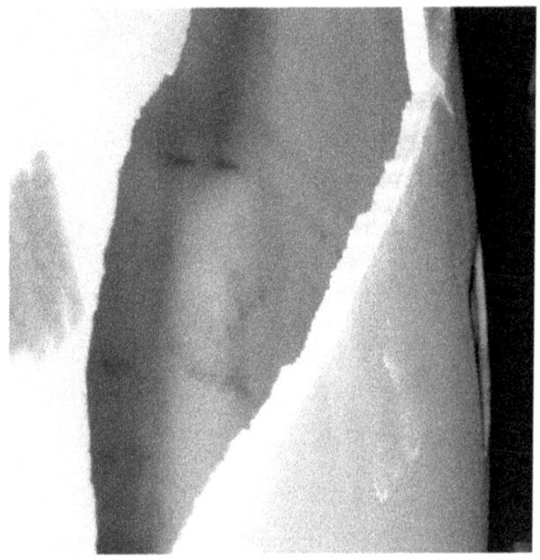

Figures 2.38–2.39. This tank showed significant UV damage; when tested using a bat, it broke open. The plastic was brittle across the width of the tank (right image).

Hitting an Empty Tank with a Baseball Bat

An empty tank showing UV cracking can be further evaluated by striking it with a baseball bat. Most people are afraid to hit their tank with a bat, fearing that they might break it, but that's just it: If it breaks, it should not be in service. Cracking an empty tank with a bat is better than having it break when filled with fertilizer or pesticide.

A good tank has the flexibility to bend outward as it is filled and inward as it is emptied. Tanks that are brittle (i.e., that exhibit excessive or advanced cracking) have lost the ability to flex under pressure and to rebound when impacted. The brittleness of an empty tank can be tested with a solid swing of a baseball bat against an area where signs of cracking were discovered during the water-soluble ink inspection. Hit the tank along the sides and top where it receives the most sunlight; then check the tank for signs of breakage. It is impossible to crack a good tank using this method because the polymer is strong and resilient; if the tank cracks or breaks open when hit by a bat, you may have saved yourself from disaster.

Repairs to Poly Tanks Unlikely

Most poly tanks with crazing or cracks, and those made from cross-linked polyethylene, cannot be repaired. Repair of polyethylene tanks is risky and not recommended. Manufacturers sell kits for repairing small surface scratches or pinholes in high-density linear polyethylene tanks. But few adhesives or caulks adhere to polyethylene, and other means such as plastic welding are best left to professionals. Even if a tank can be repaired, there is no guarantee that the repair will hold once the tank is filled. A repaired tank should be used for water only. It is best to replace a damaged tank.

Reuse or Recycling of Old Tanks

High-density linear polyethylene (HDLPE) can be recycled into other products. Conversely, high-density cross-linked polyethylene (XLPE) cannot be remolded and recycled.

All tanks, whether recycled or deposited in a landfill, should be triple rinsed prior to disposal. After rinsing, they should be broken or cut into smaller pieces before being recycled or put in a landfill. Contact your local recycling center for information on recycling HDLPE tanks.

Figures 2.40–2.41. Water tanks may be reused for other purposes instead of being cut up for disposal.

CONCLUSION

High-density polyethylene tanks have been used successfully by growers and commercial pesticide application businesses for years. They are less expensive than stainless steel and fiberglass tanks and offer ease of movement when empty, along with rust resistance. Polyethylene tanks are low maintenance and relatively reliable for storing and transporting agricultural and commercial liquids.

While the benefits of poly tanks are well established, the risk of tank failure is real. Like any piece of equipment, poly tanks need to be inspected and maintained to ensure that the benefits of use outweigh the risk of tank failure and product release.

Experience has shown that environmental, management, and design factors determine how long a tank will last and that annual or biannual inspections help prevent the unexpected and potentially catastrophic release of contents due to tank failure.

These maintenance strategies help extend a poly tank's useful life:

- Select the right tank for the intended use, based on the specific gravity rating.
- Use the tank as intended—vertical tanks for storage, horizontal tanks for storage or transportation.
- Protect the tank from UV radiation.

Never purchase a used tank without knowing its history. If you intend to buy an inexpensive used tank that appears in good operating condition, take the time to conduct a proper inspection, and secure information from the manufacturer if at all possible. Without these precautions, you may encounter sudden tank failure resulting in an expensive spill and cleanup, reporting obligations, and costly downtime.

3

Fiberglass Tanks

With contributions from Kevin Baalmann, Steve Hawkins, Jim Flanders Jr., and Kevin Leigh Smith

Growers, commercial industries, and government agencies depend on tanks to store, transport, and apply fertilizers, pesticides, and salts. Although tanks made entirely of plastic resins (poly tanks) have long been popular, interest in tanks made of fiberglass has been growing. Fiberglass tanks often last longer than poly tanks and can be repaired. This chapter examines fiberglass storage tanks: what benefits they offer, what to consider before buying them, and how to properly use, install, and maintain them.

Fiberglass tanks cost more than poly tanks, but they can store more product, which can save money. Many farmers are expanding their operations and need to purchase and store larger quantities of liquid fertilizer on the farm. For example, a 1,500-acre grain farm could use as much as 80,000 gallons of fertilizers in a year. Fluctuating fertilizer prices and product availability are two reasons influencing the trend of storing more product on the farm. More storage capacity gives growers the opportunity to buy product when costs are low. Having inventory on hand ready to use at one's convenience is another factor.

For some, the question is how long it will take for the savings of buying in bulk to pay for the more expensive fiberglass. One grower calculated it would take three years of buying bulk fertilizers to pay for the tanks, fittings, and labor for his 700-acre grain farm.

It's not just farmers who are interested in fiberglass tanks. For liability reasons, some commercial application companies have been switching from carrying pesticides in large poly tanks on their trucks and trailers to smaller, more durable fiberglass tanks. Companies often replace poly tanks that held 500 to 1,000 gallons of product with fiberglass tanks that hold 100 to 500 gallons. Carrying less product on the truck means that any spills are smaller and easier to handle. In the long run, that can mean less pollution, fewer problems, lower environmental cleanup costs, and fewer headaches.

But if truck tanks are getting smaller, why use fiberglass? Some companies see the move to fiberglass as an additional insurance policy. Many commercial insurance policies do not contain pollution coverage, may explicitly exclude pollution liability, or have very limited coverage. The more durable fiberglass tanks may be more likely to remain intact in an accident than poly tanks—the fiberglass may crack but not break. That added peace of mind is one reason some are paying more for fiberglass.

HOW FIBERGLASS TANKS ARE MADE

Fiberglass tanks are actually fiberglass-reinforced plastic (FRP)—that is, they are plastic tanks that have been reinforced with fiberglass, similar to the way that rebar strengthens concrete. There are two methods for making these tanks: the chopper gun method and the filament winding method. Understanding how fiberglass tanks are made will give you some insight about the benefits they can provide.

Figure 3.1. A manufacturer prepares the mold for the top of a fiberglass tank.

Figure 3.2. This tank bottom has just been popped off its mold.

Chopper Gun Method

With the chopper gun method, the manufacturer uses metal molds to form the sides, top, and bottom of a tank. They do this by passing a continuous strand of glass through a chopper gun, which cuts the glass into 3- to 5-inch pieces. Simultaneously, the gun mixes liquid resin and a hardener with the glass as it leaves the gun. The mixture is sprayed around the outside of the mold.

Figure 3.3. This chopper gun mixes resin with extruded glass as it leaves the gun.

The resin a manufacturer uses for a fiberglass tank is quite important. The resin is the liquid, plastic-like material that will hold the tank's contents and prevent it from seeping through the tank's walls.

Figure 3.4. A bale of glass used in a chopper gun. The glass strand is actually several strands of glass, much like the many fibers in a rope.

Not all fiberglass resins are the same. There are resins that are compatible with salts and oils, with fertilizers and pesticides, or with acids. Depending on the intended application of the finished tank, the manufacturer mixes the appropriate resin with the fiberglass. Always store products in tanks made with compatible resins—placing incompatible products in the tanks could cause the tanks to fail prematurely.

The fiberglass is made of several layers, and it can take up to six passes with the chopper gun to form a single layer. The uniformity of the wall depends on the gun's operator. Depending on the resin system, it can take 10 to 40 minutes for each layer to harden. The number of layers in a finished tank depends on the size and specific area of the tank. As a general rule, tank sidewalls are 0.5 to 0.75 inches thick, and tank bottoms are up to 1.125 inches thick.

Figures 3.5–3.7. A newly formed fiberglass tank (foreground) is being pulled off the mold (background). After the liner is removed, the tank will be cut to length, and the top and bottom pieces will be put on.

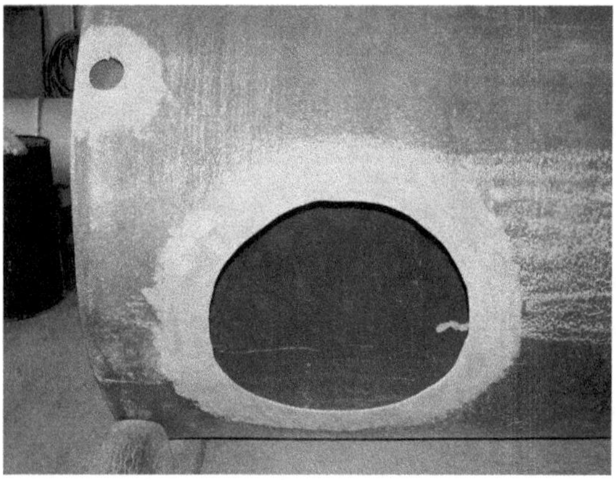

Figure 3.8. The openings for the manway and fittings have been cut in this tank.

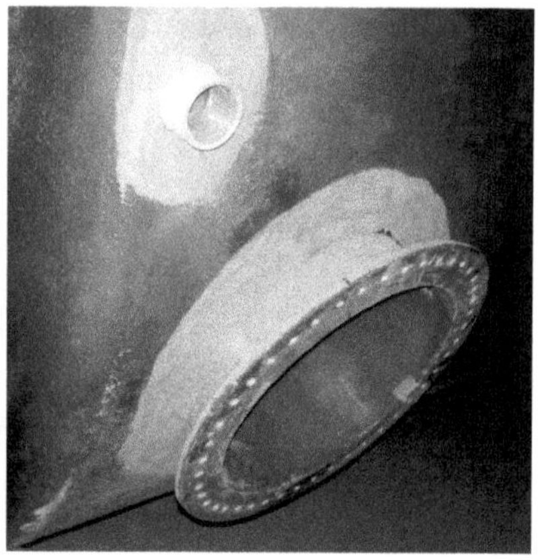

Figure 3.9. The fitting and manway are in place and ready to be glassed in on this tank. The white areas around each opening are where grinding was done to help the fiberglass and resin adhere, or bond, to the tank.

Figure 3.10. Once the tank is completed, the fiberglass seals the gap between the tank shell and flanged manway fitting. The darker circle in this image is a cosmetic effect from the use of different batches of resin.

Figures 3.11–3.12. These photos illustrate how manufacturers attach flanges for the cleanouts/manways. They take a piece of chopped glass mat soaked in resin, place it in position, roll out the air bubbles, and apply a woven mat for an additional layer of strength.

The tops, bottoms, and middles of tanks are fabricated separately. It can take an hour to fabricate the tops and bottoms of tanks that hold 5,000 to 10,000 gallons, and up to two hours to make the sidewalls.

A finished top and bottom are then attached to a sidewall with rivets or other fasteners. The manufacturer then sprays the final layers over the exterior of the tank so that the rivets and individual pieces are no longer distinguishable from each other.

Figure 3.13. A tank that has been gel coated. The manufacturer closed off the fitting with masking tape so the threads in the fitting were not harmed when the gel coat was sprayed. Manufacturing a new fiberglass tank can take up to 10 hours, depending on its size and the customer's specifications.

Openings (such as a large 24-inch hole for a cleanout/manway and a smaller 3-inch hole for a threaded fitting) are added after the tank is completely assembled and glassed.

The last layer applied to the tank is a paint or gel coat that helps protect the resin from ultraviolet radiation. Gel coat is a resin paint that is thicker than conventional paint. The tank's final color depends on the color of the paint or gel coat. Without gel coating, ultraviolet rays degrade fiberglass tanks, just as they do poly tanks.

Filament Winding Method

A second option for making fiberglass tanks is called the *filament winding method.* With this method, the glass is wound around a rotating tank mold as a mat of long, continuous glass from a spool. Some companies hand-apply a fiberglass mat around the seam between the two pieces. Instead of mixing resin and hardeners with the glass, the filament winding method only adds resin to the bottom of each layer. Rollers press the fiberglass mat in place and remove air bubbles. The number of layers of resin and glass in a finished tank depends on its size.

Figure 3.14. In this photo fiberglass is being wound onto a mold. The part to the left has already been fiberglassed (brown), while the right end shows the mold (blue). When complete, this tank will measure 10 feet wide and 15 feet tall and hold 8,800 gallons.

Figure 3.15. This photo shows the inside of the mold (foreground). Fiberglass has already been applied to the far end of the mold (in the background). Once the tank dries, the mold will contract and allow the tank to be removed.

Next, the manufacturer overlays individual strands of fiberglass on top of the mat—this is called the *chop hoop layer*. The fiberglass strands look like ropes, but each rope is made up of about 60 individual threads. The rotating mold then pulls the strands and further wraps the tank.

Once the mats and resin dry, the mold will contract and allow the tank to be removed. The manufacturer places a steel ring near the top opening to prevent the open end from collapsing.

The last step in this process is to put the top on the tank, which is held in place with pop rivets. A chopper gun adds the remaining fiberglass layers. Once this is completed, the manufacturer cuts any openings for attachments.

Figures 3.16–3.19. (Top left) The manufacturer places a tank bottom on a mold (also called a *mandrel*). (Top right) Next to the mold, spools of glass strands are lined up on racks. (Bottom left) When the tank is built, glass strands are wrapped around the spinning mandrel. (Bottom right) After that, the manufacturer applies the chop hoop layer, laying down "ropes" of glass made of about 60 strands each.

BENEFITS OF FIBERGLASS TANKS

Fiberglass tanks have many positive attributes:

- They can be custom built.
- They have good longevity and durability.
- They can resist ultraviolet radiation longer than poly tanks if they have a gel coating.
- They can be used with salts, fertilizers, or pesticides.
- They can be sanded and repainted.
- They can be repaired and put back into service.
- Baffles can be installed in transport tanks to reduce sloshing of product.
- They can have heavier valves installed because the fiberglass tank frame is more rigid.
- They can be equipped with mechanically driven agitator paddles to keep mixtures suspended.
- When they fail, they generally seep rather than burst.
- Under normal use, high and low temperatures have little impact on their service life.
- They are able to withstand rugged off-road conditions.
- The opaqueness of the walls reduces algae growth.

On the downside, fiberglass tanks cost more than poly tanks, they typically have to be ordered rather than purchased off-the-shelf (and that often means there are shipping costs too), and the walls are opaque (so you can't see what's inside).

Fiberglass tanks that hold less than 10,000 gallons are more expensive than comparable poly tanks. However, as the tanks get larger, the price differential is not as dramatic. For smaller tanks commonly used to transport chemicals, the higher cost of fiberglass is more dramatic.

If you need a tank immediately, that may be difficult for a fiberglass tank. It can take two to five weeks between order and delivery.

The multiple layers of fiberglass that prevent sunlight from penetrating the tank also prevent you from seeing through the tank walls. That can be a problem if you can't see sediment and sludge accumulating on the bottom, or if you can't see whether a fertilizer (such as 10-34-0) has salted out during the winter.

QUESTIONS TO ANSWER BEFORE BUYING

Fiberglass tanks are engineered to hold the pressures exerted by the fluids inside them. The heavier the fluid stored in a tank, the thicker the fiberglass shell will be in high-pressure areas. For example, on a 10-foot-tall tank that has 12-foot diameter, a manufacturer may add an extra piece of woven glass material to the first 6 feet from the bottom. This adds strength to the portion of the tank that is likely to experience more pressure.

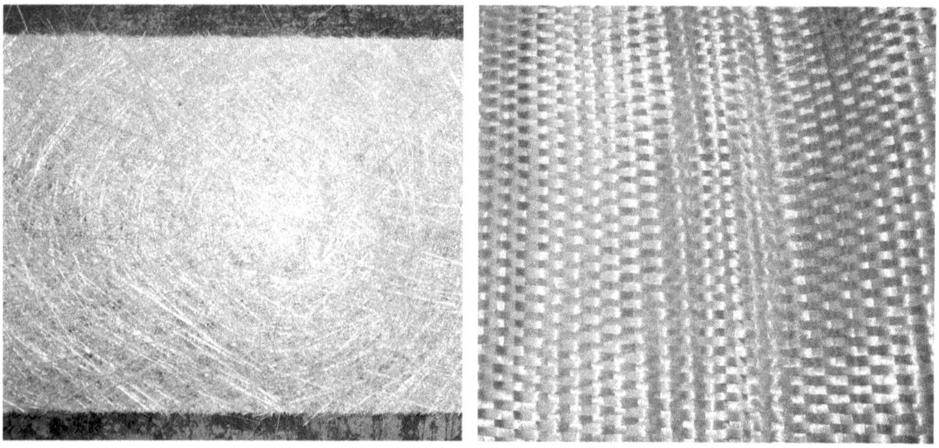

Figures 3.20–3.21. A chopped mat (left) is made of glass that has been chopped and laid into a mat. Woven mats (right) are made of glass strands that have been woven together.

Because fiberglass tanks are made to order, customers must make a few decisions before ordering. This section provides 12 questions you should answer when deciding whether fiberglass tanks are appropriate for your operation. See also chapter 1 for a fuller discussion on choosing and installing tanks.

1. What Tank Design Will Work Best for Me?

There are three common fiberglass tank designs: flat-bottom, cone-bottom, and dish-bottom. Flat-bottom tanks are harder to drain, so are best used if you plan to store just one product in them. Cone-bottom tanks are better if you need to hold different products during a growing season—such as 28 percent fertilizer, starter fertilizer, water, and pesticide. Cone-bottom tanks taper toward the center bottom, which allows for full drainage and lets you more easily clean tanks between fillings. Dish-bottom tanks are a hybrid of flat- and cone-bottom tanks. See chapter 1 for more on tank shape.

2. How Much Storage Should I Plan For?

Often, the amount of storage capacity you should have depends on whether your fertilizer or product vendors can offer you price breaks for bulk purchases. Another factor is how much inventory you want on hand at the start of the season and how much you will need over the growing season.

3. Is Bulk Containment Required?

State and federal regulations require containment structures if you store certain quantities of pesticides or fertilizers. This means that you must include the cost of containment when you analyze the costs of storing large volumes of product.

Figure 3.22. Actual measurements will help determine how many tanks can fit within a specific space.

4. What Size Tanks Do I Need?

The area of your flat space and the size of the tanks will determine how many tanks you can place on the site. One large tank may take up less surface area than two smaller tanks. For indoor tanks, the amount of storage may be limited by the size of your building. Size constraints are tighter for indoor tanks than for outdoor tanks.

5. What Materials Will the Tank Hold?

The resins used to manufacture a fiberglass tank are designed for use with a group of specific chemicals. For instance, fertilizer requires a different resin than those used with pesticides and salts. The resin, corrosion liner, and wall structure may vary depending on the product the tank is designed to hold.

It is important to consider all the possible products the tank may hold during its life, which allows the manufacturer to select the correct resins to fit your needs. When examining whether fiberglass tanks are feasible, be sure to discuss with the dealer what you will store in them.

6. Is the Expensive Gel Coat Necessary?

A final coating on the outside of a storage tank that protects it from the elements does add additional expense. It can cost $200 to $400 to gel coat a 10,000-gallon tank. Some growers skip the coating if they will store their tanks indoors. However, if you store a tank outdoors where it will be exposed to ultraviolet light, the gel coat is a must.

7. How Much Will the Fully Loaded Tank Weigh?

When tanks are full of product, that places a significant load in pounds per square foot on the concrete pad. That load will not be an issue if the concrete pad is designed properly and its bedding has the proper gravel base. In addition, the area under the concrete must be well drained to prevent water from freezing underneath it, which can cause it to crack and fail. If placing tanks on existing concrete, make sure the concrete is strong enough to handle the load before ordering tanks.

8. How Will the Tanks Be Unloaded?

Delivery trucks may come equipped with knuckle booms to unload the tanks, but don't assume this. Some companies do not have this capability, which requires the customer to have someone else bring out their equipment. If you must hire someone to unload the tanks, that will add significant cost.

9. What Accessories Are Important to Me?

Manufacturer catalogs list many different options you can purchase for a fiberglass tank. One of the more important considerations is where to place the manway, since it will allow access to the inside of the tank for cleaning.

Don't forget to get the proper-size vent for your tank. Tanks without vents (or with undersized vents) collapse when adding or removing fluids, or even during heat expansion in the spring.

10. Where Do I Want the Plumbing Fittings Placed?

Since fiberglass tanks are custom-made, you can place plumbing fittings where you want them. Consider installing two separate fittings that will allow the tank to be filled and unloaded at the same time. Adding a second flange or fitting to your custom tank will also allow recirculation of the contents. Remember that if you place fittings too low, you may not have enough clearance from the ground to fit your valves.

11. What Type of Inventory Monitoring Gauges Do I Want?

The choices for inventory monitoring gauges may include a sight gauge, a float, or even an ultrasonic system. State or federal law may regulate the gauge you select, so be sure to ask plenty of questions.

12. What Tie-Downs Do I Need?

Today there is more emphasis on securing outdoor tanks. Strong winds can knock tanks over, which can crack tanks or cause them to fail. Anchoring is also important to prevent tanks from floating in water.

Figure 3.23. A fiberglass tank provides a stronger wall that allows heavy valves to be used.

Figure 3.24. Outdoor tanks need to be secured against high winds, which could easily push them over, resulting in a spill and cleanup.

QUESTIONS TO ANSWER ABOUT VEHICLE TANKS

Fiberglass tanks for vehicles are specifically engineered for highway use. The end caps are thicker to handle the force that fluids sloshing back and forth exert on the tank walls. Before buying a fiberglass tank for a vehicle, be sure to answer the questions below.

Do I Want Baffles?

You should have baffles installed in tanks that will be used for transport. Liquid loads will shift during transport, especially when tanks are partially loaded. If a tank does not have baffles, the fluid will move to the front of the tank, which can push the truck forward and make it more difficult to stop. The sloshing liquid also places tremendous pressure on the tank walls. You can add baffles to any size tank as a way of slowing down that sloshing fluid in the tank. Generally, the larger the tank, the more important baffles are.

Figure 3.25. This vehicle-mounted tank is being outfitted with full-length and -width baffles.

What Tank Shape and Dimensions Do I Need?

Fiberglass tanks for highway transportation come in different heights, widths, and lengths. Calculating the available space on a truck or trailer will help narrow down which tank to purchase. Dimensions are available from tank manufacturers.

Elliptical tanks have lower centers of gravity than round tanks. When mounting a tank on a truck with a frame that is higher off the ground, consider an elliptical tank. This will keep the vehicle's center of gravity lower to the ground and make it less likely to roll over. This is why it is important to custom-make the transportation tank so that it has a low profile on the truck.

How Much Weight Can I Haul?

A tank's weight can be an important (and often-overlooked) piece of information. Fiberglass tanks are heavier than poly tanks. Not only are fiberglass tanks heavier, but the skids they are mounted to also add weight to the vehicle. Ask the manufacturer to provide you with the weight of the tank, skid, and any attachments you are considering.

Make sure the weight of a full tank, its mounting skid, pumping equipment, and the empty truck weight do not exceed the truck's gross vehicle weight rating. If the tank is on a trailer, the total weight of the load and trailer should not exceed the capacity of any of the hitch devices, the trailer's gross vehicle weight rating, or the truck's maximum towing capacity. More information about trailer safety is available in volume 2.

Do Multiple Compartments Offer Any Advantages?

A split tank can be divided into completely separate compartments. For example, one compartment can hold an herbicide, while another holds an insecticide. This allows you to use one tank with multiple hose reels to deliver different products at the same job site. The additional cost of multiple compartments in the same tank might be offset by not having to buy two tanks and two pumps to deliver the same service. Consider how products will actually be delivered in your operation.

How Many Accessories Can Be Added?

You can add many important accessories to fiberglass transportation tanks, including baffles, sight tubes, threaded flanges, skids, sparge tubes, sumps, lift brackets, ladders, steps, tool boxes, manways, fill ports, fittings, and lights. One of the more important accessories to consider is an agitation system. Without agitation in the tank, some products will ultimately settle out on the bottom of the tank. Manufacturers can outfit tanks with fittings, pipes, and agitators (for example, paddles) at the time of construction.

Another important question is how to estimate the volume of fluid in the tank. Because fiberglass tank walls are opaque, it is nearly impossible to see how much product it contains. Therefore, some tanks have clear sight tubes, along with graduated marks that indicate how much fluid remains in the tank. However, this clear tube will break down over time or break off. Another problem is that the clear tubes can discolor and stain, which makes it impossible to see the fluid level in them. Some manufacturers offer tanks that have built-in clear windows with gallon markings along their lengths.

Figure 3.26. A translucent area with graduated markings allows you to estimate a tank's contents.

INSTALLING STORAGE TANKS

Unloading storage tanks and moving them to their final locations can be major operations. Knuckle booms that unload tanks from semitrucks make your job much easier. Each tank will have lift points near its top and along the sides; it can be physically lifted straight up and then placed on the ground. The boom operator then switches the hooking to the top and sets the tank up vertically. A forklift then moves the tank indoors.

Valves and other plumbing fixtures are installed after tanks are delivered. While there are many choices for valves and fittings, be sure to choose materials that are compatible with the materials that will be stored in the tank. In addition, ask the manufacturer how much compression force (torque) you can apply on the fiberglass as you tighten the fittings and valves.

Figure 3.27. Fiberglass tanks allow better valves to be used rather than the lighter valves used with poly tanks.

Likewise, be careful not to overtighten the tie-downs on fiberglass tanks. These are glassed in, like all other additions to the tank, and it's possible to tighten the cables so much that they pull the metal hooks away from the tank. Tighten the cables just enough to keep the tanks in place during a storm. Check the cables annually for tension, and retighten as needed.

Figure 3.28. Tie-downs can secure tanks in the event of flooding or high winds.

INSTALLING AND INSPECTING VEHICLE TANKS

Normally, fiberglass tanks have skids that have been glassed to the tank. You can bolt the skids to a truck or trailer frame for a more permanent attachment. If you secure the tank to the frame, use stainless steel bolts of grade 5 or greater. Securing a tank to the frame prevents the tank from moving and keeps the tank from rubbing against other objects.

If you will occasionally remove a tank during the season, secure the tank with web tie-downs. Be aware the tank may still move, which may allow it to come in contact with sharp objects. For more information about tie-downs, see volume 2.

Figures 3.29–3.30. (Left) A soft insert separates this tank from metal. (Right) The soft insert for this tank fell out and the tank was damaged when it rubbed against the metal frame.

Another securement option is to place fiberglass tanks in saddles. If you use a saddle, it is important to make sure the tank does not rock back and forth against the metal frame during transport, which would allow the fiberglass tank to rub against the metal. As an extra precaution, you can insert a piece of rubber liner (or some other soft material) between the tank and the metal frame.

You need to inspect the tank for leaks each day. If a fiberglass tank has a crack, it will seep product very slowly. Remove and thoroughly inspect every tank (and the pumps and hoses) each offseason.

REPAIRING FIBERGLASS TANKS

While fiberglass tanks can withstand extreme temperatures, over time ultraviolet light will degrade the resins that hold the glass in place. As the paint or gel coat breaks down and gets thinner, small glass particles that once were embedded in resin will begin to stick out.

Ultraviolet damage is a surface phenomenon. Simply applying a coat of paint or gel coat will refasten the fibers into the paint or gel coat. Gel coatings are more expensive than regular paints but are preferred because they are resin based. When you recoat a tank, it is best to completely restore the entire tank instead of just a section that needs attention. Lightly sanding the entire tank before applying the paint or gel coat will help the coating adhere to the surface better.

Another sign that a tank may need attention is when the walls on the outside of the tank show a spiderweb pattern. Such patterns are often caused when something hits or bumps a tank (such as a forklift). You will need to repair these cracks from the outside.

If you observe seepage through a fiberglass tank or wet areas on an outside wall, that indicates that the liquid from the inside has broken through the resin. The liquid may be following a small crack in the tank wall.

Figure 3.31. This storage tank was not anchored and blew over when a storm passed through. The fiberglass tank struck the concrete containment wall, which produced a 4-foot-long crack. In the brown sanded and fiberglassed area, the lighter-colored crack has been repaired but can still be seen.

Repairing an internal leak or external crack takes more patience than just painting over a tank's surface. If you see a crack from the outside, fix it from the outside. If you see seepage in an area, you'll want to fix that from both the inside (if you can access that area) and the outside. The best repairs are done both inside and outside.

Fiberglass repair kits often contain three basic materials: resin, hardener, and fiberglass patches—the same materials used in the tank's original construction. The resin will be a thick liquid, while the hardener has the consistency and color of water.

Pay attention to the amount of hardener you mix with the resin. One gallon of resin takes only a few tablespoons of hardener to set up. When combined, the hardener and resin undergo a chemical reaction that turns the resin into a gel-like material. As it gels, the resin gets hotter. Too much hardener can create a burn spot, which in turn can create leaks. Carefully read and closely follow all repair kit directions.

How should you go about repairing a tank? The rule of thumb when you patch an area is to make a patch that's at least twice the size of the cracked or seeping area. For example, if you have a 4-inch by 4-inch spiderweb or wet spot, sand and patch an 8-inch by 8-inch area around it.

Before patching a tank, completely drain the tank and let it air-dry. This is a crucial step—fiberglass applied to wet surfaces will not cure properly (and you'll have to repair your tank again). Cut the fiberglass to the proper dimension to cover the sanded area. Once you mix the resin and hardener, they immediately begin to react with each other. Add the precut fiberglass repair strip to the resin until it is completely saturated. Remove the fiberglass and squeeze it out like a wet towel.

Then apply the repair piece of fiberglass to the sanded area, then roll out any air bubbles while the area dries. The patch will be tacky to the touch and harden over time. It can take 10 to 40 minutes for the repaired area to dry. The actual curing time depends on the temperature in the shop and the amount of hardener used with the resin. Once the fiberglass dries, apply a good-quality paint or gel coat over the surface. A general rule of thumb is to allow new fiberglass to cure two to four hours before allowing fluid to touch the repair. Again, always follow the instructions in the repair kit.

Figure 3.32. A simple fiberglass repair. Painting over the fiberglass patch would have helped protect it against ultraviolet light.

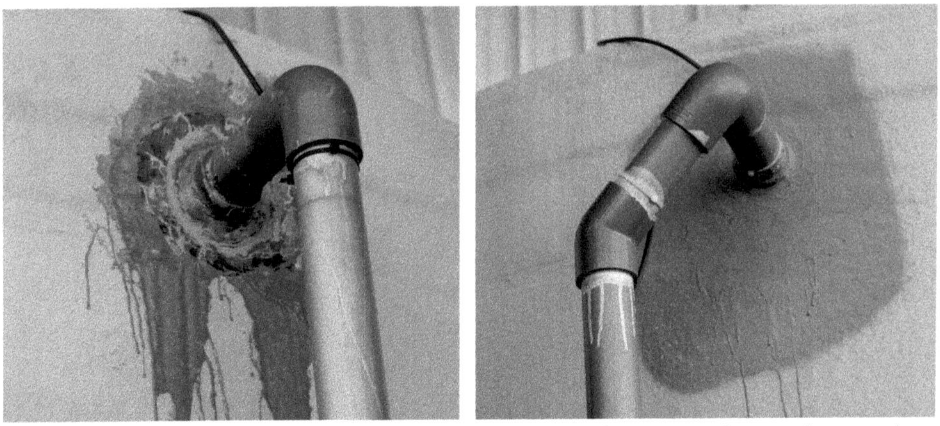

Figures 3.33–3.34. (Left) A sloppy tank repair job. (Right) This tank was properly ground, repaired, and repainted.

Cracks in the corners of transport tanks can be difficult to repair. Generally, corner cracks are caused when the protective bumpers between the skids and tanks have come out, slid down, or worn out. These conditions allow fiberglass tanks to rub against metal, which causes a weak spot or crack. Weakened areas will look like they have been sanded and will often have a dull, worn appearance.

For such cracks, try to fix the tank from both sides. If the crack is on the end of the tank where the fill is located, you will be able to fix it from both the inside and outside. However, because of the baffles, the crack may be inaccessible from the inside. If that is the case, you can choose to only fix the tank from the outside, remove the baffles so you can repair it from the inside, or have a tank repair shop fix the tank.

BUYING USED FIBERGLASS TANKS

Because of their durability and long service life, you can buy used fiberglass tanks, usually for a fraction of the cost of new tanks. But if you buy a used tank, be sure to do your homework.

Most used tanks are sold at auction. Before you bid, know what material you want to put in the tank and what resin was used to build the tank. Determining the resin can make buying a used tank difficult. You can't tell by looking at a tank what resin was used, so you cannot know what the tank was meant to hold. If you can't find out what resin material is in a used tank, then it is not a good idea to purchase it.

Examine used tanks and look for manufacturer stickers. These stickers may include serial numbers. A call to the manufacturer with an identification number may provide important information about the tank's age, its construction, and the types of resins it uses. It will be difficult to know if a tank will work in a specific application without the manufacturer's tag.

Figures 3.35–3.37. Make sure to inspect any used tanks you are considering purchasing. The top of this tank has an obvious crack. Also notice that the top is beginning to separate from the sidewall, and the fiberglass is easily seen on the tanks. More important, this tank has been sitting outside, and it does not have a gel coating to protect it against UV damage. If you were to buy this tank, you would have to consider the costs to have the top seams resealed, the crack repaired, and a gel coat applied. While these are rather easy fixes, there is still an expense associated with these repairs.

Also consider that some of the previously stored material may have penetrated into the tank itself. If the used tank stored fertilizer at a bulk fertilizer plant, then it's probably a good tank to store your fertilizer. Or if a turf company is selling a fiberglass tank that stored a pesticide, then buying the tank for your turf operation is probably fine. In other words, find used tanks that have contained the same materials you will be using. The odds of a used tank being suitable for your application decrease when you buy a tank from another industry. For instance, it would be risky to buy a used tank that stored 2,4-D if you plan to use the tank in your horticultural or nursery business.

When possible, take a strong flashlight and look inside the tank to make sure it is in good working condition and free of residue or any defects.

CONCLUSION

Fiberglass tanks provide an excellent alternative to poly tanks. While they cost more to purchase, fiberglass tanks over time provide a long-term solution to storing products on the farm and at the retail facility.

4

Steel Tanks

With contributions from Dennis Nowaskie, Brad Peas, Dean Sink, Ernie Fuller, Gerry Bunn, Chuck Kominski, Todd Cobbs, Gerry Blasé, Andy Schmerge, and Kevin Leigh Smith

Steel tanks have long been used to both store and transport pesticides and fertilizers. Their availability, cost, and durability are reasons why they remain popular. But the choice of stainless steel or mild steel tanks often depends on the products being stored and transported. This chapter explores purchasing and using mild steel and stainless steel tanks for farms, businesses, and public agencies. It provides important questions you should ask before you buy, offers tips to properly install tanks, and establishes guidelines for inspection and maintenance.

MILD STEEL TANKS

You will find steel in many products: from bolts, screws, and nuts to high-strength cables that hold up suspension bridges. Steel is composed of iron ore blended with a small percentage of carbon. There are many different processes for manufacturing and engineering carbon steel. That is why there is a grading system for steel that differentiates closely related products. The percentage of carbon in the final product determines the steel's grade.

The grades of carbon steel are mild, medium, high, and ultra. Mild grade is commonly used for tanks that store fertilizers and fuels. Mild steel contains 0.2 to 2 percent carbon by weight.

Mild steel is preferred for tanks because of its cost, durability, and versatility, which allow it to be used across diverse industries. All steel tanks are made up of steel rings stacked on top of one another and welded together. The strength of each ring and weld is important in tank construction.

However, mild steel does have a drawback: Some materials can easily corrode it. Over time, corrosive materials initially cause small, shallow pits in the tank that then become wider and deeper. Eventually, these corrosive materials eat their way through the tank and result in an unwanted release of the material.

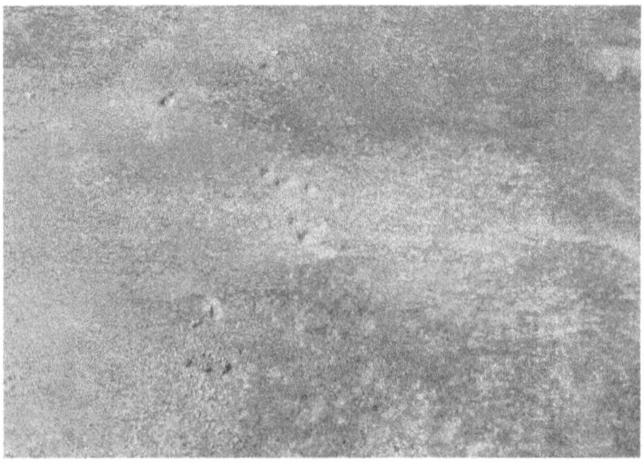

Figure 4.1. If a tank just has surface pitting (pits that aren't too deep), a good coat of paint can generally allow you to get a lot more service from the tank.

Rust is a form of corrosion commonly associated with mild steel tanks. Rust occurs when unprotected steel reacts with water and oxygen. It leaves an unsightly orange stain on the tank's exterior and interior. Left unabated, rust damage will weaken mild steel and result in leaks, tank failures, or even tank frame collapses.

Is Mild Steel Right for You?

When you purchase a mild steel tank, you should always consider how to delay the development and spread of pitting and rust. Here are 16 questions you should always ask before choosing a tank. The first six address pitting and rust.

1. What is the pH of the product you will store in the tank? This is one of the most important questions to ask. Acidic agents like battery acid (with a pH value of 2) are very destructive to steel. The lower the product's pH, the more acidic and destructive the liquid is to mild steel. Corrosion will be much worse if the intended product has a pH value of 6.5 or less. Many commonly stored liquids have a pH value just above or just below 7. For instance, the pH of liquid nitrogen (28-0-0) is 7.2, that of ammonium phosphate (10-34-0) is 6.0, and that of diesel fuel is 5.6. However, pH isn't the only indicator: Ammonium thiosulfate (12-0-0, 26% sulfur) is very corrosive to carbon steel, and its pH is 7.2 to 8.0.

Always determine the pH of the product you plan to store before buying any tank. Ask the tank manufacturer about problems that could occur by storing a product with that pH in the tank.

2. How thick do you want the walls and floors to be? A standard mild steel tank has walls and floors that are ¼ inch thick, but the thickness depends on the tank's volume and height. Bigger tanks have thicker walls. For example, a 30,000-gallon mild steel tank with a diameter of 12 feet and height of 35 feet may have a standard ¼ inch floor but have first and second rings with walls that are ⅜ inch thick. In other words, the lower parts of the tank wall need to be thicker to support the weight of the rest of the tank. On very large storage tanks, the lower-course rings are ½ inch or more thick.

Increasing the thickness of the wall and floor can extend the life of a tank. Manufacturers have guidelines for corrosion allowance based on the types of materials and the volumes to be stored in each of their tanks.

Figure 4.2. The entire outside seam of this tank is welded.

Figure 4.3. Some manufacturers only stitch weld the internal seam, which is not recommended.

3. How are the internal seams welded? You can easily determine whether the exterior seams on a tank are all welded, but you need to look more closely to see how the tanks are welded from the inside.

Tank manufacturers have different ways to weld internal seams. Some manufacturers weld the entire length of the seam, similar to the outside. Other manufacturers keep costs down by placing spot welds every so often around the seam. A tank that has a fully welded seam inside and out better resists corrosive materials and tank failure.

4. Do you want protective coatings added to the interior surface? The inside of a mild steel tank is generally uncoated unless the customer specifically requests it. There are many different coatings that manufacturers can apply to protect the inner surface, including fiberglass and phenolic epoxy. Coatings protect the inner surface of a tank against the corrosive effects of the stored material. If you plan to store a product for a long time, it may be beneficial to coat the inside of the tank.

When you regularly use a product, its level is changing inside the tank all the time. But when you use a product infrequently, problems can arise. For example, if you have a fertilizer tank that sits almost empty all winter, the fertilizer may salt out, which can cause problems (see question 5). Or if your tank is full for long periods, that could create problems high in the shell.

If you do not plan to diligently manage tank levels, it may be worthwhile to apply a coating to the entire wall surface of the tank. In other cases, you may find it worthwhile to simply coat the tank floor and 2 feet up the sides—that would save you money and protect against some corrosion.

Before you have any coating applied, ask the tank manufacturer if the coating will peel or separate. Some coatings may be incompatible with the product you plan to store. The coating could slough off and get into and plug supply lines, pumps, and screens.

Figure 4.4. This device agitates the contents of a tank to help prevent buildup of salts and other deposits.

5. Does the product you plan to store require agitation? Some products require agitation to remain effective. When a material salts out, it deposits residue on the tank floor. These solids tend to build up around the perimeter of the tank more than in the center. Unless you have an agitation system, the residue on the perimeter of the tank will not become resuspended in the liquid. Solid material can accumulate against the walls, floors, and welds and speed up corrosion.

6. Should you paint the outside of the tank? It is a good idea to paint the tank even if it will be indoors. Steel sweats during changing environmental conditions, so it will rust unless it is in a controlled environment. At a minimum, you should apply a coat of primer to every mild steel tank you place indoors. You should apply a primer coat and an additional coat of paint to every tank that will be exposed to rain. These are important steps in preventing rust.

Don't skimp: Use a high-quality paint. Metal expands as temperatures rise in the summer and contracts as temperatures fall. Over time, an inflexible paint will crack and peel as the tank expands and contracts. The tank manufacturer will be able to recommend specific paints for their mild steel tanks.

7. How much does the product weigh per gallon? When designing a tank, engineers calculate part of their requirements for wall and floor thicknesses based on the weight of the product it will store or transport. A heavier liquid will apply more pressure on the lower sections of a tank, so those lower sections will require thicker walls.

Most pesticides weigh about the same as water, which weighs 8.3 pounds per gallon. Liquid fertilizers can vary from 10.7 pounds per gallon for 28-0-0 to 11.7 pounds per gallon for 10-34-0. Phosphoric acid weighs approximately 14 pounds per gallon; it is highly corrosive and should not be stored in a mild steel tank.

Increasing the thickness of the sidewall or floor by a mere ⅛ inch can have other benefits. First, if you have a thicker tank, you can store a heavier product in the future without having to order another tank. Second, a thicker-walled tank creates a built-in corrosion allowance. Check with the tank manufacturer to be sure the walls are thick enough for the materials you want to store.

8. Will a larger tank save you money? Having the storage capacity to fully unload a tanker truck or railcar can save you money on your fertilizer and pesticide purchases. You can also take advantage of lower prices by buying in bulk during the offseason. If you buy in bulk, you may need to be able to store up to 3,200 gallons from a straight truck, 5,000 gallons from a semi-tanker, or 18,000 gallons from a railcar.

Figure 4.5. Make sure how much space you have before purchasing tanks. This is especially true when replacing a tank that is already in containment.

9. Will you place the new tank in an existing containment structure? If you are buying a tank to replace an old one, be certain the diameter of your new tank will fit in the existing containment structure. If the new tank is larger than your old one, make sure the new containment dike has enough capacity for the new tank.

Figure 4.6. Dikes can pay for themselves, especially if environmental cleanup is involved. A 2-inch steel plug blew out of one of the tanks shown here.

Don't forget to leave enough room around each storage tank to allow you to work around them to paint, inspect, and make any necessary repairs. Be sure that adding a larger tank will not make the dike out of compliance with the state's containment regulations. Follow any applicable local and state fire codes.

10. If you are placing a tank in a building, will it fit inside? Take into account that the door and space inside will need to be higher than the tank's length. Tilting the tank increases the amount of space needed to get it through the door and to stand it up indoors (see chapter 1 for how to calculate tipping height).

11. What is the condition of the concrete you'll place the tank on? Inspect the concrete pad where you will place the new tank. Look for signs of cracking and chipping. Replace any aging dike with one that has a thicker slab or reinforced concrete if you see signs of fatigue or excessive cracking. (See chapter 1 for more.)

Figure 4.7. This tank has a custom second fitting, which gives the operator more flexibility for loading and unloading the tank at the same time.

Figure 4.8. State regulations may require commercial agricultural dealers to provide regulators with a fertilizer sample to analyze and verify the product they are selling. This tank has a small tap off the sight gauge, which provides an easy port to withdraw samples.

12. Do you need a second access fitting on the tank? Manufacturers typically install only one 3-inch fitting 12 inches from the bottom of the tank to load and unload fertilizer. The 3-inch fitting normally is high enough to prevent any product that settles on the bottom of the tank from being withdrawn with the liquid. Steel tank manufacturers can custom-build your tank to suit the sizes and locations of the fittings you desire.

13. Do the pump and vent opening match? When a pump pulls liquid out of a tank, air will be sucked in from outside of the tank. When the amount of air pulled through a tank vent is less than the amount of liquid withdrawn, it creates a vacuum that can cause the tank walls to collapse inward. The vents should always be as large as or larger than the fill openings and product withdrawal lines.

The strength of the vacuum largely depends on the size of the pump and vent. A 2-inch pump requires a smaller vent opening than a 3-inch pump does. Vent size also depends on the characteristics of the pump. Some 2-inch pumps may have a higher pumping volume than some 3-inch pumps. Pumping volume depends on motor and impeller size. Choosing the correct vent size, which can range from 3 to 8 inches in diameter, will provide added pumping flexibility to your operation.

Figure 4.9. Having steps allows for the tank to be fully inspected from its top.

14. Does the tank require a ladder? The main reason for having a ladder is to be able to look inside the tank and inspect the top. If you add a ladder to a tank, you may also require a cage or railing around the ladder to prevent serious accidents. You may also need a gate or chain to keep unauthorized people off the tank. Always refer to Occupational Safety and Health Administration (OSHA) regulations for their specific requirements.

15. Do you want an internal or external sight gauge? A tank is usually fitted with a 1-inch external plastic sight gauge. Another type of gauge uses a float inside the tank and a flag on the outside to indicate the number of gallons of product in the tank. If you use the second type, a stainless steel float or cable is best to prevent corrosion. Over time, external tubes may stain, making them difficult to read. Having a shutoff valve on the bottom of the tube allows you to easily replace a stained tube.

16. How much will it cost to deliver and install the tank? A tank with a diameter larger than 12 feet will require special transportation and permits. Some delivery trucks are equipped with knuckle booms to unload tanks, but the size of the tank and the location where you place it may require a crane. When you move a tank, always use the fabricator-applied lift lugs.

Always consider delivery and setup costs when purchasing a tank. A wide-load permit and crane rental will add to your overall cost.

Installing

See chapter 1 for guidelines on choosing a location, assessing containment needs, preparing a site, and having tanks delivered and installed. You should decide where you want the manway and fittings to be located and provide this information to the manufacturer when ordering a steel tank. When your tank is delivered, inspect it before it's unloaded. Look for buckles, flat spots, or distortions on the sides of the tank that may indicate more than cosmetic damage.

Figure 4.10. When comparing cost between tank manufacturers, ask if they have a crane to unload at the site. This could make a significant difference in price between those who have cranes and those who do not.

Many horizontal storage tanks are placed in a concrete or steel saddle or in a supporting frame. In such a configuration, make sure to place a piece of rubber (or a similar product) between the tank and the saddle. The rubber will help reduce corrosion and abrasion to the tank when the steel shifts as temperatures rise and fall.

Inspecting and Maintaining

Mild steel tanks require periodic inspections and routine maintenance just like tanks made of other materials. You need to paint steel tanks to minimize rust. While the first appearance of rust may only be cosmetic, it could indicate a hidden problem that needs further inspection. Bubbling or peeling paint may be the start of more serious problems underneath the paint.

The paint on the outside of a steel tank may look acceptable, but you may not be able to see the condition of the top surface. Adding a ladder to the tank will allow you to inspect and paint the top. Always wear fall protection equipment when you inspect or paint the top. A drone can also be used to monitor the top of tanks.

Removing rust is important. Not only can rust damage the structure of the tank; it can also affect the supports. At the first sign of rust, use a wire brush or wire wheel to remove the loose or flaking paint on the affected surface to better inspect the area. You will need to thoroughly clean the steel surface, remove all rust, and paint over the area. This will prevent small problems from becoming larger and more significant.

Figures 4.11–4.12. This mild steel frame shows signs of serious rust corrosion. The owner repaired it by placing larger square pieces of steel over the rusted portion of the tank frame. If you make such a repair, be sure to seal the welded plates to avoid future problems with rust between the plate and saddle structure.

Here are eight other maintenance tasks you'll need to perform on a steel tank:

1. Apply a coat of primer before applying the top coat of paint. A primer allows the top coat of paint to properly adhere to the tank and last longer.

2. Maintain proper drainage around tanks. Keep water pumped out of dikes. Standing water promotes rust.

3. Replace sight tubes regularly. Replace external sight tubes yearly because sunlight and chemicals will weaken the plastic used in these tubes.

Figure 4.13. When first attached, a plastic hose is straight. Wear will cause it to bend, and chemicals will discolor it. The hoses shown here need to be replaced.

Figure 4.14. Cut sight hoses a little long to allow for stretching. If a sight hose is too tight, it can lead to problems during strong winds. Always close the sight gauge when unattended.

4. Regularly inspect tanks. Inspect the interior of each steel tank for pitting every one to three years. Before any inspection, remember that the confined space of a tank's interior can have dangerous fumes, making the air within the tank unfit for breathing. Before anyone enters any tank, make sure it is empty, and then remove the manhole cover and leave it open for at least a week to allow noxious vapors to escape.

Figure 4.15. This tank has a sign prohibiting entering the tank without permission. Entering a tank can be dangerous if vapors have not dissipated and oxygen levels are low.

If there are any openings or covers on top of the tank, remove them too. A fan will help clear vapors from the tank. If you need to make repairs or welds, the tank has an opening on top. Place an exhaust fan over this top opening to draw air out. This allows fresh air to come in the bottom opening and draws the welding smoke out the top.

For safety, ask your local agricultural retailer if you can use their atmospheric monitoring device to check the toxicity levels of a tank before you enter it.

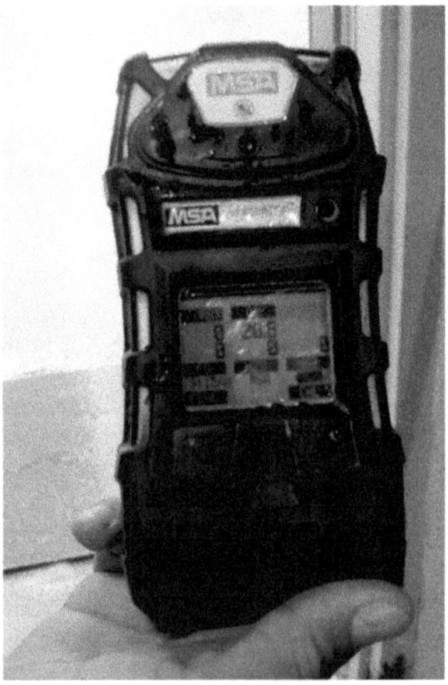

Figure 4.16. Atmospheric monitoring device used to check for vapors.

Figure 4.17. Never enter a tank unless you are a trained professional with proper safety equipment, environmental monitoring equipment, and confined space training and equipment.

Figure 4.18. A broom helps clean the bottom of a tank.

Figure 4.19. A water supply will be necessary to rinse away remaining residue.

Figure 4.20. This tank is now nearly clean enough to be inspected.

After the tank has been cleared to enter, you first want to remove any residues from the floor. Use a stiff broom to loosen any solids stuck to the bottom of the tank, and then get them into a solution by adding water. Then completely drain the tank, moving the product into a holding tank or sprayer.

Next, rinse the lower rings and floor with water to remove excess fertilizer. Empty the water remaining on the floor using the drain outlet. Closely examine the tank's floor and sides for rust, pits, cracks, and compromised welds. Make repairs as required.

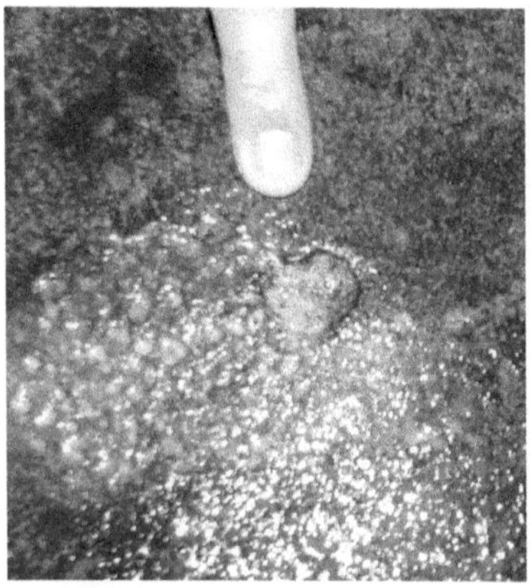
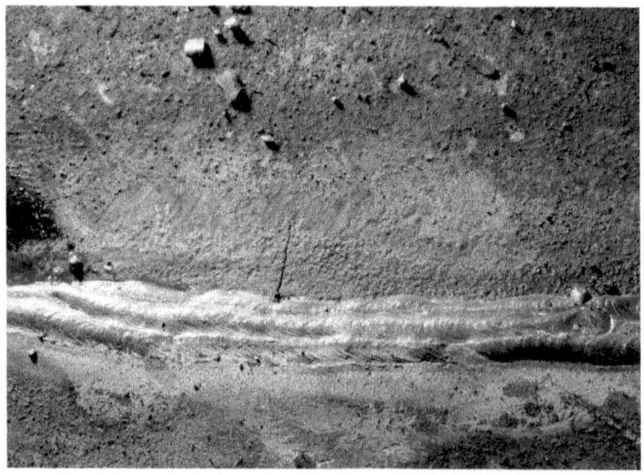

Figures 4.21–4.22. (Left) This tank has a pit. (Right) A close inspection turns up a small crack above a weld.

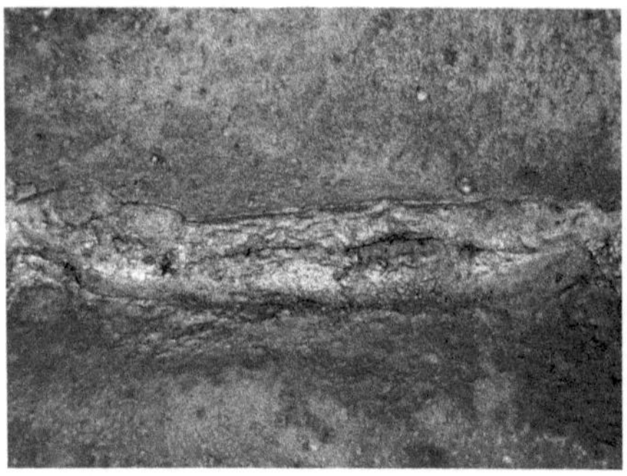

Figures 4.23–4.24. These welds show signs of corrosion damage. The most vulnerable part of a mild steel tank is where the quality of the welds is insufficient, and the steel closest to those welds is exposed to pockets of stagnant water. Always use qualified or certified welders when repairing a tank.

If you find a pit, use a pit gauge to measure how much steel has corroded. A rule of thumb says that the tank is still acceptable if the pit is no deeper than 50 to 60 percent of the wall's original thickness. For example, if your tank has a wall that is ¼ inch thick, it is still usable if the pit is less than ⅛ inch deep. Deep pits require metal plates to be welded over the damaged sections. If the sides and bottom are extremely pitted, you should replace the tank.

If you need to repair a large area on the floor of a tank, that can be difficult because welding in new plates can weaken the floor. There can also be problems if the metal piece you need for the repair is larger than the manhole opening—how will you get it inside the tank?

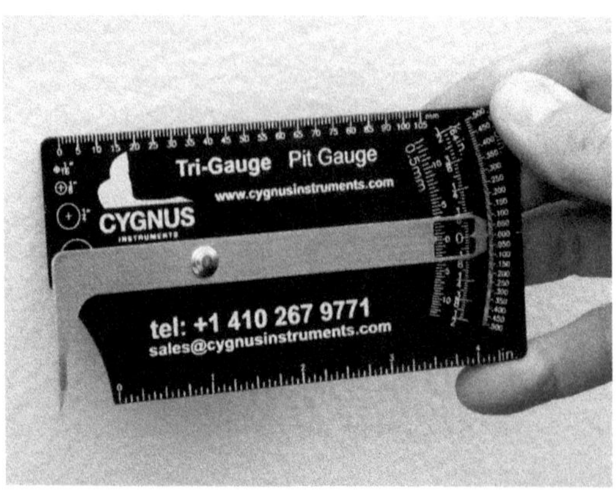

Figure 4.25. Pit gauge.

5. Replace gaskets. Pay close attention to gaskets on any frequently used manhole cover. Be sure to use the right size of gasket, and check that the gasket material is compatible with the product that will be stored in that tank. Wait 24 hours before refilling the tank to let the silicone cure.

6. Inspect tank lids. Be extra careful to inspect the lids on transport tanks to ensure they have a tight seal.

7. Inspect all attachments and clamps for corrosion. Use stainless steel clamps and attachments when possible.

Figures 4.26–4.29. Replacing a gasket. A light film of grease on both sides of a gasket may help prevent it from being torn when you take off the lid. Using anti-seize on the bolts will help the next time you need to remove the manhole lid.

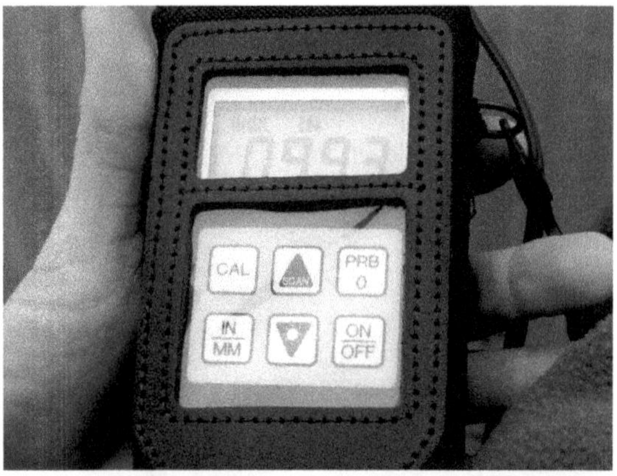

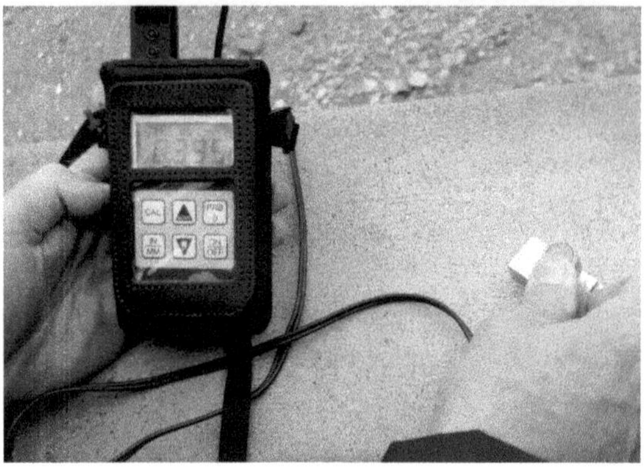

Figures 4.30–4.32. There are good instruments that measure the thickness of tank walls, but the devices should complement a careful visual inspection of the area. Instruments can give you a good reading in one spot, but just 1 inch away there could be a hole nearly all the way through the wall.

8. Consider investing in detection tools. There are a number of electronic tools that use ultrasound to measure the thickness of the tank wall from the outside. You can compare a reading from such a device to the original wall thickness. If you use a device and detect that the wall's thickness has been compromised by corrosion, examine that area inside the tank.

STAINLESS STEEL TANKS

Basic stainless steel begins with iron and carbon just like mild steel, but then manufacturers add chromium and nickel to improve corrosion resistance. Like mild steel, there are various grades of stainless steel. The stainless steel used most in the pesticide and fertilizer industry is 304 (or 18-8) stainless. Stainless steel used in tanks contains 18 percent chromium and 8 percent nickel. Tanks made with 300-series stainless steel offer excellent corrosion resistance, so they can store a very diverse range of products: from fuels, to pesticides, to fertilizers.

The two most common 300-series grades are 304 and 316 stainless. Of the two, 304 stainless is used the most. It has excellent corrosion protection and costs less than 316 stainless (which is superior in quality). Steel manufacturers further designate a 304L stainless, which indicates the carbon content is no more than 0.02 percent, further enhancing its corrosion resistance.

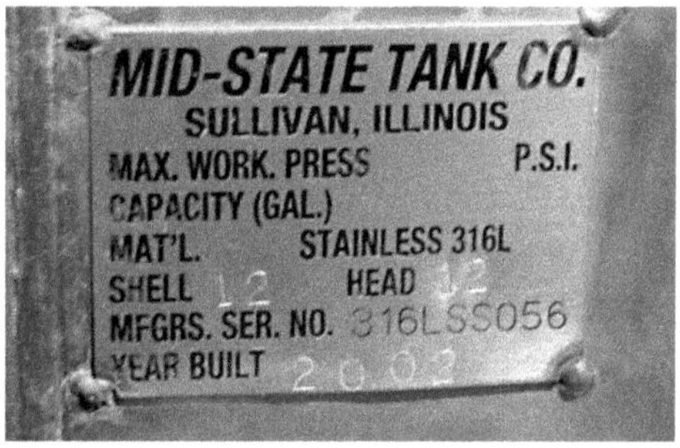

Figures 4.33–4.34. The stainless steel grade is marked on these tanks: 304 (left) and 316L (right).

However, even 304 stainless can still be degraded by extremely acidic liquids or product that remains in the tank for extended periods. Calcium chloride is one product that will corrode 304 stainless. Grade 316 stainless steel is used in tanks that store very acidic liquids such as calcium chloride and some micronutrients. Grade 316 stainless may also have the L designation.

Figure 4.35. Nipples, elbows, and pipe fittings made from stainless steel have information about their grade stamped on them.

The major advantage of using stainless steel for storage or transport tanks is that liquids seldom interact with, adhere to, or penetrate the inside walls of the tank. Stainless steel also allows you to completely rinse the tank with little chance of cross-contamination or chemical residue. These attributes make stainless steel the more desirable metal tank material for transporting fertilizers and pesticides to and from businesses and farms.

There are other advantages to using stainless steel:

- It is not affected by sunlight.
- It is not subject to rust.
- It can be painted.
- It can have decals attached.
- It can remain in service longer.
- It will hold its value longer.

Of course, the major drawback for stainless steel tanks is the initial cost compared to poly, fiberglass, or mild steel models. This higher initial cost is the major reason large fertilizer storage tanks are built of mild steel or fiberglass.

Although stainless steel resists corrosion better than mild steel, you still need to regularly inspect and maintain these tanks.

Questions for Stainless Transport Tanks

There are several options to consider when you choose a stainless steel tank. Here are 17 questions that will guide you in customizing a stainless tank to meet the needs of your operation.

1. What grade of stainless is best for the products you will store? Contact the manufacturer of the product you plan to store to determine which stainless grade (304 or 316) is most beneficial.

2. Will the tank be full most of the time? The less time a product stays in the tank, the easier it is to justify 304 stainless. If the product remains in the tank for an extended period, then 316 stainless will offer better resistance against corrosion.

3. How thick do the walls need to be? The wall thickness of a truck-mounted tank is measured by its gauge. Table 4.1 shows the relationship between gauge and inches. Note that the same gauge designation has a different wall thickness depending on whether the tank is made of mild or stainless steel.

Gauge and wall thickness are inversely related. The lower the gauge value, the thicker the wall. For example, 12-gauge stainless is 0.1094 (a little less than ⅛) inch thick, while 16-gauge stainless is 0.0625 (roughly 1/16) inch thick. For each gauge number there is about a 10 percent difference in thickness compared to the gauge directly above or below it.

Table 4.1. Wall Gauges and Thicknesses for Mild and Stainless Steel

Gauge	MILD STEEL THICKNESS (inch)	STAINLESS STEEL THICKNESS (inch)
3	0.2391	0.2500
4	0.2242	0.2344
5	0.2092	0.2187
6	0.1943	0.2031
7	0.1793	0.1875
8	0.1644	0.1719
9	0.1495	0.1562
10	0.1345	0.1406
11	0.1196	0.1250
12	0.1046	0.1094
13	0.0897	0.0937
14	0.0747	0.0781
15	0.0673	0.0703
16	0.0598	0.0625

Figure 4.36. The gauge is indicated on this label. This tank was constructed from 304 stainless steel.

A tank made of 10-gauge stainless instead of a higher gauge may be important if the tank will carry heavy liquids, will be used primarily for delivery, or will carry a large volume (such as 3,500 gallons). Heavier-gauge steel is recommended when purchasing a transport tank because the liquid inside it constantly moves back and forth during travel. A heavier gauge may reduce the likelihood of a serious environmental mishap, hefty environmental fine, or serious long-term legal issues.

4. How are the internal seams welded? The entire seam should be welded on transport tanks. This will protect the seams from weakening or leaking from the liquid shifting back and forth.

5. How many gallons will the tank store? How you use the tank could determine its size. For example, if you will use the tank to transport product to an applicator, try to match the transport tank size to the receiving tank size. So, if the pesticide sprayer has a 1,200-gallon tank, a 1,000-gallon stainless nurse tank will never quite fill the sprayer, while a 1,500-gallon nurse tank will always have some product left over. Try to match supply and use tanks as closely as possible.

6. Will the tank be mounted on a truck or trailer frame? The height of truck-mounted tanks tends to be limited so the driver can see though the back window. However, you have more choices for trailer-mounted tanks.

Figure 4.37. A truck-mounted tank.

7. How much weight is the truck or trailer rated to carry? How much weight a truck or trailer can transport depends on how much weight the truck, axles, and tires are rated to carry (see volume 2). If you want to mount the tank on a truck frame, check the distance from the truck's cab to the back axles, and check the wheel base length to make sure the weight over the axles does not exceed its ratings.

8. Does the size of the vent opening match the pump? As with standing tanks, the vent's diameter needs to be large enough to prevent the tank from collapsing when you remove fluids. Know how many gallons per minute the pump is rated for.

Standard vent openings will easily handle a 2-inch pump pulling 150 to 180 gallons a minute. In contrast, a 3-inch pump can withdraw more than 325 gallons per minute, which can quickly create a vacuum strong enough to collapse the tank if the vent is undersized. Consider adding a larger vent opening, or multiple openings, to allow you to use a larger pump in the future.

Figures 4.38–4.39. These tanks imploded because of inadequate venting. Tanks implode when you pull liquid out of the tank faster than air can enter the tank. Under the right conditions, even storage tanks can implode.

Figure 4.40. Keeping the fill port open while you fill the tank allows enough air to escape and prevents a seam or weld failure.

9. Does the tank come with baffles? All larger transport tanks should have baffles. Permanent baffles are important, because they help reduce the forward and backward movement of the fluid during transit.

10. Does the tank come with an attached frame? Stainless steel tanks usually come separate from the supporting frame. Adding a frame will increase your cost but will provide added securement.

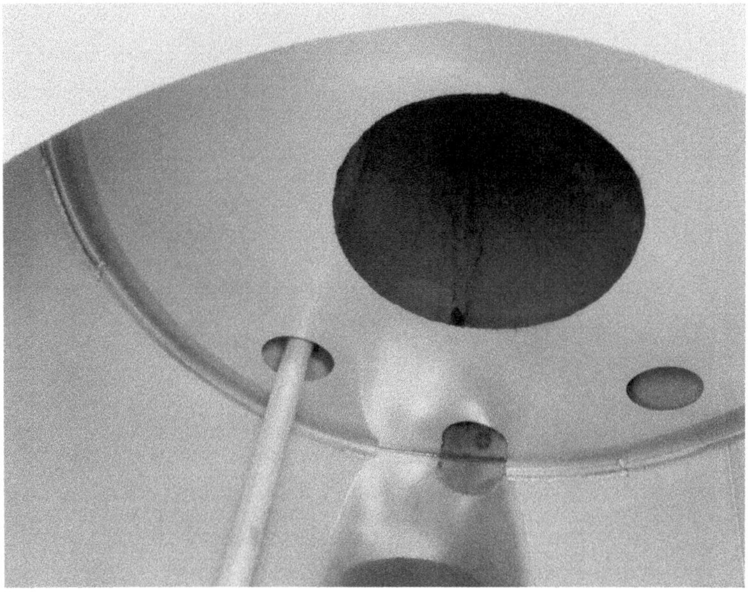

Figure 4.41. Baffles in a steel transport tank.

Figures 4.42–4.43. Normally, the number of lids determines how many compartments are in a tank–a stainless tank with a single compartment (left) and one with two compartments (right). Some manufacturers will add a second lid to a single tank for better access to the inside of the tank when cleaning.

11. Does the tank need single or multiple compartments? You can order tanks with multiple compartments. For example, you can divide a 3,200-gallon tank into two separate compartments, each holding 1,600 gallons.

12. Does the tank need a collection sump? You can fit stainless tanks with sumps and troughs to collect and remove liquids. You can position an 8-inch by 8-inch box sump to help drain the entire tank. A trough sump runs the entire length of the truck or compartment and slopes to a collection point where the material can be pumped out of the tank and cleaned easier.

13. Does the tank need a sparger tube for agitation? Many liquid products require agitation to keep them suspended within the tank. Most applicators agree that a tank used for spraying should have a sparger tube.

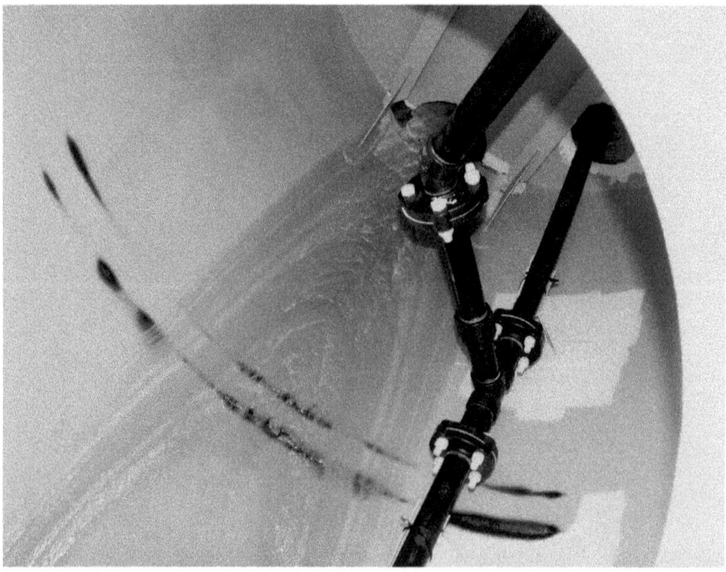

Figure 4.44. Sparger tube added to allow for rinsing remaining residue from the tank.

14. Does the tank need a rinse line? Rinse lines installed at the top of tanks vary from 1 to 1.5 inches in diameter. The lines make it easier to rinse the tank by applying clean water from an external water tank.

15. Does the tank need a ladder, platform, or railing? A ladder or platform provides easy access to the top of the tank, and a railing may be required for a work platform. Make sure that the ladder, platform, pump, valves, and hoses are on the driver's side of the vehicle. This should help the driver react quickly to an emergency.

16. Does the tank have an area for a freshwater auxiliary tank? While you can purchase one separately, the manufacturer also can add a tank to hold clean water.

17. Is the final finish of the tank important? Most stainless steel tanks have a standard flat finish. While the finish is mostly a cosmetic consideration, some companies prefer a brighter finish to help advertise their operations.

Figure 4.45. A stainless tank with a ladder to allow easier access to the opening at the top.

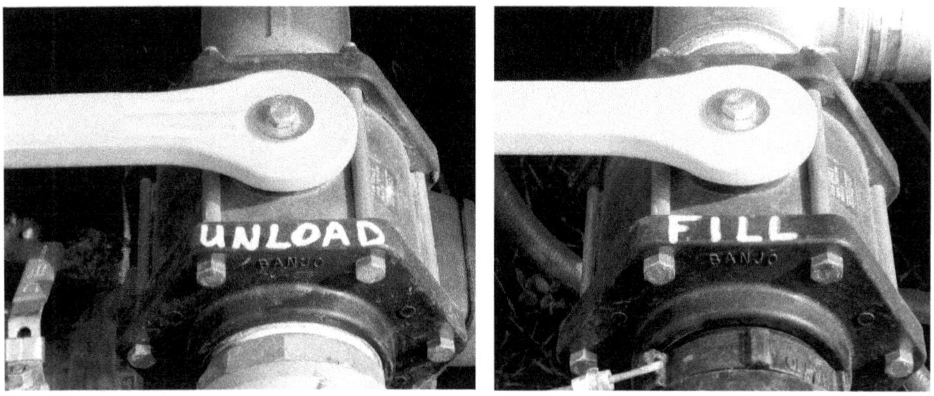

Figures 4.46–4.47. Clearly mark valves to prevent mistakes.

Figure 4.48–4.49. (Left) This tank has chains directly on it, which could lead to problems. (Right) On this tank the chains are run through hoses that help to protect the tank's surface.

Installing Tanks on Trucks and Trailers

Manufacturers should provide the installation procedures their tanks require. Follow these procedures closely. If you do not mount it properly, a stainless steel tank will likely develop stress cracks from the twisting and turning of the truck frame or trailer.

Have you ever held both ends of an aluminum pop can and twisted each end in opposite directions? You can feel the can weaken as you continue to twist until the surface of the aluminum creases. Without the ability to flex, a stainless tank will develop stress cracks just like the can, and it will lose its strength and integrity.

Many methods are used to reduce the stress on tanks. Some companies place a 1-inch composite board or piece of wood between the tank and the frame of the truck or trailer. Another method is to mount the tank frame to the trailer or truck with springs, which act much like shock absorbers. The springs will absorb the shock instead of the tank.

You can add stop blocks to either end of the trailer or truck bed to keep the tank from sliding back and forth when the tank is not bolted to the frame. You may also use straps, bands, and chains to hold the tank to the trailer or truck. Make sure that rubber pads or similar materials are placed between the tank frame and the chains to prevent metal-to-metal contact.

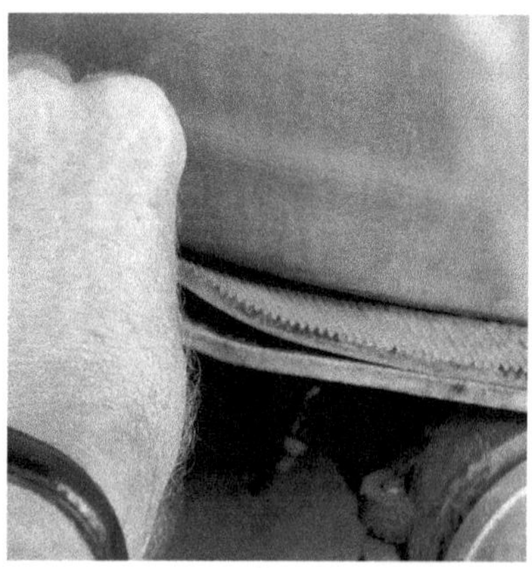

Figure 4.50. Rubber strips placed between the tank and metal frame help protect the tank from getting holes rubbed in it.

Inspecting and Maintaining

A common misconception is that stainless steel is impervious to corrosion. While it's true that stainless resists corrosion better than mild steel, chemicals left in any tank for an extended period can damage the metal—even stainless steel. Empty and thoroughly clean the inside surface of a stainless steel tank occasionally to help protect against corrosion.

You should inspect tanks as part of any maintenance schedule. During inspections look for:

- Areas that contain a manufacturing flaw
- Tanks that may be incompatible with the products they store
- Points of contact between the tank and other metal
- Exterior sight gauges that are stained or broken
- Worn and corroded valves, hoses, and fittings
- Stress cracks from improper mounting

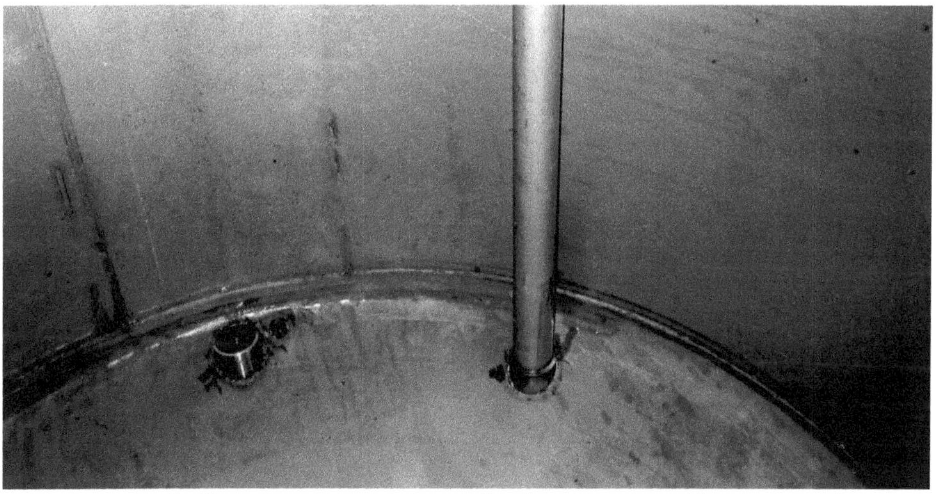

Figure 4.51. Periodically washing out a stainless steel tank can increase its lifespan. This tank is still in service even though it was manufactured in the 1970s. Notice how little pitting there is on the inside of the tank.

Figure 4.52. Chemicals spilled on the exterior of a stainless tank can cause reactions with unsightly effects. The backs of these stainless tanks are beginning to show signs of corrosion.

Figure 4.53. Properly storing hoses helps maintain their integrity during transit.

REPAIRING TANKS

The tanks described in this chapter are atmospheric tanks (unpressurized tanks) and store commonly used products such as liquid fertilizers, diesel, gasoline, and pesticides. Another group of tanks, known as *code tanks* (pressure vessels), hold products such as anhydrous ammonia, propane, and fumigants (see chapter 8). While code tanks and atmospheric tanks are both designed to hold products, their similarities end there.

Repairing Unpressurized Mild Steel Tanks

Most leaks in unpressurized mild steel tanks occur through a corroded weld rather than through the side or floor. Any leak will eventually get worse and must be repaired. Repair the damage as soon as possible after you discover a leak. While you can make some repairs while the tank is full by welding a steel plate over the outside, it is best to repair it when the tank is empty.

When making a repair, put a 3- or 4-inch plate over the area that's been breached. Try to cover the leak area with as large a patch as possible, since you don't initially know the extent of the damage. If you weld a patch that is too small, you could create more leaks if the area being repaired is weak and thin. However, be careful that the flat plate is not so large that it sticks out over the curvature of the tank.

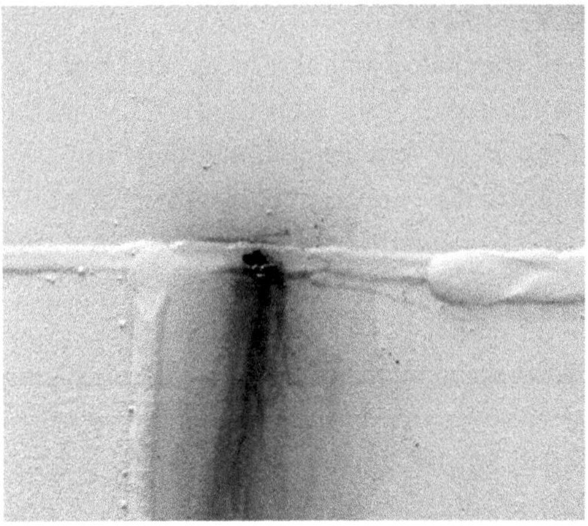

Figure 4.54. A permanent repair to this tank would involve removing the damaged area by grinding and replacing the weld with a full-penetration weld.

Figure 4.55. A plate used to repair a tank.

EMPTY THE TANK FIRST

Remember: Patching a tank that is full of product could lead to more problems because you won't know the extent of the weakened areas without inspecting the inside. When possible, pump the tank out at least below the leak. If the area around a hole is weak, you may end up making the leak larger. Sometimes the liquid in the tank can keep the metal too cool to get a good weld in the repair area.

While you can make repairs to stop the leak, it's important to consider these repairs as only temporary. Depending on the damage, you may also need to inspect and repair the inside of the tank. When you do, make sure to check the air quality and properly vent the tank prior to entering (see p. 78). Carefully inspect the area on the inside that was repaired on the outside.

Welding rods known as 6013, 7018, or 6010 are often used to repair mild steel tanks. A 6013 welding rod is easy to work with, versatile, and readily available in most shops.

Repairing Unpressurized Stainless Steel Tanks

It's been said that welding stainless steel is as much art as science. You must weld stainless steel in a manner that repairs the problem without altering the physical and chemical composition of the steel. The best approach is to hire a welder who is experienced on stainless tanks and knows which technique, rod, speed, and heat are most effective.

Be Considerate of Commercial Welders

Many times, you don't discover a leak until the tank is full and in use and must be fixed quickly. In some rare instances, you can repair leaks with product still in the tank, but you'll have to drain the tank in most cases. Very few welding shops will have a way of properly disposing of any chemical you leave in the tank or of working on a loaded tank. In other words, you must remove all pesticides and fertilizers from the tank before it can be repaired.

Have the welding shop test for vapors in the tank to ensure it is safe to weld. To expedite repairs, consider preparing the tank before bringing the welder on-site or bringing the tank to a shop. Here are some things you can do to prepare your tank for repairs:

- Completely empty the tank and hoses of any liquids.
- Pressure wash or steam clean the inside of the tank.
- Force air through the system to flush out any remaining liquids.

PURCHASING USED TANKS

You can often find used steel tanks at acceptable prices. But just like buying a used car, tractor, or wagon, it's buyer beware when you purchase a used tank. Thoroughly inspect any used tank to ensure you don't end up buying one that should have been sold for scrap.

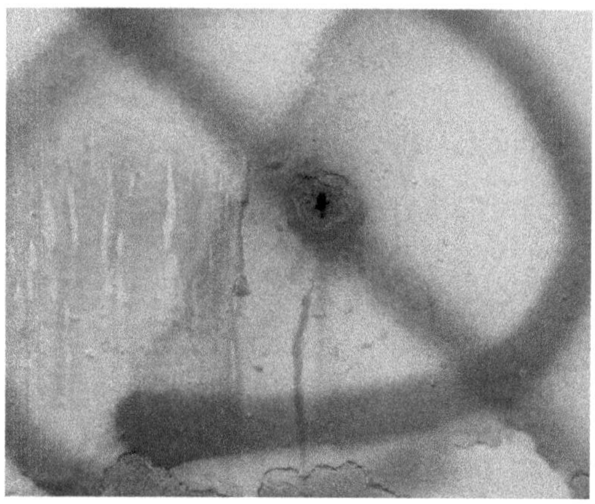

Figures 4.56–4.58. Here's what the owner of this tank had to say: "This 25,000-gallon tank was set in a dike in either 1997 or 1998. It was a used tank when we put it there. The circle and X is where there was a small bird's eye (top right). It was not a hole. The real problem was on the throat of the manhole (bottom left). After we put two loads of 28-0-0 in the tank, a small drip developed. The tank was emptied right away (there was no loss except for the tank). After scratching at the paint we found it was punctured on the throat. I was glad we caught it when we did."

If you can locate the identification plate attached to a used tank, it will provide you with a tremendous amount of information. The identification tag can include:

- The manufacturer's name
- The grade of steel
- The thickness of the steel
- The tank's unique serial number
- The date the tank was built

What to Look for When Buying a Used Mild Steel Tank

Thoroughly inspect a used mild steel tank (or any used tank) before buying it. Walk around the tank and look for any signs of repairs, rusting, or weak spots. Don't buy a tank that's been repaired without looking at the state of the repair from inside the tank.

A fresh coat of paint will make a tank look good from the outside, but it's the inside that really counts. When you inspect a mild steel tank, use a strong flashlight inside the tank and scan the floor and walls for pitting. The severity of the pitting will be a key to knowing whether you should buy the tank. You should not purchase a tank that has been sitting on the ground unless you can inspect the bottom, because it may have a serious rust problem.

Figure 4.59. A tank with rivets was probably made before or during World War II. The arrow points to a missing rivet. Thoroughly inspect an old tank like this before you buy it.

What to Look for When Buying a Used Stainless Steel Tank

Used stainless steel tanks are in great demand. But without the identification plate attached to the tank, it may be difficult to distinguish a stainless steel tank from a mild steel one. Distinguishing stainless from mild steel is even more complicated if the previous owner painted the tank.

Here are a few tips that can help you distinguish stainless steel from mild steel:

- Upright storage tanks that hold more than 10,000 gallons are usually made of mild steel.
- The tank's data plate should list what steel it is made of.
- A magnet is strongly attracted to mild steel. A magnet will not adhere to 304 stainless. There may be a weak attraction for 316 stainless steel but nothing compared to mild steel.
- Depending on the finish and age of a stainless steel tank, galvanized steel can look similar. Steel that has been galvanized has a zinc coating that reduces corrosion to the underlying steel. A magnet will firmly attach itself to galvanized steel, much like it would to mild steel.

Before you buy a used stainless tank, consider the shell thickness. Many of the used tanks you'll see on the market were originally built for the food industry. Food industry tanks are often built with lighter-gauge metal (such as 12-gauge). Lighter-gauge steel may work well for fluids that weigh about the same as water. For instance, milk weighs 8.2 pounds per gallon, and beer weighs 8.4 pounds. But heavier materials like fertilizers weigh about 11 pounds per gallon and may create a problem in light-gauge tanks—especially if used for transport. Check the plate on the tank to see if it provides information about the gauge. If not, call the manufacturer and use the serial number to determine what gauge steel was used on the tank when new.

Figure 4.60. A magnet is attracted to mild steel but less so to stainless steel. It is an easy tool to separate mild steel from stainless.

Never assume that a stainless tank wrapped with insulation is in good shape. Insulation wrapped around a tank can hide significant problems. In some cases, insulation can react with water to produce hydrochloric acid, which in turn cracks the tank's surface. Without pulling the insulation away from the tank, it is nearly impossible to thoroughly inspect the tank.

If there are stress cracks, a dye penetrant tank kit will help you determine how deep the cracks may be. Dye penetrant also may expose cracks that you cannot see.

A vacuum box is an instrument that runs on electricity and requires you to apply soapy water to the suspicious area on the tank. If it pulls air, it will show bubbles coming from the crack. This device works on any steel or fiberglass tank. If it pulls air through the crack, there is usually a leak.

Other areas of the tank to inspect include:

- The baffles
- The manway (if accessible by a ladder)
- The sparger line or rinse line (if the tank has one)

Figure 4.61. Show caution when buying used tanks because there is a lot at stake. These two 40,000-gallon tanks failed. One of the tanks damaged a nearby building as it fell. Saving a few dollars up front might cost you a lot more in the long run.

CONCLUSION

When you purchase any tank, good judgment is the most important factor. Remember that a tank is just as important to your operation as a tractor, truck, sprayer, and other equipment. Take the time to match the tank material and size to the intended contents and application. Making a careful selection can give you years of trouble-free service. And before you consider a used steel tank, carefully inspect the tank and assess its history and current condition. This may prevent problems later.

5

Pesticide Minibulks

With contributions from Joe Becovitz, John Obermeyer, Mike Titus, Julie Lamberson, Kip Landwehr, and Kevin Leigh Smith

Minibulks—also referred to as *shuttles*, *totes*, or *intermediate bulk containers*—are routinely used to safely transport thousands of gallons of pesticides, fertilizers, and adjuvants between retailers and farms. They have become an irreplaceable part of agricultural operations. Minibulks provide quantifiable advantages to both the retailer and the farmer:

- **Bulk pricing:** Retailers that order pesticides in bulk quantities receive significant price discounts from manufacturers. In turn, the retailers can pass those discounts on to their farmer customers. This is especially important when you consider that a 2.5-gallon container typically costs much more than the same material purchased in bulk.
- **Bulk storage:** If used correctly, minibulks allow farmers to obtain the benefits of buying in bulk without building costly permanent bulk storage facilities.
- **Return for credit:** Farmers could possibly return any unused product left in minibulks to their retailers if the tank's seals remain intact and the retailers accept bulk returns.
- **Less plastic use:** Minibulks can replace hundreds of smaller plastic containers, which can save a significant amount of time and money. For instance, one 250-gallon shuttle will replace opening, pouring, rinsing, and disposing of 100 2.5-gallon containers.
- **Ability to be refilled:** Retailers can fill and refill minibulks many times over their lifespan. Compare this to the cardboard and plastic waste generated from single-use and 1-gallon and 2.5-gallon containers.
- **Fewer injuries:** Applicators can dramatically reduce their pesticide exposure when they can pump product from minibulks directly into their sprayers. Pouring out of containers requires repeated climbs up and down the sprayer, which increases the potential for slips and falls.

Despite these advantages, there are also legitimate concerns when using minibulks. Larger containers can mean bigger problems during transport, storage, and use. In fact, most minibulk spills occur during the loading, transporting, or unloading of the container. A minibulk spill can mean hundreds of gallons on the ground (which can then potentially move to environmentally sensitive sites) instead of just 1 or 2 gallons from a jug.

Figure 5.1. Minibulks are commonly transported over roads between the ag retailer and the farm, or from the farm to the site of application.

In addition to transportation concerns, there are rules governing the filling and storage of bulk and minibulk pesticide containers. These rules are administered and enforced by the US Environmental Protection Agency

Figure 5.2. Minibulks are easy to load and move with a forklift, as long as care is taken to avoid puncturing the container.

Figure 5.3. Many minibulks are refilled from larger storage tanks, saving time when filling sprayers as compared to using 2.5-gallon jugs.

Figure 5.4. Filled minibulks ready for shipment.

(EPA), US Department of Transportation (DOT), and state offices (for state contact information, see the Association of American Pesticide Control Officials' resources page: https://aapco.org/2015/07/28/resources-2/).

This chapter provides farmers and retailers guidance and best practices for managing, storing, handling, and transporting pesticide minibulks.

UNDERSTANDING *BULK* AND *MINIBULK*

We often use terms like *bulk* and *minibulk* without much thought; however, the regulations clearly identify the exact meaning of each. The EPA defines liquid bulk pesticide as an undivided quantity greater than 55 gallons. This definition is different from the DOT's, which is 119 gallons for hazardous materials. By default, 55-gallon drums and 2.5-gallon containers are not considered bulk pesticides. This distinction is more than a little important to understanding pesticide regulations.

There are two types of bulk pesticide storage tanks: larger stationary storage tanks and smaller minibulks. Stationary storage tanks are generally constructed of stainless steel for durability and ease of cleaning. They hold between 400 and 5,000 gallons. These tanks require secondary containment.

In addition, stationary tanks are usually equipped with dedicated plumbing. The dedicated plumbing is used to fill minibulks and sprayers and helps to significantly reduce cross-contamination by keeping products separate from one another. Stationary pesticide tanks (unlike bulk fertilizer tanks) are usually stored under roof.

Figure 5.5. Minibulks waiting to be cleaned and refilled at the ag dealer's site.

State and federal regulations define minibulks as tanks with a capacity of more than 55 gallons but not exceeding 400 gallons. Most minibulks in use today are somewhat standardized and hold roughly 110, 125, or 265 gallons (tank capacities of 120, 135, and 275 gallons, respectively). Minibulks may arrive at the retailer already filled by manufacturers but are most often refilled from the retailer's bulk tanks. Retailers often store minibulks in the same heated buildings with containment systems as their stationary bulk tanks.

Minibulks are designed to be mobile, shuttling product to the farm or field, where users can transfer the material directly to sprayers. Farmers and retailers also place minibulks on trailers along with water and other supplies. The trailers become mobile mixing sites and can easily follow sprayers from one field to the next.

At one time, minibulks came in dozens of shapes and colors. It was easy to identify a particular pesticide product based on the shape, color, and size of the container. Today most manufacturers place their products in caged plastic tanks. Cage tanks are basically nondescript molded plastic tanks that have supporting steel or aluminum shells around them and bladders often placed inside the tank.

Figure 5.6. A minibulk that is properly labeled according to EPA and DOT regulations.

Figure 5.7. The operational pad holds filled minibulks, while the empties are stored outside of containment.

SECONDARY CONTAINMENT REGULATIONS FOR FARMS AND RETAILERS

Although minibulks are sometimes used for semipermanent storage, they are really designed to be mobile containers that can be filled, emptied, and refilled quickly. When they are used for storage, secondary containment may be needed depending on the requirements in your state.

REQUIREMENTS FOR FILLING MINIBULKS

Federal regulations spell out specific requirements that dealers must comply with when they take product from stationary bulk tanks to fill minibulks.

EPA Registration

According to the regulations, any facility that "produces" a pesticide product is required to register with the EPA as a pesticide-producing establishment. Production can mean that the facility mixes ingredients according to a labeled formula to make a pesticide product. Or, as is usually the case with ag retailers, production can mean simply transferring pesticide product from one container to another for the purpose of resale. All pesticide-producing establishments must report their production (even if zero) to the EPA by March 1 of the following year.

Figure 5.8. Moving a filled minibulk with a forklift inside a refilling operation.

Repackaging Agreements

Retailers must have repackaging agreements from each manufacturer to repackage bulk pesticide products. Retailers must make copies of current agreements available for inspection by state officials at each repackaging location. The written contracts lay out exactly what products ag retailers can repackage, what containers the products can be repackaged in, how and when minibulks should be cleaned, and what inspection protocols should be followed.

Unique ID

Whenever users refill or repackage minibulks, they must mark each one with a unique ID or serial number. However, minibulks obtained directly from manufacturers and shipped to customers do not require unique serial numbers. If farmers bring minibulks back to retailers for refilling, all of the repackaging agreement requirements kick in, including assigning the shuttle an ID number.

Figure 5.9. Refillable minibulk with an ID number (on the green tag).

One-Way Valves

Refillable minibulks must have one-way valves that only allow product to flow out of the tank.

Figures 5.10–5.13. The top image shows the valve with its cover, and the others show it closed, partially open, and open. This setup allows the product to flow out but prevents product from being pumped into the tank.

Figures 5.14–5.15. Tamper-evident seals prevent cross-contamination.

Tamper-Evident Seals

Retailers and manufacturers must place tamper-evident seals on or around openings before they ship minibulks. These external seals provide important visual clues when users return minibulks. The seals help protect the product's integrity while in service. If the seals are intact, the product in the tank has not been tampered with. If any of the seals are broken, retailers must follow the specific cleaning instructions found in their repackaging agreements before they can refill the minibulks—even if they refill minibulks with the same product.

New Labels

New labels must be attached to minibulks each time you fill them. The labels should include the EPA establishment number and net contents. The establishment number on the minibulk is the same one the EPA assigned to the facility. This holds true even if a facility fills a minibulk for a sister facility. The establishment number written on a minibulk is always the number for the location where the container is filled.

Figure 5.16. It is not legally required to include the date the pesticide was put into the tank

It is important to note that the tank's net contents are not the same as the tank's capacity. The net contents are the amount of product that was put into the tank at refilling. For example, let's say a customer brought back a 250-gallon tank to the retailer for refilling. There were 50 gallons remaining in the tank when the customer brought it in, so the dealer put 200 gallons in the tank. The net contents written on the label would be 200 gallons, not 250 gallons. Although it's not required, some dealers also include the date the minibulk was refilled on each label.

Recordkeeping

Each time retailers fill containers, they must record the unique ID numbers, dates, and EPA registration numbers for the products. Retailers must keep these production records for three years. If the product is a restricted-use pesticide (RUP), then retailers must keep an additional, more detailed record of sale or application for two years. Each year, retailers must submit a summary of all pesticide production (amounts filled for each bulk pesticide product) to the EPA.

Inspection and Cleaning Requirements

Retailers must determine if minibulks can be refilled by reviewing the tank labels to see if they are marked as refillable or nonrefillable. Since most minibulks are meant to be reused, inspecting them is perhaps the most important part of the refilling process.

Proper inspections can reveal faulty containers, leaking valves, tanks that need to be recertified (pressure-tested), or contamination issues. Failing to inspect returned containers before refilling them can result in contamination issues, which may cause thousands of dollars in unintentional crop damage and ruined reputations.

When you inspect minibulks:

- Ensure that all valves are functioning.
- Examine the outside of the tank for any spills, leaks, or cracks.
- Examine the pesticide label on the outside of the tank and all tamper-evident seals to ensure they are intact. If the tank labels do not match the product about to be put into them, or if any seal is cut or broken, then retailers must clean the minibulks according to the instructions in the repackaging agreement. They must clean the tanks over a wash pad where they can collect the rinse water.

PRESSURE-TESTING MINIBULKS

EPA regulations defer to the DOT when it comes to testing the integrity of minibulks. The DOT requires pesticide manufacturers to build their shuttles to specific standards that depend on the classification of the pesticides they will transport. DOT regulations require that some, but not all, minibulks used to transport pesticides must be pressure-tested every 2.5 years (30 months).

For the purpose of this chapter, we have arbitrarily sorted minibulks into four major categories under DOT rules:

1. **Minibulks that hold materials not classified by the DOT.** While some of these containers are still in use, these tanks were some of the first minibulks to be used and did not require pressure testing. These older minibulks include green and white shuttles for glyphosate-based herbicides.

2. **Single-use, nonrefillable tanks.** These tanks are more commonly used for adjuvants. Pesticides are seldom packaged in one-way shuttles. Since these are one-way, disposable tanks, they do not require pressure testing. Check labels to confirm tanks are nonrefillable.
3. **Limited-use refillable and reusable minibulk tanks.** Refillable minibulk tanks (or cage tanks) are containers that the retailer will dispose of at the end of 2.5 years. Consequently, these tanks do not require pressure testing.
4. **Long-term refillable and reusable minibulk tanks.** These are the tanks that require pressure testing every 2.5 years. Check labels to confirm tanks are refillable.

Figure 5.17. Minibulk containing materials not classified by the DOT.

One of the quickest ways to determine if a minibulk requires pressure testing is to look at the UN (United Nations) code assigned and affixed to the minibulk. The testing requirements for any tank are based on the UN specification mark on the tank itself. The specifications include the UN symbol:

Figure 5.18. A single-use tank.

UN containers have an X, Y, or Z marked on the tank. Refillable UN containers require pressure testing every 2.5 years.

At a minimum, employees who pressure-test minibulks should wear eye protection, but be sure to follow label directions relative to personal protective equipment. To begin the pressure test, make sure that all openings (like caps, one-way valves, and vents) are completely closed. Pressurize the container to 3 pounds per square inch (psi) and wait 5 minutes to see if the tank maintains pressure. The pressure added to the tank may cause the container to slightly bulge out. Do not apply too much pressure—that can permanently damage the container and/or injure the tester.

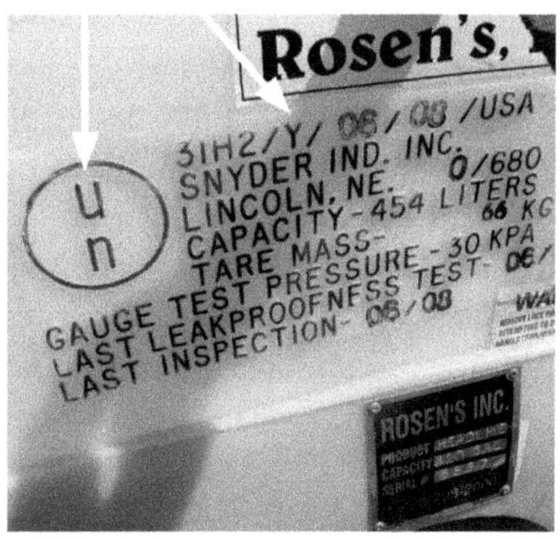

Figure 5.19. This tank is refillable.

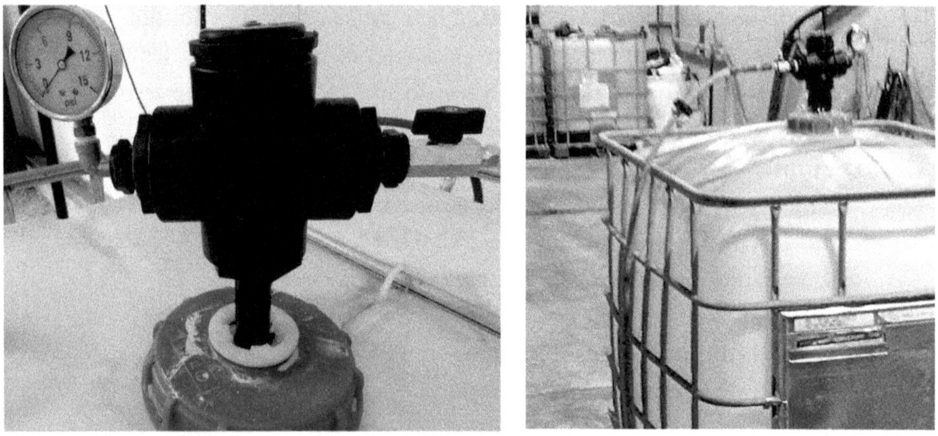

Figures 5.20–5.21. (Left) A close-up view of preparing to pressure-test a minibulk. (Right) This tank was pressurized to almost 2 psi. "We stopped the test, because the two metal straps on top were about to break," the operator said. Always use caution when pressure-testing cages!

If a tank does not maintain the pressure, it means there is a leak. Take either soapy water or glass cleaner and spray it around the lids, valves, and other areas to pinpoint the leak's source. The solution will bubble up, revealing the exact location of the leak. Repairing leaks is often as easy as replacing an O-ring in a cap or tightening a loose valve.

If the container holds 3 psi for 5 minutes with no signs of leaks, then it has passed the pressure test. Permanently mark the month and year of the inspection on the container near the UN marking.

In addition, you must create a record to document the testing protocols. These documents must be kept for three years. That documentation must include:

- Tester's name
- Name and address of the testing site
- Date of the test
- The container's identification number based on its serial number or another number permanently marked on the tank
- Conclusion statement of pass or fail
- Description of the testing procedure used

There are times when tanks fail pressure tests and must be taken out of service. When that happens, permanently and prominently mark the minibulk as "out of service" or "failed pressure test." Make sure to note in your records when you took a container out of service and then destroyed or recycled it. The Pesticide Stewardship Alliance lists minibulk recyclers on their website: https://tpsalliance.org/mini-bulk-ibc mgmt/container-collection-and-recycling-resources.

FARM-OWNED MINIBULKS

As end users, farmers are not required to follow the same production rules as ag retailers. This means you can transfer material for your own use from a retailer's minibulk into your own minibulk without worrying about production reports, UN codes, or even pressure testing. However, just because you are allowed to skip something doesn't mean it's a good idea to do so! Consider the following problems that can be caused by ignoring industry requirements, common sense, and a few best-management practices.

Spills

If you never inspect your own minibulks, you almost guarantee a tank failure in the future. In addition, insurance policies do not normally cover spills along highways or ditches unless you added special coverage. The potential economic losses (not to mention the environmental impact) from a pesticide spill could force a farm out of business.

Tank Contamination

Failing to dedicate minibulks to a single product or failing to properly clean minibulks can cause product contamination. Contamination usually results in substantial crop damage and injury, which your insurance may not cover.

Retailers Won't Fill Your Minibulk

Because retailers are required to follow repackaging agreements, they will not fill minibulks that do not meet the standards set out in their contracts. Farmer-owned tanks that don't have one-way valves, have no tamper-evident seals, or, worse, still contain product represent huge liability risks for retailers. Likewise, retailer-owned minibulks that are returned with broken seals and leftover product also represent, at the very least, a dilemma for retailers: Do they take a chance by taking the product back? Or do they possibly lose customers by telling them they will not accept the return? The simplest way to avoid these possible problems is to use only retailer-maintained tanks that are returned in good working order with seals intact.

Figure 5.22. Retailer warning label on a minibulk that it will cost $300 to clean a tank.

The most serious of the issues is the retailer's response to material left in a shuttle when an inspection shows that the seals have been broken. The farmer may be asked to take the product home, since the retailer cannot know for certain whether the product has been adulterated.

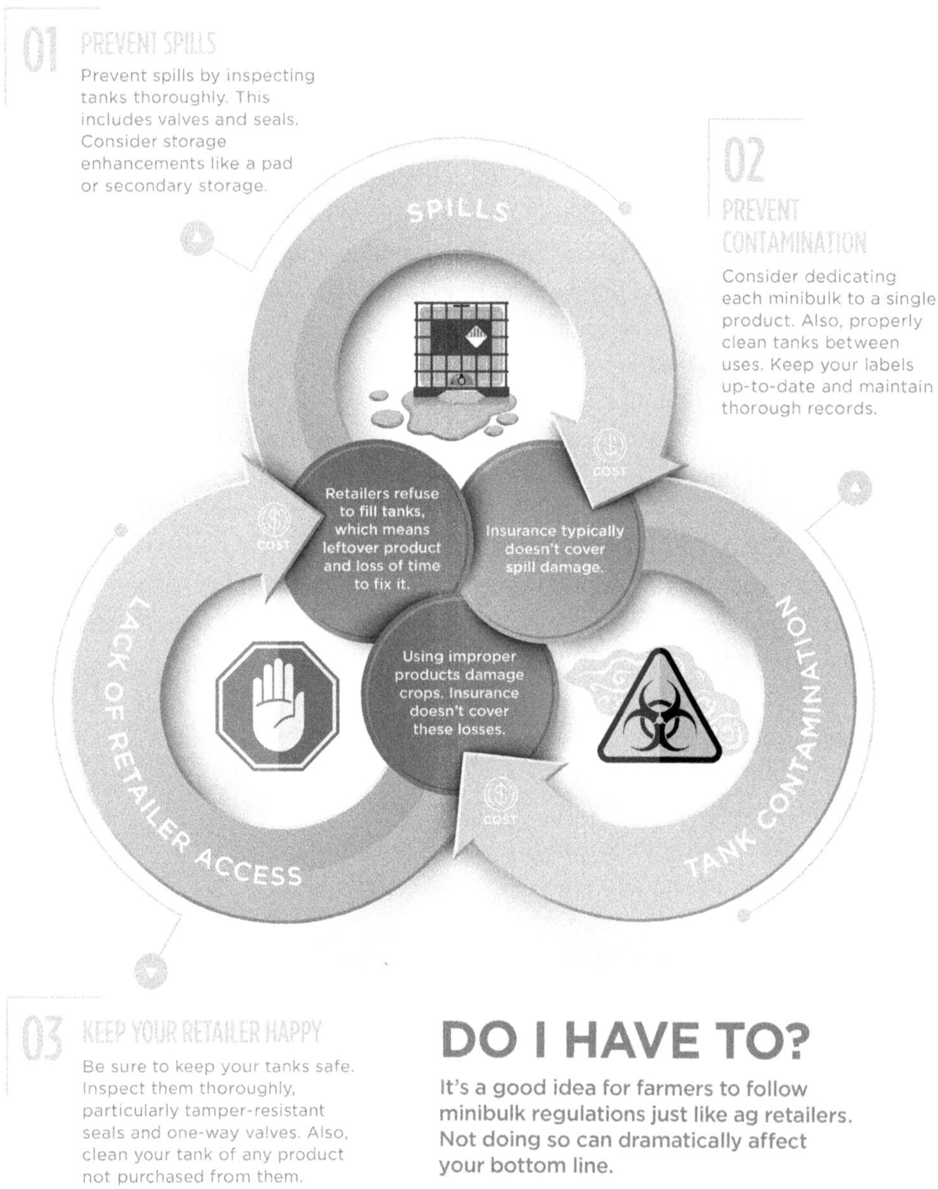

CONCLUSION

Agriculture has embraced minibulk use as a very cost-effective and efficient way of moving pesticides from retailers to farms. The convenience of bypassing many smaller containers and the lower price of bulk products make minibulk use an important part of agricultural production.

Farmers and retailers know the benefits of minibulks well, but they also need to know about the concerns of spills and tank contamination. Retailers and farmers agree that the added cost of providing secondary containment and following repackaging agreements is significant. However, they also agree that following the regulations helps prevent spills, lowers the frequency of tank cross-contamination, and improves environmental quality. When farmers and retailers work together as part of a minibulk system, they both can reap the benefits of lower product cost and liability and increased safety and profit.

Figure 5.23. A minibulk provides so many benefits, including quicker filling and cleanup compared to 2.5-gallon jugs.

6

Tanks for Micronutrients

With contributions from Brad Peas, Darrell Oltman, John Obermeyer, and Kevin Leigh Smith

Progressive farmers have long been early adopters of new technologies to help them get the most economic return from their crops. One practice receiving attention today is applying micronutrients. Farmers are asking whether it is worth the additional cost per acre to apply secondary nutrients and micronutrients like sulfur, boron, copper, zinc, and manganese.

Like for any new technology, many of the questions farmers and retailers are asking about micronutrients are being investigated by university and industry scientists. However, it seems that a number of commercial retailers and farmers are convinced they are getting economically significant yield gains from using micronutrients. And micronutrient use is increasing.

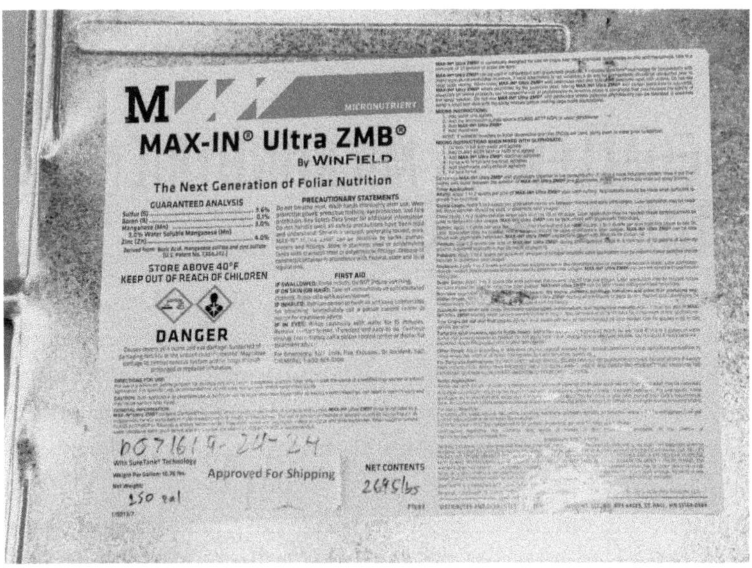

Figure 6.1. A label for a micronutrient blend contained in a minibulk.

MY $20,000 TANK IS READY FOR THE JUNK HEAP?

Never order a tank to store micronutrients without first doing your homework and asking lots of questions. Your effort might just save you from the need to purchase a second tank to replace one that failed prematurely. Questions you should answer before purchasing any tank include:

- **How are the micronutrients formulated?**
- **Is the tank material compatible with the specific micronutrient(s) you intend to use?**
- **Is the tank designed to let you completely remove settled micronutrients?**
- **Can you prevent micronutrient solutions from freezing?**
- **What size tank will you need to store enough for the entire season?**
- **When will you need the tank delivered by?**
- **How much lead time will crane operators need to install large tanks?**
- **Will adding micronutrient tanks put you over the limit for bulk storage regulations?**

With micronutrient use increasing, and with projections that the market will continue to expand, the question arises of which tank material is best suited for storing these products. Micronutrient formulations may be different enough that the tanks and conditions that store urea ammonium nitrate (UAN; 28%–32% liquid N) may not be suitable for micronutrient fertilizers.

BASIC CHALLENGES MICRONUTRIENTS POSE FOR STORAGE TANKS

Plant nutrients include macronutrients, secondary nutrients, and micronutrients. Macronutrients include nitrogen, phosphorus, and potassium. Secondary nutrients include sulfur, calcium, and magnesium. Micronutrients include zinc, copper, boron, manganese, iron, and others. While we often consider all of these as fertilizers, sulfur and micronutrients form a diverse group of chemistries that can interact in various ways with the materials from which storage tanks are made.

If you're browsing online, it might seem that there are plenty of tanks you can purchase to store micronutrients. However, the reality is that the material used to construct the tank, its design and shape, the cost to transport and set it up, and its availability will quickly narrow down the choices to just a handful of tank products. If you want to purchase a tank to store micronutrients at a retail facility or on the farm, consider the eight points outlined in this chapter to help you narrow down your options to one or two tanks that will provide the best value over the long term.

Tank Compatibility: Working from the Inside Out

Most manufacturers and distributors that sell micronutrients offer different formulations. But the main products, such as boron and zinc, are the same because they are elements. What differentiates micronutrient products is the way they are formulated. That should not come as a surprise. After all, nitrogen can be formulated as anhydrous ammonia, UAN (both 28% and 32%), aqua ammonia, and urea.

Just like the different forms of nitrogen, manufacturers formulate their micronutrient products differently. The chemical makeup of the unique micronutrient formula can create compatibility problems with some materials used in constructing storage tanks.

Formulations are important. Plants cannot take up many micronutrients in their pure form, so manufacturers have to place the micronutrient in an acid or alkaline solution that makes it available for crop uptake. The acidity or alkalinity of the micronutrient solution can range from a pH of 1.2 (equivalent to battery acid) to a pH of 11 (equivalent to bleach). Knowing the micronutrient product's pH is critical when determining the types of tanks you can use.

Four general tank types have found wide use in agriculture: poly, fiberglass, carbon (mild) steel, and stainless steel. When you select a storage tank, your goal is to make sure the tank material is compatible with the chemical and physical characteristics of the micronutrient product.

Poly Tanks

The pH of a solution does not affect poly tanks, which is why people often use them to store highly corrosive materials such as sulfuric acid. When ordering a poly tank, you must consider that micronutrients typically weigh 10–11 pounds per gallon. Therefore, they must be stored in a tank with at least a 1.5 specific gravity rating (a 1.9 rating would be even better). If you store them in a tank that wasn't designed to hold that weight, the tank can fail prematurely. (See chapter 1 for more on specific gravity.)

Figure 6.2. New tanks provide their specific gravity ratings (e.g., 1.5 or 1.9) on a label. Contact manufacturers to determine tank densities when the labels are missing.

Figure 6.3. Make sure the tank's rating matches or exceeds the weight per gallon of the stored micronutrient product, or else the tank may fail.

In general, poly tanks are made from high-density linear polyethylene (HDLPE) but can also be made with cross-linked polyethylene (XLPE). HDLPE is generally compatible with most micronutrients and is the most common tank material. XLPE is generally more expensive but can be beneficial if surfactants are part of the micronutrient blend. If the tag identifying the tank material is missing, then call the manufacturer and ask them to confirm its composition by providing the serial number, which is often etched around the lid of the tank. See chapter 2 for more on poly tanks, including maintenance and testing.

Fiberglass Tanks

Fiberglass tanks are also a good choice for storing micronutrients. On the outside, one fiberglass tank looks the same as the next. But the resins used to manufacture fiberglass tanks make them different. For corrosive materials like micronutrients, manufacturers will add an additional protection layer known as a *veil.* This is a more expensive (about 15 percent) vinyl lining. Once the tank has been ordered and made, the tank can't be upgraded. See chapter 3 for more on fiberglass tanks.

Carbon Steel Tanks

Carbon (mild) steel is unsuitable for storing most micronutrients because of the corrosiveness of the materials. At a lower pH, the acidity of products like copper and zinc can degrade carbon (mild) steel quickly. Because of these concerns, you should not use tanks made of mild steel to store micronutrients.

Stainless Steel Tanks

Most retailers and farmers believe that stainless tanks are compatible with all liquids. That's not necessarily true when you store more corrosive micronutrients. Stainless steel's suitability depends on the grade used in the tank's construction. Traditionally, agriculture has used the less expensive 304 stainless, which contains a higher percentage of carbon in its construction than 316 stainless. But some micronutrient products attack carbon, which makes 304 unsuitable for storing them. A 316 stainless steel tank may be able to store most micronutrients, but check with tank manufacturers about the products you plan to store. See chapter 4 for more on steel tanks.

Figures 6.4–6.5. The plates on stainless steel tanks indicate the type of steel used: 304 (top) and 316 (bottom).

Micronutrients can affect pumps. You should expect cast-iron pumps to last fewer years when using some micronutrients. Stainless steel pumps are available but are about four times the cost of standard pumps.

DON'T FORGET THE STAINLESS STEEL GRADE OF FITTINGS

A common mistake is to forget that many fittings and connectors are made from 304 stainless. As these photos suggest, many micronutrient products can rapidly degrade 304 stainless. When upgrading tanks, let installers know you will need 316 fittings and connectors. In general, polypropylene fittings are also compatible with most micronutrients.

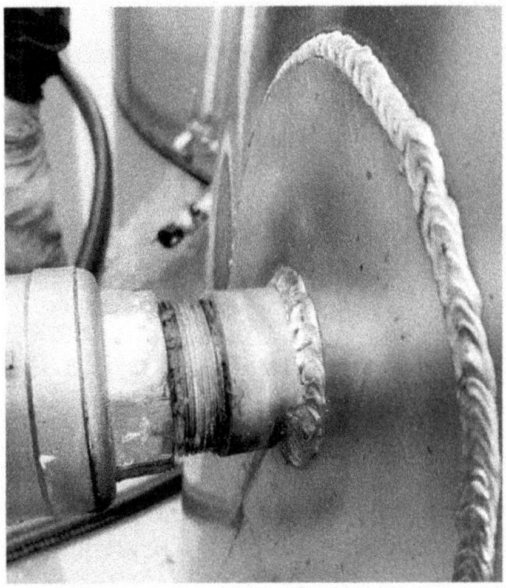

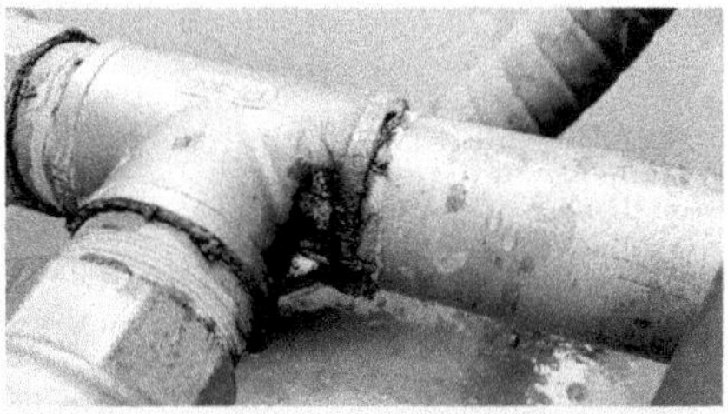

Figure 6.6. This pump was degraded by micronutrients. You can help reduce corrosion by occasionally washing off the exterior to remove spilled material.

CHECK THE SAFETY DATA SHEET (SDS)

Safety data sheets can provide important information about micronutrients and the materials they are incompatible with. Some will caution you to avoid fittings made from copper and its alloys, aluminum, and nickel. When these fittings react with micronutrient products, this can be very destructive to 304 and 316 stainless. Check with the stainless steel manufacturer to see if their tanks and fittings are compatible with the micronutrient products you plan to store.

Tank Design: Emptying Tanks of Settled Product

Some micronutrients are formulated as suspensions, which means that the products "float" in the liquid instead of being dissolved in a solution. Without agitation, these suspended micronutrients tend to settle on the bottom of the tank over time. Therefore, we recommend that you purchase tanks that have a cone-shaped or sloped bottom that drains to a common outlet. Such designs allow you to easily remove settled materials from tanks.

It is common to add recirculation pumps to tanks to help keep micronutrient products in suspension. You can add additional fittings to tanks to aid in recirculating the product. You can also add eductor nozzles that will help move the chemicals faster and more efficiently. A recirculation pump pushes material through the eductor nozzle. The nozzle's shape pulls in additional volumes of material, which causes much greater flow in the tank. Eductor nozzles commonly have a 4:1 or 5:1 mixing ratio. So, flowing 50 gallons per minute (gpm) through the eductor causes a mixing flow in the tank of 200 (4:1) or 250 (5:1) gpm.

Another option for agitating products is available for larger fiberglass and 316 stainless steel tanks. Some of these tanks have a side manway that allows access to the inside. Once inside, you can sweep the bottom debris back into the liquid as a suspension.

Gaskets and Seals: Keeping Product Flowing Instead of Leaking

It is also important to review the materials used to make the gaskets and seals that will come in contact with the micronutrients. Gaskets are generally constructed of either EPDM or Viton. EPDM generally works well with high-pH (caustic) products, while Viton generally works well with low-pH (acidic) products. You can see why it is important to know the pH values of the micronutrients you plan to store.

Figures 6.7–6.9. These photos show what micronutrients can do to seals.

WHAT ABOUT PUMPS?

Micronutrients should not affect pumps that are used for multiple products. Pumps are often used to move water, fertilizer, pesticides, and micronutrients. This mixed use allows the micronutrients to be flushed. Just make sure to flush pumps with clean water and antifreeze before storing them for the winter. **Pumps that are dedicated to micronutrient products can have problems if they remain bathed in the products.**

Availability and Pricing: Having Enough Storage Capacity

While micronutrients are available in convenient 2.5-gallon containers, larger operations prefer to purchase micronutrients in less expensive bulk quantities, which lowers their cost per acre. Note that micronutrients are, like other ag products, subject to market disruptions and availability. If you have enough storage capacity to offload a year's supply of these products, that can guarantee you will have the micronutrient products on hand, especially in the spring.

If you want larger storage tanks, that may eliminate poly as a choice. If you decide to use a cone-bottom poly tank, manufacturers may limit the larger sizes to around 11,000 gallons. Going above this capacity would require a fiberglass tank or a 316 stainless steel tank with sloped floors and manways. In that case, you will need to weigh the increased cost for fiberglass or stainless against the benefits of storing these larger amounts for lengthy periods.

Storage Conditions: Preventing Micronutrients from Freezing

Do not assume all liquids freeze at 32°F. Some micronutrient products will freeze at different temperatures than water. Be sure to ask the micronutrient manufacturers if their products will freeze, whether warming their products will resuspend the products, and whether the micronutrient remains effective after thawing out.

Figure 6.10. This manganese label specifies that it should be stored above 40° F.

If possible, store micronutrients that you must keep during the winter in a temperature-controlled building or barn. If that is not an option, you can order poly tanks that have heat tracing and insulation. You can keep these custom poly tanks above freezing with electrical heating systems, and these systems can be covered with insulation to help keep the heat inside the tanks. Don't forget that you will also have to heat and insulate any hoses or pipes connected to the tanks.

Figures 6.11–6.12. These photos show zinc and sulfur products in heated storage rooms.

Figures 6.13–6.14. Another option for tanks that must remain outdoors is to order poly tanks with insulation and heat tracing like the ones shown here.

Schedule: Giving Yourself Time to Order

The tanks we've described may not be available from your local vendors. It could take months to get your specific tank made and delivered. This will require you to plan ahead before the growing season starts, so you can get your tanks delivered and filled with micronutrients. You will also have to work with local contractors who can be standing by to offload the semis.

It's important to understand the micronutrient product you intend to store. If you plan to store a product that contains zinc in a formulation that largely stays suspended, then you can order a flat-bottom tank from a local vendor instead of waiting for a cone-shaped poly tank to be made and delivered. However, if you plan to store a different zinc formulation or a different micronutrient product, then the flat-bottom tank might not work as well without agitation or recirculation. The flexibility of cone-bottom tanks may be the best option for your operation, which is why they may be worth the wait to have one custom-made.

Containment: Planning Ahead

Some states consider micronutrients a fertilizer, so storing them could throw you over the limits allowed under your state's containment regulations.

CONCLUSION

Farmers and ag retailers are always looking for ways to get more bushels out of every acre. In recent years, Corn Belt farmers have begun using more micronutrients as part of their fertilizer programs to try to improve their crop's yield potential. University and industry scientists are searching for answers to many of the questions about micronutrients. That said, many commercial retailers and farmers are convinced that micronutrients return more in yields than what they cost to purchase, store, and use.

With more farmers using micronutrient products, questions about storing them in larger quantities are becoming more common. Do your homework, so you don't end up like the person who spent $31,400 only to find out that the corrosive materials turned expensive tanks into scrap metal (see "A Costly Mistake!"). After you narrow down your tank choices based on what you read in this chapter, talk with your tank supplier and micronutrient supplier to get the answers you need about compatibility and other features that may be important in the final selection. Ask before you spend!

A COSTLY MISTAKE!

Storing a micronutrient in the wrong tank can be an expensive mistake. These photos show a horizontal 304 stainless steel tank that stored a corrosive material. You can see where the caustic material penetrated the welds. The owner had two tanks like this, each costing $15,700, so the two tanks heading to the scrap heap was a $31,400 mistake. If a picture is worth a thousand words, then it should be evident that asking a few questions before purchasing a tank for micronutrients can be worth tens of thousands of dollars.

7

Aboveground Petroleum Tanks

With contributions from Steve Hawkins, Lysa Holland, Barbara Carr, Dennis Meredith, Klein Ileleji, Don Campbell, Larry Gentry, Larry Hamby, Max Roach, and Arlene Blessing

Petroleum products such as gasoline, diesel fuel, biofuels, lube oils, and engine oils are essential to farm operations, commercial pesticide application businesses, and American commerce. It is widely accepted that the lower (bulk) cost and operational convenience of storing gasoline, diesel fuel, and oil offset any potential problems associated with on-site storage. Quick and easy access alone is valuable, as are savings realized by maintaining productivity, that is, not pulling employees off the job and sending them to pick up fuel. On-site storage also assures timely access to bio-based fuels and off-road diesel fuels that may be unavailable from local retail sources.

But there are risks related to the potential for soil contamination, surface water and groundwater contamination, theft, and vandalism. Zoning restrictions, increased insurance costs, reduced cash flow, and significant regulatory controls also must be considered. The potential risk of on-site fuel storage can be reduced by following a set of safety, security, environmental, and regulatory guidelines. These include choosing the best location for aboveground storage tanks (ASTs), incorporating good engineering practices, and observing industry standards.

Poorly managed aboveground storage of petroleum products can be costly. Spills and other consequences of poor management can involve years of cleanup efforts and lengthy legal battles. Issues of concern include the following:

- Soil and water contamination
- Decreased property value due to spills and releases
- Cleanup costs
- Evacuation of nearby neighborhoods in the event of a sizable release
- Theft and vandalism
- Potential for terrorist activity
- Insurance specifications
- High cost of insurance coverage for environmental pollution
- Local, state, and federal regulations, inspections, and enforcement actions
- Adverse environmental audit (may prevent sale of the property)

Figure 7.1. While storing fuel on the farm has many advantages, making sure it is stored properly is critical.

This chapter addresses these potential problems and consequences. It describes field-tested management procedures and security measures to use when storing, dispensing, and disposing of petroleum products. It will raise your awareness of federal and state petroleum regulations that impact your facility. Hopefully, it will motivate you to upgrade management practices at your facility by implementing our suggestions. Your mishap probability—and your liability potential—will decrease as a result.

BEST MANAGEMENT PRACTICES FOR ASTS

The use of aboveground (versus buried) fuel storage tanks has increased over the past two decades, both on the farm and at commercial businesses. This is due largely to the cost—to property owners and insurance companies—of cleanup associated with leakage from buried tanks. When private wells and community water supplies are contaminated, cleanup efforts are extensive and costly, often reaching six figures. Stricter federal and state regulations governing underground fuel storage have resulted.

Properly engineered and installed aboveground tanks have many advantages over those placed in the ground. Leaks are easily detectable and quick to contain. Aboveground tanks facilitate visual inspection and can be painted to prevent corrosion, making leaks less likely. Aboveground tanks can be safely relocated to a different part of the property as circumstances change (e.g., additional building construction, altered traffic flow). Last, tanks can be removed when farms or businesses are sold, enhancing property value and eliminating risk to the new owner.

Despite these advantages, aboveground fuel tanks have drawbacks. Vehicles can back into them, vandals can deface or damage them, and trespassers can steal their contents. They are vulnerable to high winds, flooding, and lightning. Adjacent property owners may consider the pipes and hose apparatus—even the concrete slab on which the tanks rest—unsightly.

Figure 7.2. Shrubs have been planted and mulch applied for aesthetic purposes near these aboveground fuel storage tanks.

Figure 7.3. These tanks have been painted dark green to blend into the landscape. The bollards keep vehicles from hitting them.

Formerly, such concerns guided a trend toward burying petroleum fuel tanks, but in the long run, tank leakage created environmental problems far greater than any posed by aboveground tanks. More stringent federal regulations subsequently forced many farmers and business owners to switch to ASTs.

Proper management of the risks associated with ASTs is essential. Train your personnel to identify and eliminate risks; to conduct routine inspections of fuel storage containers; to dispense fuel and operate pump shutoffs properly; to contain spills; and to conduct cleanup procedures, including the safe operation of equipment. Involve your employees in scheduled reviews of your operation, identifying steps you can take to minimize spills.

Selecting Tanks

Selection of a proper storage tank for your operation prevents problems. Here are some things you should know up front:

- A tank intended for outdoor use should have an Underwriters Laboratories (UL) outdoor-use designation.
- Indoor basement tanks are not intended for outdoor use.
- Tanks designed for flammable liquids such as gasoline should be fireguarded or vaulted to prevent fuel ignition for two hours during a fire.
- Diesel fuel tanks do not have to carry a two-hour fire rating if the flash point is greater than 100°F.
- A secondary containment area must have the capacity to hold the contents of the primary tank, should it fail.
- Double-walled tanks satisfy secondary containment requirements, as do dikes or "bathtubs" in which the primary tank is situated.
- Double-walled tanks are preferable because rainwater is not a problem; removing accumulated rainwater from dikes can be a hassle.

Every tank fill-up poses a spill risk, so limit fill-ups to once a month by selecting a tank that will hold a month's supply of fuel. In selecting a tank size, calculate its usable volume; note that the pump intake is situated several inches above the bottom of the tank, which reduces the volume of usable fuel. Also, to allow for expansion, tanks should be filled to no more than 95 percent capacity.

When your tank is installed, you will receive a chart for converting inches of fuel in the tank to gallons. Measure the depth of fuel by carefully lowering a marked dipstick into the tank and removing it. Note the depth of fuel indicated on the stick and use the tank chart to convert inches to gallons. Tank charts are specific to the tank's configuration, so always use the chart that corresponds to your specific tank.

Figures 7.4–7.5. This double-walled unit is a tank within a tank. The space between the internal tank and the outer tank is monitored by a leak detector. In this case, the internal tank is divided: Half the tank holds gasoline, and the other half holds diesel. A divided tank is a good option when space is a concern.

Figure 7.6. This tank without containment is located beside a pond, leaving little doubt where a spill would end up.

Figure 7.7. These uncontained tanks are located too near a well house. If a spill were to contaminate the well, years of expensive treatments would be required.

Selecting Tank Sites

Tanks should never be placed near lakes, streams, ditches, drains, or well. Uncontained spills could quickly pollute surface water, impact drinking water supplies, and harm wildlife.

Strategies and Recommendations

The placement of ASTs is critical, and the safety of people and property is your primary consideration. Tanks have to be accessible to employees, fuel delivery drivers, and emergency responders but must not be near traffic patterns. They should be located a minimum of 50 feet away from buildings, creeks, roads, wells, power lines, grain bins, and property lines. Do not place tanks where ignition sources such as welding and cutting torches are likely to be used. Avoid roof drip lines; water may freeze as it flows from the eave of the building onto the tank, or snow may fall from the roof onto the tank. Flat ground is a must so that a spill or leak will not run downhill toward creeks, ditches, tiles, or drains before you can deal with it.

Figures 7.8–7.9. The area where delivery trucks park to fill the ASTs—the off-loading area—should be close to and within clear sight of the tanks. Trucks should not be required to back up or make sharp turns. Tanks should be placed in an area accessible to employees and fuel delivery drivers, but away from the main traffic pattern and not visible from the highway. Bollards should be installed to prevent trucks from backing into the tanks pictured here. Notice in the second photo that tanks are not placed above underground utilities nor directly beneath overhead power lines.

Figure 7.10. This tank is situated in flat, open fields. If an accidental release were to occur, it is unlikely that fuel would enter surface water, drains, or ditches.

Placing Tanks on Secure Bases

The importance of placing a tank on a secure base is illustrated in figures 7.11–7.13. An aboveground tank was supported for many years on a wooden base. The wood eventually rotted, tipping the tank over and releasing its 118 gallons of fuel onto the ground. Notice that the tank would have landed in a creek had it fallen to the left. This is an example of poor tank placement (near a creek) on a poor base (wood instead of a concrete pad). Only luck prevented this spill from becoming an environmental disaster.

Figures 7.11–7.13. A wooden base that supported a petroleum tank collapsed, leading to a release near a creek.

Strategies and Recommendations

Ideally, tanks should be placed on a reinforced concrete pad at least 6 inches thick; large tanks may require stronger support. Packed stone will support some tanks, but tanks never should be placed on wood or hollow concrete blocks. Tanks should be at least 6 inches off the ground to reduce corrosion, allow for inspection, and permit painting of the undersides. Tanks should rest on steel supports manufactured for their weight plus the maximum contents. The area under and around tanks should be sufficiently impermeable to retain a discharge until cleanup can occur.

Figure 7.14. This tank is leaning because of soft ground underneath.

Figure 7.15. These tanks were inappropriately mounted on concrete blocks. Notice that the tank on the left is leaning, with the main part of it barely resting on the blocks. There is no containment. Hoses left in contact with the ground tear and deteriorate quickly.

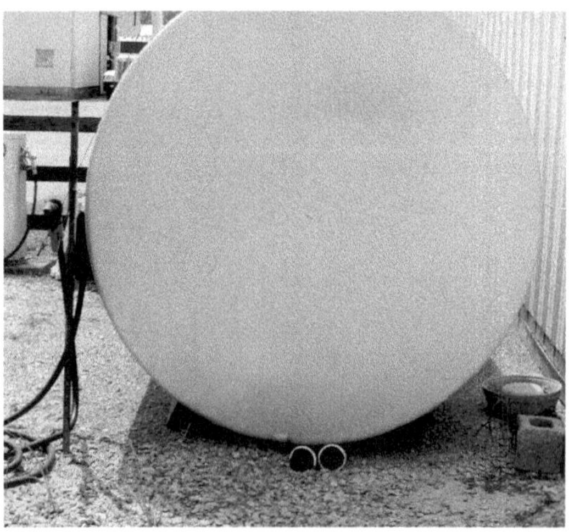

Figure 7.16. This silver tank rests on gravel. The drainage pipe underneath it prevents erosion: Rainwater flows through the pipes and away from the tank.

Identifying Tank Contents

Improperly labeled tanks can be a serious issue when an emergency occurs or when employees place the wrong fuel in trucks and equipment.

Strategies and Recommendations

Proper labeling includes identification of contents, tank volumes, hazards, and warnings that ensure the correct fuel is delivered to the appropriate tank, that equipment is filled with the proper fuel (e.g., gasoline versus diesel), that dyed and undyed fuels are used in the appropriate vehicles, and that emergency responders can easily determine what product they are dealing with in an emergency.

Figure 7.17. Each tank should have clearly visible labels indicating its contents and associated hazards. The Department of Transportation's diamond-shaped placard and the National Fire Protection Association label are good choices. Each fuel tank should bear the words "Flammable: No Smoking." Post danger signs in both English and Spanish.

Product signage should be 5 x 14 inches, with black lettering on a white background. "Flammable: Keep Fire and Flame Away" placards should be 10 x 14 inches, with red letters on a white background. "Dyed Fuel" and "Undyed Fuel" should be 10 x 10 inches, black on white.

Figures 7.18–7.19. These tanks are marked with large lettering indicating the product type and warning the user about hazards associated with the product.

Preventing Vehicular Contact with Tanks

Sometimes aboveground fuel storage tanks must be situated in high-traffic areas at the farm or business due to limited space or for easy access. But tanks can rupture upon contact with vehicles, leading to expensive remedial action. Bollards become one of your most valuable assets when prevent vehicles from hitting tanks.

Figure 7.20. Vehicles are parked head-on to these tanks.

Strategies and Recommendations

Farm managers and business owners all want their aboveground tanks near their base of operations to save time going back and forth for fuel. In most cases, some type of barrier has to be installed to prevent vehicular contact with the tanks. Barriers can include piping, wood framing, concrete walls or pillars, concrete wheel stops, and guardrails (figures 7.21–7.24). Curbing is useful for controlling vehicle placement near tanks. Pipe bollards can be installed around fuel tanks. They normally are 4–6 inches in diameter; common spacing is 3–4 feet apart and at least 3 feet from the nearest tank.

Figures 7.21 -7.24. Examples of barriers that protect fuel tanks from vehicles.

Grounding Tanks

Sparks can ignite built-up vapors, with disastrous consequences. An equipment grounding conductor should be installed if electrical equipment is allowed within the fuel storage area. It is important that all electrical connections to petroleum equipment be made with explosion-proof fittings and petroleum-resistant wiring. Explosion-proof fittings also should be used at the electrical panel. The conduits at each end must be sealed to prevent vapor release.

Strategies and Recommendations

Tanks should be grounded. Some are manufactured with a specific location for attaching the grounding wire. Explosion-proof pumps offer additional protection.

All fittings, whether the union is ½ inch or ¾ inch, must be vapor proof. An explosion-proof seal is required where the power enters the dispenser and also at the end of the conduit leading to the circuit box. All electrical elements such as conduits and fittings should be explosion proof, and all wiring should be fuel resistant. The emergency shutoff should be at least 15 feet from the tank but no more than 120 feet.

The National Fire Protection Association requires explosion-proof wiring methods, including whips to connect motors. Grounding conductors for equipment must have branch circuits. Do not use a metallic conduit for grounding equipment in hazardous locations. Grounding conductors that are not tightly fitted can lose their ground, creating a shock hazard.

Venting Tanks

Venting is a safety requirement to keep vapors from collecting in the tank as the temperature rises; venting allows vapors to escape, decreasing the likelihood of explosion. ASTs should have both primary and emergency vents. The primary vent should be at least as wide as the fuel fill pipe. It should extend above the normal snow level for combustible fuels, or 12 feet aboveground for flammable fuels such as gasoline. The primary vent pipe should be checked regularly for obstructions such as bird and insect nests.

An emergency vent also should be installed on the tank. The most common type is called a long bolt manhole vent; in case of fire, the increased tank pressure lifts the weight, venting the tank. The weight should periodically be lifted manually to ensure that it remains free-moving, that is, has not rusted to the tank. Double-walled tanks require two vents, one for the inner tank and one for the outer tank.

Figure 7.25. Tall primary vent pipes, compared to short tank vents, allow vapors to dissipate over a larger area, preventing the buildup of vapors around the tanks. Some insurance companies require the taller vents.

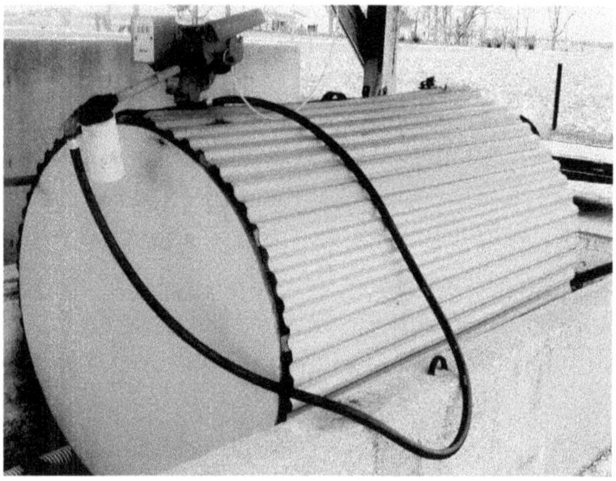

Figure 7.26. This tank under roof is wrapped to deflect heat, helping to reduce vapors.

Preventing Contamination Around ASTs

Small amounts of fuel spilled when filling equipment, during careless fuel delivery, or as a result of poor management can slowly contaminate the soil. A leak of only one drop per second can release about 400 gallons of petroleum into the environment in one year, resulting in significant soil and groundwater pollution. Many insurance policies will not pay for remediation of contaminated soil and drinking water supplies. Review your policy to see what coverage you have. In addition, you must consider that 400 gallons at \$3 per gallon would represent a loss of \$1,200 from your farm or business budget!

Figure 7.27. Notice the soil staining in the fuel loading area. Over time, fuel spilled from hoses can lead to significant problems.

Figure 7.28. Here the nozzle has been left on the containment wall, risking a spill from any fuel left in the hose.

Figure 7.29. These fuel tanks are contained in a portable metal dike, which simplifies the cleanup of spills.

Figure 7.30. Concrete containment can have short or tall walls. It must be designed to hold at least 110 percent of the volume of the largest tank.

Figure 7.31. This nozzle is hung on a steel fence post to prevent dripping.

Figure 7.32. Although the nozzle is hung within the containment system, the hose remains outside. A leak in the hose would spill fuel onto the ground outside the containment walls.

Figure 7.33. This tank has a sign reminding employees to place the hose and nozzle inside containment walls. A plastic sleeve protects the hose in case the sign goes unheeded.

Strategies and Recommendations

Prevention of fuel spills and leaks is the most important management tactic in minimizing pollution liability at your farm or business. You can purchase materials to absorb spilled fuel. Some types absorb petroleum only, while others absorb water and coolants as well. Always keep a spill kit at each fuel tank location. Contents should include absorbent pads, pillows, and socks to contain and recover spilled fuel.

Managing Rainwater Within Containment Systems

Managing rainwater that falls into containment structures presents a difficult challenge. Clean water may be released into the environment, but contaminated water may not. An oily sheen on the surface of contained water indicates the presence of fuel, and separation is necessary before the water may be returned to the environment.

Figure 7.34. Your spill kit should contain enough materials to handle a typical spill. Commercial spill kits often contain materials of different shapes: pads, socks, pillows, etc. Each product is color-coded. Blue/gray is universal for absorbing water, coolants, and oil. White absorbs oil and fuel only. Yellow and pink absorbent pads are designed for hazardous chemicals. Instead of a commercial kit, you may also use a covered drum or tote with a couple of bags of oil dry, a dust pan, and a broom.

Strategies and Recommendations

It is important to release clean water from containment as soon as possible after a rain. The longer it stays contained, the greater the opportunity for it to become contaminated with fuel (e.g., from a nozzle). In addition, corrosion may begin on the tank's surface or supporting structure, compromising its integrity. And the volume of rainwater decreases the containment capacity for petroleum in the event of a spill.

In case rainwater does become contaminated with petroleum, it is important to have spill pads that absorb oil only. These float on the surface of contained water, absorbing spilled fuels before you release the water.

Figures 7.35–7.36. (Top) The rolled-curve loading pad in front of the tanks contains spills when unloading delivery trucks or filling equipment. (Bottom) Clean rainwater that collects on the pad is released through a pipe. The pipe's valve remains closed and locked when not in use. Some businesses pass all collected rainwater from containment into an oil-and-water separator.

Figure 7.37. This metal containment system had its plug removed, allowing water to drain. However, since the plug was never put back, any spill would be released onto the ground.

Figure 7.38. The concrete containment system shown here is well built, the tank contents are identified, and the nozzles are locked when not in use. Nevertheless, there is a major problem: The drain valve in the back is open. This allows water—and any spill that might occur—to drain immediately, which defeats the purpose of the containment facility. Also, the paint on the right-hand tank has eroded down to the primer.

Preplanning for a spill reduces its environmental impact—and probably the cost of cleanup. The product you choose should be available at the fuel storage site. After the absorbent is saturated with the spilled product, store it on plastic and keep it out of the rain. Put plastic beneath the saturated absorbent even if it is stored on concrete; otherwise, the petroleum could be absorbed by the concrete, causing additional contamination and cleanup. Vapors escaping from the saturated absorbent can be a fire hazard, so be sure to properly dispose of the material as soon as possible. Small amounts of used absorbent can be double bagged in plastic trash bags to await disposal, but always check local regulations.

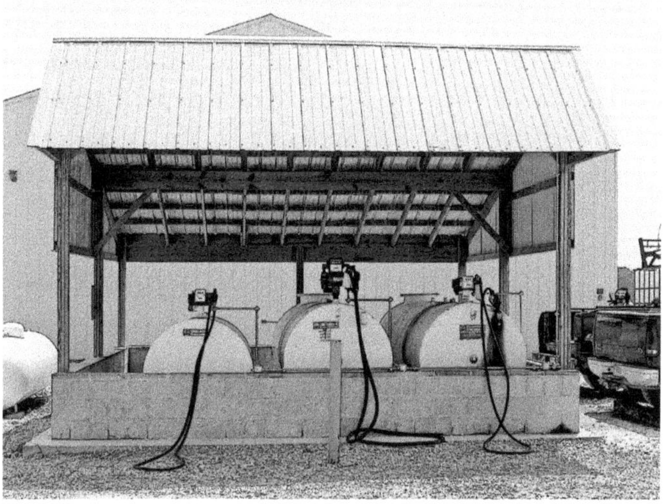

Figures 7.39–7.41. Covering tanks eliminates most rainwater concerns and reduces fuel loss due to evaporation. The cover in the top photo should have a vent at least 3 feet above the roof and at least 12 feet above grade. The open-air, shed-type cover in the middle photo is better because it does not trap potentially explosive vapors.

General Maintenance

Simple maintenance items often get forgotten. Tanks need to be repainted to prevent corrosion on the surface, and weeds need to be removed to reduce fire hazards.

Monitoring Fueling Operations

You are responsible for spills made during the delivery of fuel to your facility. It is advisable to be there or assign an employee to monitor fuel delivery to ensure the following:

- The correct fuel is placed in the tank.
- The delivery person knows how much fuel is needed to fill the tank to 90 percent capacity (calculate this beforehand).
- The correct amount of fuel is dispensed.
- The spill basket is emptied and cleaned prior to fuel delivery.
- Spill response materials such as socks, absorbent pads and pillows, and drain covers are within reach.
- The delivery person is aware of drains and other surface water features prior to filling the tank.
- The delivery person knows what type of overfill device is on the tank and what action to take if it activates.
- No one smokes within 50 feet of the fuel delivery truck, its hoses, and the tank vents.
- Spills are cleaned up immediately.

Use this formula to calculate the amount needed to fill the tank to 90 percent (0.90) capacity:

$$\textbf{(tank capacity in gallons} \times \textbf{0.90)} - \begin{array}{c}\textbf{gallons of fuel}\\ \textbf{in the tank}\end{array} = \begin{array}{c}\textbf{maximum amount}\\ \textbf{of fuel to add}\end{array}$$

Facility personnel who monitor fuel delivery must observe the entire process and remain attentive. They are responsible for directing the delivery person to stop the flow of fuel in the event of a leak, spill, or unusual occurrence. Fuel delivery personnel are responsible for proper hookup and for stopping the fuel flow instantly if necessary.

Once the tank is filled, delivery personnel are responsible for disconnecting the hose, securing all ports, and verifying that the correct amount of fuel has been dispensed (by remeasuring). Facility personnel should then verify that the spill basket is empty and return all spill control materials to their storage location.

Figure 7.42. A fuel delivery without personnel visibly present. In case of an emergency, the unattended spill would be worse than if the delivery driver had been present.

Employee Training

All farm or business personnel who pump fuel should be trained on good fueling procedures. They should know how to use all spill kit items, how to prevent overfills, and how to react if an emergency occurs. Hands-on training usually is better received and remembered than training from written materials.

Spill prevention is the single most important thing, and periodic refresher training should be provided. Employees need to be reminded of routine filling procedures such as turning pumps on and off, locking tanks, inspecting hoses and fittings, and completing the use log. They must be reminded to always stay at the tank when fueling.

Figure 7.43. Hoses should be inspected periodically for signs of sun rot such as seen here.

Detecting Water in Tanks

The accumulation of water in tanks can cause them to rust on the inside. It is a good idea to check stored fuel for water content by using a special paste on a dipstick. The paste changes color if water is present. If water is detected, contact your fuel company and ask them to remove the low-quality fuel.

Inspections

You should perform thorough monthly inspections of your storage tanks and keep documentation of the inspections. Deficiencies noted during the inspection should be corrected as soon as possible. Your inspection should include at least the following:

- A visual examination of the tank system for deterioration, including but not limited to the tank and its coating, hoses and fittings, pipes, the foundation, and the drainage mechanism
- A dipstick inspection of tank contents, at the lowest point, for the presence of water
- A check of the interstitial space (i.e., the space between the walls) of a double-walled tank for accumulation of fuel or water
- Confirmation that all drain valves are securely closed when not in use
- An inspection for accumulated water or fuel in the containment area
- A check of the spill basket to make sure it is clean and functional
- A check of normal and emergency vents for obstructions or restrictions that could interfere with proper operation
- A check of auxiliary equipment for operational malfunctions
- An investigation of conditions that may pose a fire, safety hazard, or environmental hazard
- A search for evidence of a release from the tank system
- Confirmation that spill kits are immediately accessible

Plan a Walk-Through with Firefighters

Contact your local fire department and first responders to schedule a tour of your fuel storage facility. Develop an emergency response plan that you can give to the local fire department. See volume 2 for more details.

Indoor Gasoline Storage

Many companies use 5-gallon gasoline containers to fill their equipment. Such containers must be managed to reduce the threats of fire and theft.

Strategies and Recommendations

Limit the quantity of gasoline stored indoors. Consult your insurance agent to determine if your coverage hinges on a specified quantity.

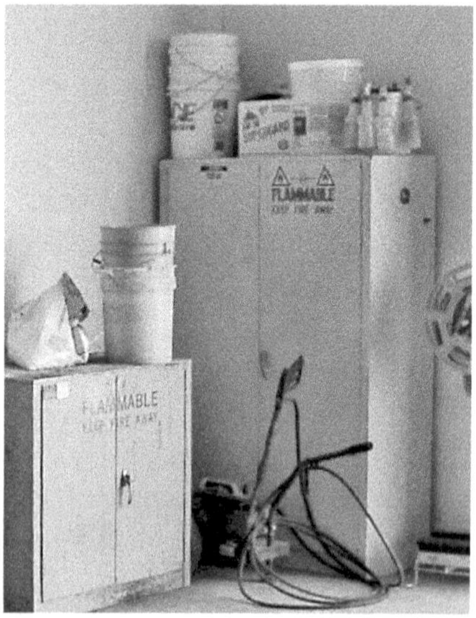

Figure 7.44. These fireproof cabinets are ideal for storing gasoline indoors.

Handling Oil Products and Filters

A clean work area means a safer environment for employees. Your efforts to maintain a good working environment will result in fewer employee injuries and fewer worker's compensation claims.

Strategies and Recommendations

Many farmers and business owners maintain and repair their own vehicles and equipment. Their attention to detail in preventing spills, storing and ultimately disposing of used oil, and cleaning up after any mishap along the way is extremely important. Good housekeeping is a vital component of good oil management.

Figure 7.45. A good example of a bad situation. Always drain oil filters and containers before throwing them in the trash.

Figures 7.46–7.47. In these photos oil containers are placed along the wall, out of the way. (Left) The drip pan has holes that allow oil to drain into buckets, and oil barrels are placed on racks. (Right) Drip buckets catch any oil that drips from the spigots.

Figure 7.48. Oil barrels, used oil, and equipment are stored over containment. The funnels hanging on the wall have drip cans below them.

Figure 7.49. The drums are stored over containment, and the contents are clearly identified.

Figure 7.50. Oil storage containers of all sizes must be labeled, whether a 5-gallon bucket, 55-gallon drum, or large tank.

Securing Tanks from Attack by Vandals and Thieves

Storage tanks that are visible from the road or highway are vulnerable to vandals and thieves. Vandals may slash the hoses, releasing fuel onto the ground, or pollute the tank contents; thieves may simply help themselves to a tank of gas.

Strategies and Recommendations

There are various strategies for protecting unattended tanks, starting with a good padlock. Expect to pay a good price for a superior lock that can withstand a blade or bullet. A series of locks and methods used in combination can greatly deter a thief. Nozzles as well as outside pumps and switches should be locked. Indoor switches should be turned off when not in use. The fuel pump breaker can be turned off at the main box; if your breaker box has a hinged lid, it also can be locked.

Timers can be used to control electricity to your fuel pumps. If you place them inside a building, choose a secure location and padlock the cover. They may be placed inside a vandal-resistant enclosure located a safe distance from the tank.

Figure 7.51. Thieves cut this switch lever to turn on the pumps. This theft took place before it became routine to shut off the power to pumps inside buildings. A padlock designed to resist bolt cutters is recommended.

Figure 7.52. Thieves broke into a building, turned on the power at two locations, and then broke the switch housing to get power to the pump.

Figure 7.53. A good padlock is a starting point for security. Locking the handle slows down a thief, giving you more time to detect them.

Figure 7.54. Turning off fuel pumps at the circuit breaker box is an additional security mechanism.

Figure 7.55. People are moving fuel switches indoors, so a would-be thief has to break in and turn the switch on to steal fuel.

Figures 7.56–7.57. Security lighting around your fuel tanks deters thieves; it also can aid in the discovery and cleanup of spills. When considering lighting, request vandal-proof glass. The product stored and tank location also may enter into your decision.

Prepare for a Petroleum Emergency

Emergencies do happen. Extreme weather conditions and human interference—intentional or accidental—are the two main causes. In any case, you want to be ready. The chances of minimizing your loss increase with preparedness and are enhanced by the quick, appropriate response of your local emergency personnel. Think about what could happen, and what might happen. Think about how to react and whom to call after 911. Write things down. Develop an emergency plan. Include telephone numbers for reputable spill cleanup contractors so that cleanup can begin as soon as possible.

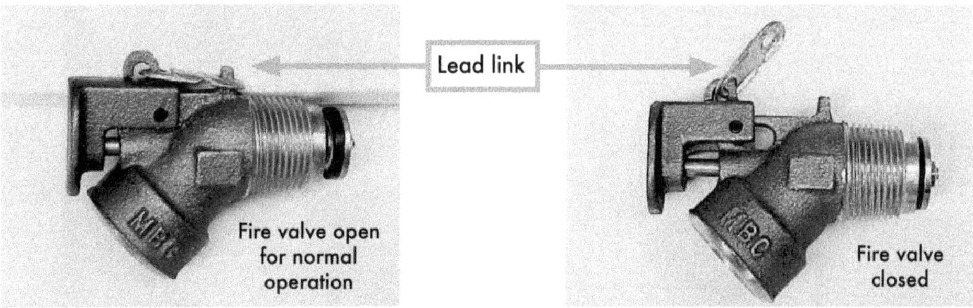

Figures 7.58–7.59. Fire valves are required. During a fire the lead link melts at a low temperature, allowing the valve to close; if the tank hose burns off, the valve will be closed. After fueling a vehicle, the link can be manually released, and the valve will close. A padlock can be used to secure it.

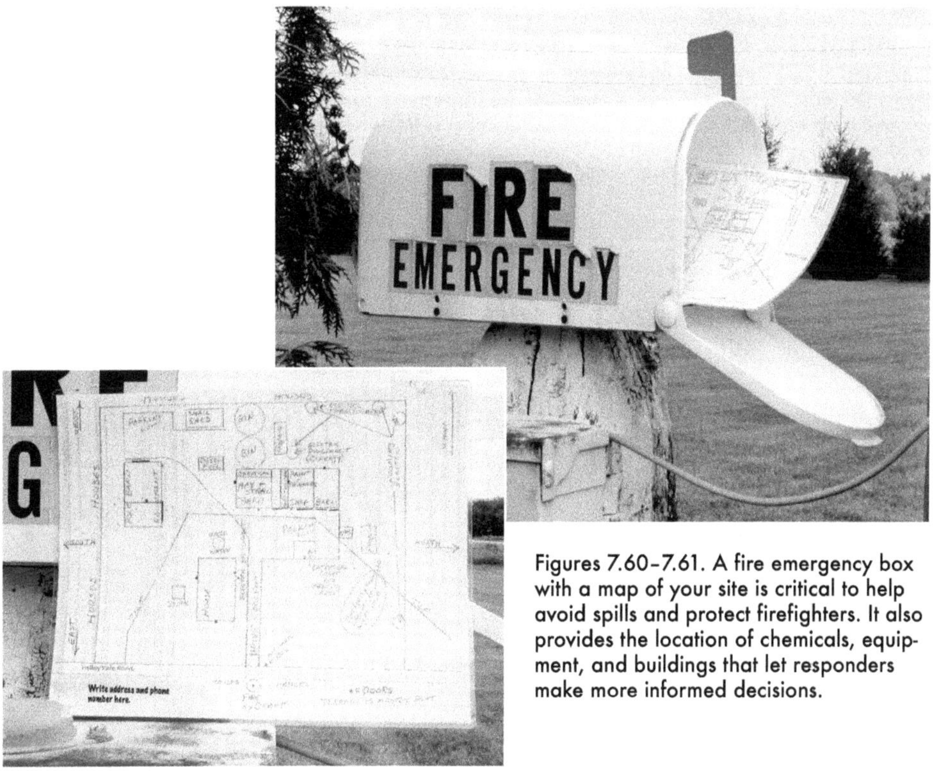

Figures 7.60–7.61. A fire emergency box with a map of your site is critical to help avoid spills and protect firefighters. It also provides the location of chemicals, equipment, and buildings that let responders make more informed decisions.

Figures 7.62–7.63. Placement of all-purpose ABC fire extinguishers near aboveground storage tanks and in buildings can help prevent minor fires from becoming major. Fire extinguishers must be inspected monthly, and the importance of these inspections cannot be overemphasized. Fires often occur during busy farming seasons when equipment is hot while refueling. The risk of a fire caused by overfilling is greatest at this time. A well-maintained fire extinguisher can save your machinery—and possibly your life!

Figures 7.64–7.65. Clearly marked emergency shutoffs are essential; once a release is in progress, it is important to stop the fuel flow as soon as possible to prevent further contamination. Shutoffs should be at least 15 feet from the fueling area but not more than 120 feet away.

Figure 7.66. This alarm system is triggered when a spill occurs.

Figure 7.67. Posting spill procedures is essential if the spill is to be quickly dealt with.

ABOVEGROUND PETROLEUM TANK REGULATIONS

Petroleum regulations are written to protect people, property, and the environment. The regulations often require the following:

- Tanks must be constructed according to engineering specifications and industry standards.
- Tanks must bear product identification and warning signs.
- Tanks must be secured and in containment.

Petroleum regulations not only protect private property and enhance personal safety but are designed to augment public protection measures at all levels of government. For instance, a business may be required to notify various government agencies if its petroleum storage exceeds a specified threshold. These individual reports alert emergency responders to prepare a plan for accidental petroleum spills, natural disasters, and fires; they allow fire departments and other officials to establish plans for handling intentional petroleum releases by vandals or terrorists. Preplanning facilitates a more coordinated effort to protect the public. Many regulations also are best management practices that you should follow even if not required to do so. You should review all of these to make your ASTs as safe as possible.

Key Terms

Vapor: a visible exhalation of gas, like a mist, suspended in air

Flash point: the lowest temperature at which a liquid petroleum product can release sufficient vapors to ignite

Flammables: products capable of catching fire at flash points below 100°F

Combustibles: products capable of catching fire at a flash point of 100°F or above

The correct interpretation of petroleum regulations hinges on understanding the four terms broadly defined above. All petroleum products release vapors, some sooner (i.e., at lower temperatures) than others. The amount of vapor released into the atmosphere is temperature driven: More vapor is released at higher temperatures, less at lower temperatures. The amount also depends on the chemical and physical properties of the petroleum product.

Flash points matter to emergency responders. Gasoline has a flash point of –45°F, while the flash point of diesel fuel is 105°F. So gasoline is more flammable than diesel at normal temperatures. Petroleum products fall into one of two groups categorized by flash points. Products with flash points below 100°F are categorized as *flammables*, while those with flash points of 100°F or above are categorized as *combustibles*.

Flammables are Class I materials, subdivided into Class IA, Class IB (gasoline), and Class IC. Combustibles are Class II and Class III materials. Class II materials such as heating oil and diesel fuel have lower flash points (100°F to 140°F), while those of Class III combustibles range from 140°F to 200°F. The differences between flammable and combustible liquids determine the regulations with regard to tank construction and ventilation, the maximum fuel volume (to allow for vapor expansion), the separation distance between aboveground tanks and buildings, and identification tags and warning signs on tanks.

Petroleum vapors are heavier than air and tend to settle near the ground on a calm day, so the regulations address the reduction of vapor accumulation. They define how tall the tank vents need to be to help dissipate vapors, and they state management practices that can reduce ignition sources such as sparks from metal-to-metal contact from a nozzle being inserted into a tank, contact with electricity, or contact with live ash or a cigarette butt.

Spill Prevention

The federal Spill Prevention, Control, and Countermeasure (SPCC) Regulation (40 CFR Part 112) addresses farms and businesses that store more than 1,320 gallons of oil or oil products in aboveground storage tanks/containers. *Oil* includes, but is not limited to, the following products: waste oil, used oil, heating oil, diesel fuel, kerosene, gasoline, crop oil, adjuvant oil, lubricating oils, naphtha, mineral spirits, vegetable oils, animal fats, and blends such as biodiesels; synthetic oils are also regulated as oil or oil products. Some of the things required by the SPCC regulation include appropriate secondary containment and/or diversionary structures, security measures, inspections and recordkeeping, and employee training. Its primary goal is to prevent oil from being discharged into or upon surface waters or adjoining shorelines of the United States.

The SPCC regulation is based on the total capacity of all on-site oil storage containers that can hold 55 gallons or more. All aboveground containers (tanks, drums, totes, etc.) 55 gallons or greater are included, even if empty. The threshold capacity is 1,320 gallons; that is, 1,321 gallons would trigger the regulation requirement at your facility.

For instance, a farmer has the capacity to store 2,110 gallons of oil on the farm: two 55-gallon drums of oil in the shop, a 500-gallon diesel tank, a 500-gallon gasoline tank, and an empty 1,000-gallon tank for storing off-road diesel during the planting and harvest seasons. The total capacity of 2,110 gallons exceeds the 1,320-gallon threshold, so the farmer must comply with the federal SPCC requirements.

A farm starting operation on or before August 16, 2002, must maintain its existing SPCC plan based on the SPCC requirements prior to the new 2002 amendments. A farm starting operation after August 16, 2002, must prepare and implement an SPCC plan when the EPA promulgates a rule specific to farms and specifies a compliance date for farms.

- A written SPCC plan. (1) If you store 10,000 gallons or less at your facility and have not had a discharge (see "Spill Reporting" in the next section) into or upon waters of the United States in the past three years, you may write your SPCC plan and certify it. (2) A licensed professional engineer (PE) may write your SPCC plan after visiting and examining your facility and certify that the written plan satisfies the requirements. The plan is kept on-site; it is not submitted to the EPA unless specifically requested.
- If a farm has oil-filled operational equipment (e.g., hydraulic and/or lubricating systems associated with a cotton gin, or transformers) and has not had a discharge (see "Spill Reporting") into or upon waters of the United States in the past three years, it may institute an inspection and monitoring program (along with some other requirements), in lieu of secondary containment for oil-filled operational equipment.
- Sample plans are available through the Oil Information Center at (800) 424-9346 or at https://www.epa.gov/oil-spills-prevention-and-preparedness-regulations/spill-prevention-control-and-countermeasure-spcc.
- Oil tanks and products must be contained. The regulation does not specify the type of containment or diversionary structure required. Good engineering practices and a risk analysis (to determine the path a spill would take) should be used to determine the best type of equipment or structure for containment at your facility.
- You must train employees who work in the oil-handling area on preventing, controlling, and responding to an oil spill at your facility. The training must include a discussion and review of the SPCC plan, the procedures in place to prevent releases, applicable pollution control laws and regulations, equipment operation and maintenance, procedures to safely load and unload oil products, and actions to be taken during an emergency. Spill prevention briefings are required at least once a year to promote adequate understanding of the SPCC regulation, reemphasize the plan requirements, and address any discharges or failures that have occurred at the facility. For example, if a spill has occurred, the discussion should address the cause and containment of the spill, any equipment malfunction related to the spill, and new measures in place to prevent a recurrence.

Reporting On-Site Petroleum Storage to Local and State Agencies

The Federal Emergency Planning and Community Right-to-Know Act (EPCRA) requires businesses and farmers who store hazardous chemicals (e.g., gasoline) above threshold levels to report to government agencies on an annual basis. The main objective of this regulation is to allow local authorities to preplan for fires, spills, accidents, vandalism, natural disasters, and acts of terrorism (40 CFR Part 370; https://www.epa.gov/epcra).

Farmers and business owners must report on-site storage of gasoline, fuel oil, or diesel totaling 10,000 pounds or more. Any substance, to the extent it is used in routine agricultural operations, or any fertilizer held for sale by a retailer for ultimate delivery to the customer, is exempt from this reporting requirement. However, farmers lose the exemption if they operate another business from the farm (e.g., trucking or commercial pesticide application). The Tier II forms must be submitted annually by March 1, reporting the amount of fuel stored for the previous year.

A gallon of diesel fuel or gasoline weighs about 7.5 pounds. Divide the 10,000-pound threshold by 7.5 pounds (one gallon) to determine that 10,000 pounds equals 1,333 gallons. This is the quantity of fuel that requires EPCRA reporting. If you have a commercial business, you must report storage above the threshold quantity, using the Tier II Inventory Form (https://www.epa.gov/epcra/tier-ii-forms-and-instructions). Submit the completed Tier II form to your state's Emergency Response Commission (https://www.epa.gov/epcra/state-emergency-response-commissions-contacts), the county's Local Emergency Planning Committee, and the fire department(s) responsible for protecting your facility.

The list of all hazardous substances covered under this regulation is found in the Title III List of Lists at https://www.epa.gov/epcra/consolidated-list-lists. The list also is available through your Local Emergency Planning Committee, and the state Emergency Response Commission.

Even if your storage capabilities fall below the reporting threshold, it is a good idea to establish an emergency plan. Work with your fire department, providing a map of your property and all pertinent information. See volume 2.

Spill Reporting

Knowing whom to call during an oil or chemical spill is essential, but this can be confusing. You may need to notify multiple local, state, and federal agencies. Who is called, how soon, and under what circumstances depend on whether the petroleum spill is on-site or in transport.

Whenever an oil discharge reaches surface waters of the United States or adjoining shorelines, you must notify the National Response Center (800-424-8802) and potentially the US EPA. If the amount discharged in a single release exceeds 1,000 gallons or if there are two releases exceeding 42 gallons (each) within a 12-month period, you must send a written report to the appropriate EPA regional office within 60 days of either release. Contact information for EPA's regional offices and the states assigned to each office is available at https://www.epa.gov/aboutepa/regional-and-geographic-offices. A copy of the report also must be sent to your state's emergency response department. The written information must include the following:

- Your name
- The name of the facility
- The location of the facility
- The maximum oil storage capacity of the facility

- Corrective action and countermeasures taken, including equipment repair and replacement
- A facility description, including site maps, flow diagrams, and topographical maps pertinent to the spill
- The cause of the discharge, including a failure analysis of the fuel storage system
- Preventive measures incorporated or currently under consideration

Emergency Planning and Community Right-to-Know Act

The regulation that covers the storage of hazardous chemicals also includes spill reporting requirements, which are based on the Reportable Quantity (RQ). The RQ sets the amount of an active ingredient that is reportable in the event of a release. There are no RQs for petroleum products, and they are excluded from EPCRA spill reporting. EPCRA does require you to call 911 when gasoline or diesel fuel is spilled during transportation, and the call must be made immediately.

US Department of Transportation (DOT)

If you are registered with the DOT, you have to keep records at your primary place of business for three years following an accident involving a commercial motor vehicle. Your records must include the date of the accident; the city, town, and state where the accident occurred; the driver's name; the number of injuries; the number of fatalities; a statement on any hazardous materials released (except from the tanks of vehicles involved in the accident); and copies of all accident reports filed with the state or the insurers.

As a carrier of hazardous materials (e.g., diesel fuel, gasoline), you are required to notify the National Response Center (800-424-8802) immediately after any incident that involves death, injury requiring emergency hospitalization, damage exceeding $50,000 to the vehicle or property, towing of a vehicle, evacuation of the general public for one or more hours, or closure of a major road for an hour or more. You must submit a written report to the DOT within 30 days, using Form F 5800.1, available at https://www.phmsa.dot.gov/sites/phmsa.dot.gov/files/docs/IncidentForm010105.pdf.

US Occupational Safety and Health Administration

The Occupational Safety and Health Administration (OSHA) regulates all businesses as well as farms with more than 10 employees. OSHA has many regulations in the Code of Federal Regulations dealing with tank construction specifications. Title 29 CFR Part 1910.106 addresses petroleum and fuel products:

- Tanks designed for underground fuel storage may be used aboveground if they hold less than 2,500 gallons. However, reuse of underground tanks for aboveground fuel storage is not recommended.
- A minimum of 3 feet must separate any two storage tanks holding flammable or combustible liquids.

- A minimum of 20 feet must separate a liquefied petroleum gas container from a tank holding a flammable or combustible material.
- Vent pipe outlets for tanks holding Class I liquids shall be distanced from buildings so that vapors will not pose a human health threat, shall be at least 12 feet above the ground, shall be positioned to prevent vapors from accumulating under eaves or other obstructions, and shall be at least 5 feet from building openings.
- Vent pipes shall be at least as large as the largest fill or withdrawal pipe but no smaller than ¼ inch in diameter.
- Tank supports shall be placed on firm foundations constructed of concrete, masonry, or steel. Single wood timber supports (not cribbing) laid horizontally may be used for outside aboveground tanks if not more than 12 inches high at their lowest point.
- Tanks can rest on the ground or on foundations made of concrete, masonry, piling, or steel. Tank foundations shall be level and must be positioned to minimize tank corrosion.
- Portable fire extinguishers and other fire control equipment must be provided at points of operation and wherever flammable or combustible liquids are stored. Personnel expected to use this equipment must be trained in its use and operation.

CONCLUSION

On-site storage of petroleum products benefits farmers and business owners. It provides easy 24-hour access and price breaks, and storing diesel for off-road agricultural use yields a tax break. But the benefits are clearly offset by significant risks posed by storing fuel and other petroleum products. Placing fuel tanks based on employees' convenience can lead to neglect. That is, if they are not in an area that you view regularly, spills and other accidents are more likely to go unnoticed. Don't let neglect trump wise management on your farm or at your place of business.

You can implement simple procedures and best management practices that will significantly reduce the risks of on-site storage. A single fuel tank on your property carries enough risk to warrant your attention to the safety procedures and protection measures addressed in this chapter. The benefits of storing petroleum products on-site hinge on your good management of the process.

8

Anhydrous Ammonia Tanks

With contributions from John Rebholz, Jim Camberato, John Obermeyer, Mike Titus, Jamie Southard, Danny Starke, Shawn Lambert, and Kevin Leigh Smith

Anhydrous ammonia is available from hundreds of agricultural retailers for their farm clientele who use this fertilizer on their corn acreage. While driving through Midwestern farmland in late fall or spring, it is common to see retailers and farmers pulling anhydrous ammonia nurse tanks to and from fields. During peak times, retailer lots that were once filled with nurse tanks and applicators (tool bars) stand empty as every working piece of anhydrous ammonia equipment is pressed into service.

Using anhydrous ammonia offers many benefits:

- The cost per pound of actual nitrogen with anhydrous ammonia is less than with urea, urea ammonium nitrate (UAN) 28%–32%, and aqua ammonia.
- Anhydrous ammonia is 82 percent nitrogen, so transportation costs per unit of nitrogen are much lower than for other forms of nitrogen.
- It is a popular form of nitrogen recommended for fall application in Midwest corn-producing states.

Figure 8.1. On many farms, anhydrous ammonia provides the nitrogen source for crops.

However, there are major risks associated with using anhydrous ammonia. We probably have all seen, read, or heard about anhydrous ammonia releases. A "white vapor fog" sweeps across the landscape, and emergency responders are called out to deal with the escaped gas. Depending on the direction in which the release is carried by the wind, roads in and out of the area may be blocked, and the people near the release site evacuated. In other situations, local authorities may order people to shut windows and shelter in place until they give the all-clear notice.

Many things can cause releases: a ruptured hose at the filling riser or during field applications, an overturned anhydrous ammonia tank in a road accident, or improperly connected hoses.

Figure 8.2. A ruptured hose can create a very dangerous ammonia cloud that can be deadly to the farmer, firefighters, and those living around the area.

Figure 8.3. A nighttime release poses additional risks to emergency responders who are working in the dark, not knowing what is happening around them.

Firefighters responding to an incident have only a short time to spray water onto the anhydrous ammonia cloud to mitigate the vapor and reduce its movement. While most anhydrous ammonia releases are safely dealt with, in some cases local residents and firefighters have been permanently disabled or killed when engulfed in the vapor cloud. When a release is major, local hospitals can quickly become overwhelmed by injured people seeking emergency care. What's more, ammonia releases that injure or kill may lead to lawsuits that can drag through the courts for years.

Figure 8.4. When transporting ammonia tanks on the highway, any accident can create road hazards.

It's undeniable that anhydrous ammonia has what we might call Jekyll and Hyde attributes. Anhydrous ammonia is a low-cost and readily available nitrogen fertilizer. However, because of its potential hazards to humans and the environment, those who store, transport, and use anhydrous ammonia must be constantly reminded to focus on even the smallest of safety details.

Ammonia training—and retraining—needs to be provided to part-time seasonal workers, newly hired recruits, tenured veterans who work outdoors, and even those who work behind customer counters at retail facilities or on farms. A constant focus on anhydrous ammonia safety helps ensure that everyone fully appreciates their role in protecting themselves, their coworkers, and the community. Farmers and farm employees who handle ammonia also would benefit from this training. In some states, farmers and farm operators are now required to have ammonia safety training. This chapter provides you with a better understanding of what hazards anhydrous ammonia poses, how you can safely work around this product, and what you should do in the event of a release.

CHEMICAL CHARACTERISTICS OF AMMONIA WORTH KNOWING

If you want to know why there are so many regulations surrounding anhydrous ammonia, then it helps to understand its chemical and physical properties. You don't have to be a chemist or be familiar with complex chemical reactions to understand the chemical properties and reactions associated with anhydrous ammonia.

Let's start with anhydrous ammonia's structure. You may see or hear anhydrous ammonia referred to by its chemical formula: NH_3. The formula tells us that anhydrous ammonia molecules contain three parts of hydrogen (H) for every one part of nitrogen (N). But the weight of the N atom makes up 82 percent of the anhydrous ammonia compound. This makes anhydrous ammonia a concentrated nitrogen product, 82-0-0.

Liquid Under Pressure

At ambient temperatures and pressures, anhydrous ammonia is a gas. It is transformed into a liquid when it is compressed under pressure in an approved storage tank. When released into the air, anhydrous ammonia converts back to a gas (the term *vapor* is often used interchangeably with *gas*). An interesting fact is that 1 cubic foot of liquid anhydrous ammonia, contained under pressure, produces 850 cubic feet of vapor when released, an expansion rate of 850 to 1.

Easy to Detect with Its Distinctive Odor

When released into the air, anhydrous ammonia is colorless, but the odor is quite recognizable: It smells like ammonia, a common household cleaner. The odor is often described as sharp or pungent. At high concentrations, it can be described as suffocating.

Most people can detect ammonia's distinctive odor at between 5 and 50 parts per million according to OSHA. That is well below the levels (greater than 220 parts per million) that can cause irreversible harm to humans.

Strong Affinity for Water

The word *anhydrous* means a substance without water. For this reason, anhydrous ammonia has a strong affinity for water—whether in the atmosphere, in the soil, or in your body. Anhydrous ammonia takes water from anything it comes in contact with. This is why anhydrous ammonia can seriously damage your skin, which is 44 percent water. Even more serious is the potential damage to eyes (92 percent water) and lungs (72 percent water).

This is why emergency water tanks must always be available. Nurse tanks have 5-gallon water tanks, while the retail facility will have a larger emergency soak tank that an exposed person can immerse themselves in.

Figure 8.5. Emergency water tank on a nurse tank and safety information decal.

Figure 8.6. A release of ammonia vapor will move up into the atmosphere because it is lighter than air.

Lighter than Air

Anhydrous ammonia vapor is lighter than air, which means it moves up into the atmosphere from the point of its release. However, high relative humidity traps the vapor. When ammonia encounters moisture in the air, the vapor becomes heavier and hangs lower to the ground.

Emergency responders often describe what they call the upside-down J phenomenon of an anhydrous ammonia gas cloud. When released from the top of a tank, the vapor goes straight up into the atmosphere. After the vapor rises 50 to 75 feet, the anhydrous ammonia cloud starts to curl over and flatten out as it picks up atmospheric moisture (getting heavier in the process). This phenomenon has misled people into thinking it was safe to reenter release areas too soon; they have then suffered severe injuries.

Caustic

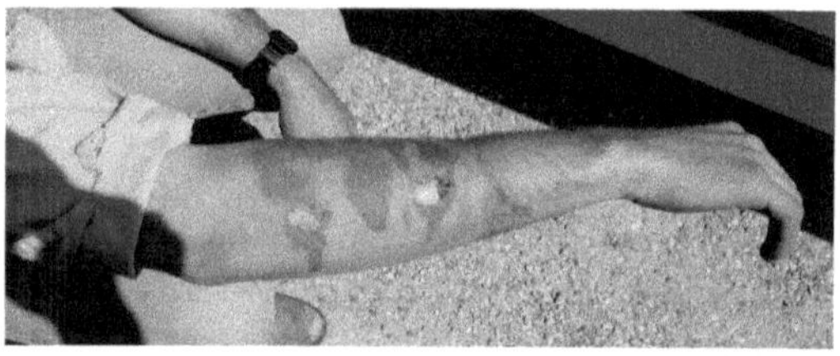

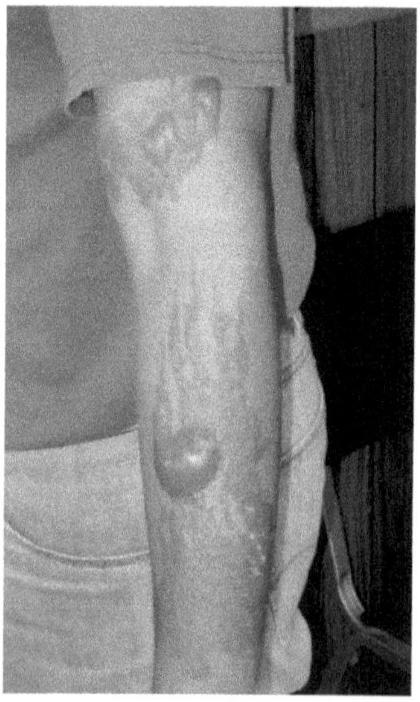

Anhydrous ammonia has a pH of 11, which is just slightly less alkaline than bleach, which has a pH of 12. Remember the chemical formula for anhydrous ammonia: NH_3. When it absorbs water (H_2O), anhydrous ammonia removes a hydrogen atom from water to take on the more stable form of NH_4+ (ammonium).

When this process takes place in the soil, this reaction is good, because the nitrogen remains in place for the plants to use. But when that reaction happens in the air because of a release near you, removing the hydrogen atom from water leaves behind a molecule of hydroxide (OH–), which is very caustic to skin, lungs, and eyes. Water is the only immediate first aid for ammonia exposure.

Figures 8.7–8.8. Anhydrous ammonia injury to the skin, causing significant burns. While these burns are on the surface, the eyes and lungs can be burned in a similar manner.

Figure 8.9. The white area on the black hose where it comes out of the valve is frost because liquid anhydrous ammonia is so cold.

Extremely Cold When Liquid (Low Freezing and Boiling Points)

Anhydrous ammonia boils at -28°F and freezes at -108°F. This is why it turns into a vapor when it is released at ambient temperatures. When anhydrous ammonia changes from liquid to a vapor, it is at first very cold, which can further damage human and plant tissues.

PROTECTING YOUR FACILITY WITH INSPECTIONS AND SIGNS

The description of its chemical properties provides ample evidence for why there is such an investment in specialized infrastructure to store, dispense, and transport anhydrous ammonia. Because of this, major anhydrous ammonia releases at retailer sites are fairly rare, even though the industry moves hundreds of thousands of tons of ammonia each fall and spring. Tanks, plumbing, and additional safeguards constructed according to national standards and government regulations are partially responsible for making releases rare.

Furthermore, initial installations undergo routine inspections by independent industry consultants, company-employed safety and risk coordinators, and government inspectors who help ensure continued compliance with regulations and company safety practices. Ultimately, the safe use of anhydrous ammonia depends on employees who pay close attention to every aspect of the process and report potential problems.

Emergency Contact Information Posted at Site Entrance

One of the first steps in that process is that every facility should place an emergency sign at the most prominent entrance that emergency responders would use (including fire, police, and medical professionals). Emergency signs must be placed at every entrance. Each emergency sign provides the name of the location and its 911-approved address. The sign also provides the names and phone numbers (including area codes) of those with the knowledge to deal with an anhydrous ammonia release at that facility.

Figure 8.10. Emergency information is often posted at the main entrance to an ammonia facility and at the bulk storage tank.

Emergency signs posted at entrances also provide legends that inform emergency responders what the pipes carry and what the valves control. Pipes painted white are carrying anhydrous ammonia as either a liquid or a vapor (the white color helps reduce heat). Valves painted orange indicate the lines are carrying liquid. Valves painted highway yellow indicate vapor lines.

DELIVERING ANHYDROUS AMMONIA TO THE FACILITY

Drivers who deliver anhydrous ammonia to retail facilities and farms are expected to conduct their business in a professional and safe manner—no different than anyone else working around ammonia. Two critically important tasks that drivers must complete occur at the beginning and at the end of the delivery process.

First, drivers should not fill storage vessels to more than 85 percent of their capacity. This allows room for the anhydrous ammonia in the tank to expand as ambient temperatures rise. Storage tanks typically hold anywhere from 12,000 to 30,000 gallons of liquid anhydrous ammonia—tanks holding 45,000 and 60,000 gallons are also common. Each storage tank has a float gauge that helps indicate how full it is (as a percentage).

A transport truck normally carries 8,000 gallons (or roughly 20 tons) of anhydrous ammonia; anhydrous ammonia weighs 5.14 pounds per gallon. Since a driver can fill a storage tank to 85 percent of its rated capacity, a 12,000-gallon storage tank could hold one delivery truck's worth of ammonia.

A MAJOR DELIVERY SCREWUP

Many retail facilities that handle anhydrous ammonia also handle propane. There are serious consequences if someone mistakenly pumps ammonia into a propane tank. The fittings on propane tanks are made of brass and copper, while the fittings on ammonia tanks are made of mild steel and stainless steel. Putting ammonia in a propane tank creates pinholes in the brass and copper fittings.

You want to know with certainty that delivery drivers and employees are putting ammonia and propane in the right tanks. If the truck in front of the propane tank has ammonia written on its side, or vice versa, then employees have the responsibility to tell the drivers to stop and make sure they are filling the correct tank. It is better to ask and be wrong than to ignore what your eyes are telling you.

Second, after drivers fill the storage tanks and shut down both their trucks and the storage tanks, they still have one critical task: Do not allow any liquid trapped in the hose to escape into the environment. There can be as much as 1 gallon of liquid left in the delivery truck's hose, which will turn from a liquid to a gas if drivers disconnect the hose at this point.

When drivers unload ammonia, the hoses are connected between the trucks and the tanks and are pressurized. When drivers complete their product transfers, they must leave the hoses connected under pressure until they relieve the pressure in the hose by opening the bleed-off valve. Doing so will drain the remaining ammonia into a nearby enclosed water tank designed specifically for this purpose.

It's common to use a 275-gallon caged minibulk tank to collect this bleed-off anhydrous ammonia. The anhydrous ammonia is absorbed into the water. This leaves little product in the hose when it is disconnected from the fitting at the bulkhead.

It is important to close the valve to the bleed-off tank when it is not being used—leaving the valve open can cause back pressure and suck water back into the ammonia piping. The bleed-off water contains nitrogen, which can be applied at agronomic rates to cropland—either by itself or along with other fertilizer or chemical applications.

Figures 8.11–8.12. Bleed-off tanks.

The bleed-off water can become saturated with ammonia. You should change the water several times during the season considering that as many as 10 transports come to a facility each day during the busy season.

Delivery drivers should be expected to:

- Know the correct traffic patterns for unloading at a storage site.
- Have a valid commercial driver's license (CDL) and hazardous materials endorsement.
- Set the brakes on their vehicles and chock the wheels.
- Know where the emergency shutoff is and how to operate it.
- Ensure clean water is available in the event of an emergency.
- Wear long-sleeved shirts and pants, gloves that are rated for anhydrous ammonia, and nonvented or indirect vented goggles (at a minimum).
- Make sure the bleed-off system is shut off before starting the transfer pump. Bleed-offs need to be vented. There have been fatalities when the bleed-off containers were not vented and became pressurized.

Figure 8.13. A delivery driver unloading anhydrous ammonia who is properly attired to handle the product.

- Connect both the vapor and liquid hoses to bulkheads before starting the transfer process.
- Open valve-side systems first to allow the vapor to equalize between cargo tank and storage tank.
- Fill storage tanks to no more than 85 percent of their capacity.
- Bleed off any remaining ammonia in the hoses.
- Stay with their trucks until the process is totally complete.
- Monitor and observe the transfer activities from the ground and not from inside their cabs.

Bulk Storage Tanks

Tanks used to store anhydrous ammonia are basically heavy, thick submarine-like tanks. Many of these tanks were built in the 1960s and 1970s. A storage tank contains the largest volume of anhydrous ammonia at the facility, which is why maintenance on these tanks is so vital. That maintenance includes keeping them painted and keeping the pipes free of corrosion.

Storage tanks must be built to withstand internal pressures up to 250 pounds per square inch (psi). However, anhydrous tank manufacturers actually construct them with a 4-to-1 safety factor, meaning they can withstand pressures higher than 250 psi.

These tanks have a series of external and internal valves. There are pressure relief valves that open in the event the pressure gets too high within the tank. There are excess flow valves that can shut down the flow of liquid anhydrous ammonia in the event of a plumbing failure or hose rupture. There are back check valves that allow product to move in only one direction. All of these built-in safety devices are yet another step to prevent major releases.

It is important to familiarize yourself with the locations of each of these safety devices and the way they operate. Each has its own specific function. Knowing what those functions are and how they operate within the system allows a quicker response in the event of an emergency.

Can you find these at your facility?

- Crash protection around the storage tank to protect tanks, piping, and valves from moving vehicles and equipment
- Pull-away protection in case hoses are not disconnected from the bulkhead or refilling station after refilling
- Color-coded valves and a piping legend that indicates whether lines contain liquid or vapor
- "Anhydrous Ammonia" or "Caution—Ammonia" painted on the side of the storage tank, as well as the National Fire Protection Association diamond (NFPA; the diamond's multiple colors gives first responders quick information about the hazards posed by the anhydrous ammonia)
- Manual valves to turn off tanks
- Liquid level fill gauge

- Bleed-off tank
- Emergency shutoff system on the transfer truck and storage tank
- Two tanks that hold at least 150 gallons of water that a person can physically jump in if that person has been exposed to ammonia (the tanks should be able to hold the largest employee working at the facility)

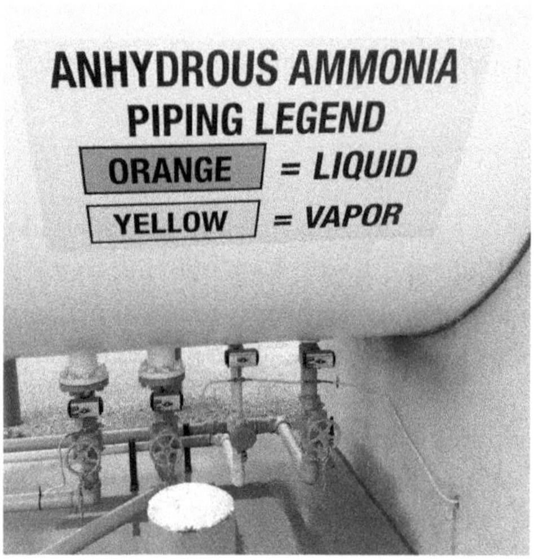

Figures 8.14–8.15. Ammonia pipes and valves are color-coded. In the right-hand photo the lighter-colored, smaller pipe in back is the yellow vapor phase line.

Figure 8.16. It is important to know the location of the emergency shutoff.

Figure 8.17. A water tank in which a person can immerse themselves if exposed to ammonia.

TRANSFERRING ANHYDROUS AMMONIA TO APPLICATION TANKS

The location where you fill application nurse tanks with anhydrous ammonia is an elevated area commonly referred to as the *riser*, *rack*, or *filling station*. Many locations have a liquid pump that transfers anhydrous ammonia from the larger storage tank via pipes coming up through the center of the riser. At the riser, the liquid anhydrous ammonia enters hoses that lead to primary shutoff valves, commonly called *hose-end valves*.

This elevated platform allows one person working on the riser to access the valves on top of the nurse tanks to connect the hoses. Most risers can fill four nurse tanks at one time—two on each side. Filling time varies, but at newer facilities, it can be as little as 10 minutes per nurse tank.

Nurse tanks are built to exact specifications dictated by federal regulations and industry standards (American Society of Mechanical Engineers; ASME). This is a good thing when you consider that these tanks are pressurized vessels on wheels carrying hazardous materials. Older nurse tanks have capacities of about 1,000 to 1,450 gallons. Today, with larger farm equipment, it is common for tanks to hold up to 3,000 gallons of anhydrous ammonia. Each nurse tank must have a clearly legible data plate. If it does not, then the tank must undergo a series of tests to ensure its safety and integrity.

Farmers are exempt from US Department of Transportation (USDOT) regulations dealing with CDLs and hazardous materials endorsements when they transport 3,000 gallons or less of anhydrous ammonia in any combination. Be aware that these rules may be more stringent when crossing state lines.

Figure 8.18. A riser for filling nurse tanks. There are a number of safety procedures that must be followed to prevent injury to the persons filling and moving the anhydrous tanks.

Understanding Nurse Tank Valves and Decals

To help you safely handle anhydrous ammonia, here are some facts about nurse tanks:

Most nurse tanks are painted white. This reflects sunlight to reduce radiant heating of the tank, thus minimizing the pressure within the nurse tank.

Each nurse tank is required to have three different types of decals: (1) a 1005 placard, (2) an "ANHYDROUS AMMONIA" decal placed on all four sides (even on side-by-side nurse tanks), and (3) an inhalation warning decal placed on each side. Some states may require additional markings. Emergency responders use the placards and decals during incidents involving anhydrous ammonia.

Figure 8.19. Decals on a nurse tank that provide important information to those responding to an ammonia release.

Nurse tanks have four legs that are secured to their running gear. The running gear needs to be adequately sized for the tanks and weight placed on them. Because they are implements of agriculture or implements of husbandry, nurse tanks are not required to have brakes. However, many newer nurse tanks are equipped with brakes.

Slow-moving vehicle (SMV) emblems or metal signs are affixed to nurse tanks. SMV signs inform those approaching that the equipment in front of them is moving slower than normal traffic.

Figure 8.20. Nurse tanks ready to go on the road.

The valves on top of nurse tanks are enclosed in a roll cage to protect them in case the tank is involved in a roll-over accident. While the order of the valves can vary, nurse tanks are constructed with similar valve configurations and are consistently labeled.

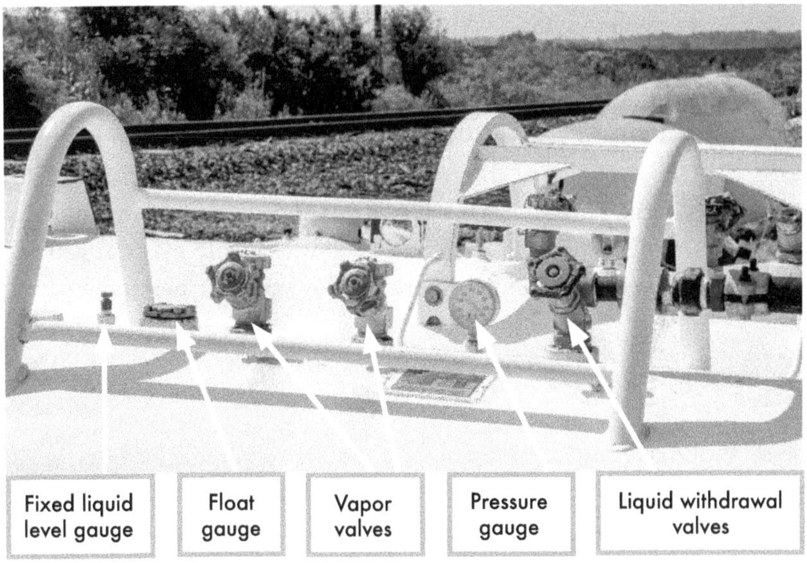

Figure 8.21. Valves outside of the tank, which are enclosed within a protective cage.

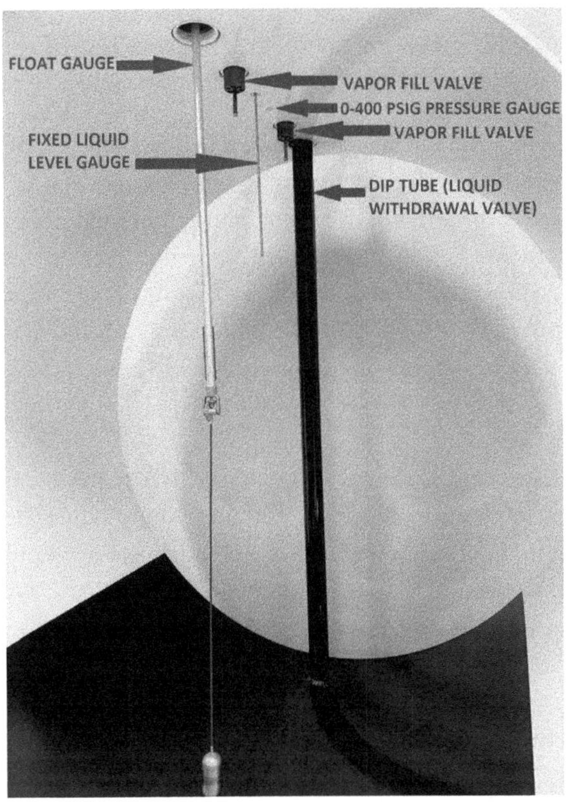

Figure 8.22. Valves and pipes inside the tank. This shows the importance of their placement and provides clues to how the valves work on top of the tank.

Figure 8.23. Excess flow valve.

Here are the six kinds of nurse tank valves and their functions:

1. The Liquid Withdrawal Valve

The liquid withdrawal valve, commonly known as the *farmer's valve*, has a tube that goes to the bottom of the tank that allows the withdrawal of liquid anhydrous ammonia. This valve is always painted orange, which indicates liquid.

The withdrawal valve incorporates an excess flow assembly that shuts the system down if the flow rate exceeds a certain number of gallons per minute (gpm). That's generally 45 to 60 gpm, depending on the size of the container's opening. In the event of a release exceeding those amounts, the spring in the excess flow assembly snaps shut. The excess flow assembly may not work if it is rusted or hampered by other debris, or has degraded O-rings. In addition, the excess flow assembly may not activate to stop a release if a hole develops in a hose where full flow is not achieved.

REPAIRS BEYOND YOUR WELDING SKILLS

Keep in mind that nurse tanks are actually pressurized vessels that carry hazardous material. Because the material is stored under pressure, any repairs to the tanks must be performed by professional welders who have special certifications and experiences. Never weld anything to the actual container unless you are a welder with an R stamp certification.

2. Vapor Fill Valves

Vapor fill valves are used to add liquid anhydrous ammonia and to equalize vapor. They are painted yellow to indicate that the valve extends only into the vapor space.

When you are filling the tanks, either valve will work for pumping liquid anhydrous into the tank. If nurse tanks are involved in an accident that results in the tanks being upside down, then the valves will operate in a different space—the orange valve will be in the vapor space and the yellow valves will be in the liquid space. When that happens, the tank must be turned upright before you can remove liquid anhydrous.

3. Fixed Liquid Level Gauges

The fixed liquid level gauge is commonly called the *spit gauge*. It is a bleeder valve that has a dip tube that extends inside the tank to the 85 percent liquid fill line. When the liquid hits the 85 percent level, the liquid ammonia pushes up through the tube and creates a "spitting" action that can be observed.

4. Float Gauges

Float gauges are visual indicators of the liquid levels in nurse tanks. However, you should use float gauges only as a reference, because internal floats can easily be damaged.

5. Pressure Gauges

Pressure gauges typically read from 0 to 400 psi. You should not rely on pressure gauges to determine the fill status of the tank because the pressure will vary depending on the temperature. Remember: These gauges only tell you pressure, not fullness.

6. Pressure Relief Valves

Pressure relief valves (commonly referred to as *PRVs* or *pop-off valves*) open if the pressure inside the tank exceeds 250 psi.

Filling Nurse Tanks from the Riser

Employees on the riser have one main job: to safely fill the nurse tanks. There are three golden rules that people charged with filling nurse tanks need to keep at the top of their minds before they open the first valve at the start of their shift.

Golden Rule Number 1

Every person on the riser should wear the appropriate safety equipment at all times. In addition to wearing long pants and long-sleeved shirts, personal protective equipment approved for anhydrous ammonia includes gloves and chemical splash goggles.

SAFETY EQUIPMENT FOR ANHYDROUS AMMONIA USE

When working with anhydrous ammonia, it isn't enough to just have gloves and goggles—be certain you use the proper safety gear. Gloves must be insulated to protect hands from freezing and must be impervious to anhydrous ammonia. In addition, the gloves should be cuffed to prevent liquid ammonia from running down the operators' arms when they lift their hands.

Goggles can be non-vented or indirect vented. Safety shields do not take the place of goggles. In addition, employees who work around anhydrous ammonia should never wear contact lenses.

Golden Rule Number 2

Always check the water supply in the emergency soak tank that someone would jump into in case of an emergency. Tanks collect algae, dead animals, and trash. Remember that water is your best friend when you have been exposed to anhydrous ammonia. Water will relieve pain and minimize skin damage from ammonia exposure. However, if a catastrophic release occurs and getting to your emergency water source puts you in harm's way, run upwind and find an alternative source of water outside the danger zone.

Figure 8.24. A clean emergency water tank.

Golden Rule Number 3

Inspect hoses for cuts, nicks, or soft spots. Hoses that come in contact with metal risers can become abraded over time. Small pinpricks are normal. Never take a chance with a damaged hose—replace it! Never use duct tape or black tape to repair a hose, not even for a short-term fix.

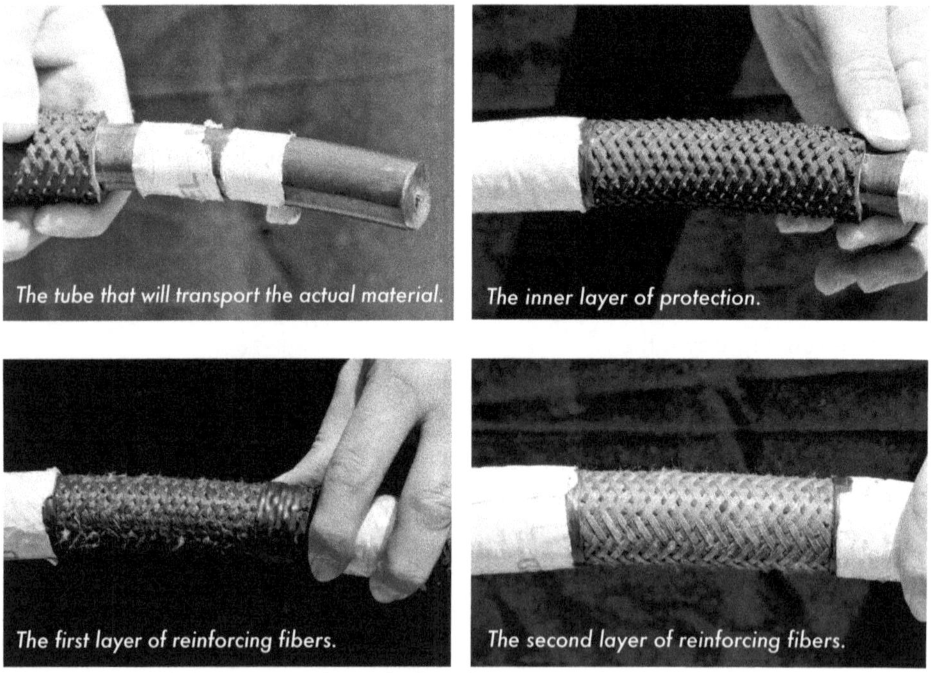

Figures 8.25–8.28. Components of an anhydrous hose.

NEVER CUT CORNERS WITH ANHYDROUS AMMONIA

Every facility has a risk management plan that lists standard operating procedures (SOPs) for how ammonia should be handled. SOPs can vary from company to company and even branch to branch within a company. To use a sports metaphor: The SOP is the playbook or game plan.

Company personnel should review their SOPs in the spring and again in the fall. The review will often cover such items as how valves are labeled, how to open valves on storage tanks, how to fill nurse tanks, how to safely disconnect hoses, what safety equipment to wear, how to shut down systems in the event of emergencies, and how to transport nurse tanks on the highway.

An SOP is more than words on paper. It is a step-by-step process that describes how to work safely around anhydrous ammonia. Cutting corners and not following procedures puts any ammonia releases squarely on the shoulders of the employees who decided to circumvent those established procedures. *Follow the procedures so you can return home at night.*

After you have donned your protective equipment, filled the emergency water tank, and checked the hoses, it is useful to take a general walk around the riser before the day's work begins. Consider the following when filling tanks with anhydrous ammonia.

- Always assume ammonia is in the hose—just like you should always assume a gun is loaded! There is always ammonia residue left in the nurse tank that turns into vapor. Make sure to bleed hoses using valves designed to do so.
- Keep the platform clear of any trip hazards such as tools and hoses left on the deck. Many risers do not have guardrails for fall protection surrounding them.
- Fill nurse tanks to no more than 85 percent using the fixed liquid level gauge as the main indicator.
- Be alert to potential problems. For instance, anhydrous ammonia fittings are designed to be hand-tightened. If you are unable to hand-tighten the valves, it could be that the O-rings are damaged, or the threads are worn. If you cannot hand-tighten the fittings, then you should stop and see what the problem is. If something doesn't appear right, stop and investigate. Attention to detail is critical. You should never use a pipe wrench to tighten fittings, because doing so can damage the threads.

- Keep unauthorized people off the risers. All ag retailers know that there are times when there is not a spare tank on the lot—everything is in the field. This means that when people are returning empty tanks, they want a quick turnaround on their fills. While it can be difficult, remember to keep people away from the risers as much as possible. The personnel filling tanks must focus on doing their jobs. Having two or three people on or around the platform can become distracting.
- Designate a specified drop-off area where clients bring their empty nurse tanks. This will help keep everyone's focus on what needs to be done.

One of the problems ag retailers face is having growers who bring in two tanks (one behind the other), drive across the scale, and then go directly to the filling station. Always tell growers why they can't go directly up to the riser. We suggest they go to the office or another location on-site while you fill the nurse tanks. That way you don't feel rushed or have to make idle talk with them that can reduce your concentration on the task at hand. More important, if an accident occurs, you don't have to worry about the safety of the growers.

This might not be followed 100 percent of the time, but it should be the goal. There is no reason for growers to be within the area where tanks are being filled. Insurance companies sometimes require buffer zones for nonemployees for certain operations as well.

Figure 8.29. Signs to keep bystanders away are essential to keep them safe if an emergency was to arise.

Moving Nurse Tanks to and from the Riser

When application seasons begin, there is normally a person filling nurse tanks and shuttling empty tanks and full tanks around the lot. As the application season gets busier, other employees may be asked to move tanks. A few safety considerations apply to the person moving nurse tanks:

- Nurse tanks are bulky and heavy. Take care not to drop the nurse tank tongue on your feet or smash your fingers at the hitch.
- Individuals who move nurse tanks should wear the same PPE as employees on the riser at the filling station. Some companies require any employee within a 25-foot radius of the riser to wear gloves and goggles.
- You should know where to activate the emergency shutoff in the event that the person on the riser is unable to do so.
- Every nurse tank should be chocked when parked at the riser. This prevents any forward or backward movement.
- When nurse tanks are being filled with anhydrous ammonia, they should not be attached to a vehicle in case someone tries to drive off. Pull the hitch pin out and unhook the safety chains. Let the tongue sit on the drawbar for a quicker reattachment.

Figure 8.30. Chocks prevent movement of the nurse tank.

- Nurse tanks need a quick visual inspection. Pay attention to the condition of the tires; make sure that the lugs appear tight; that the 5-gallon water tank is full; that there are no loose bolts; that placards are legible, not faded or folded; that there are no signs of leaks; and that the hitch pin, keeper, and chains are in good shape.

If something on a nurse tank needs repair, tell the manager or mechanic it needs attention. Put an out-of-service tag on the nurse tank to keep it from being put back into service until repairs are made.

Although anhydrous ammonia is exempt from certain regulations, it is still a hazardous material. From a retailer's standpoint, a DOT out-of-service violation is a serious matter.

Figure 8.31. Whether a tank is owned by a retailer or a farmer, a release of ammonia can have serious repercussions.

TRANSPORTING ANHYDROUS AMMONIA IN AND OUT OF THE FACILITY

A single agricultural retailer can own dozens to hundreds of nurse tanks. Whether farmers pick them up or retailers deliver them depends on the trade area. In some areas, 95 percent of the farmers pick up nurse tanks from retailers. At other facilities, 95 percent of the farmers pay to have them delivered to the fields of application. Approximately half of the equipment used to apply anhydrous ammonia is owned by retailers and is rented to their customers. Retailer-owned equipment that is rented to farmers puts retailers in the liability crosshairs.

All nurse tanks and application equipment receive a quick walk-around inspection. You don't need to be a trained mechanic to see a missing lug nut, notice the cord through the tire tread, observe that placards are partially torn, see that the SMV emblem has faded, or discover that a 5-gallon water container hose has dry-rotted from sitting in the sun all year. These are the small things that need your attention before

IF YOU OWN IT, YOU'RE RESPONSIBLE FOR IT

When retailers own the nurse tanks and applicators, equipment maintenance is their responsibility. Poor maintenance subjects companies and responsible employees to increased scrutiny, questioning, and liability.

Generally, maintenance on tanks and applicators happens in late winter or early spring, right before they are placed in service. The employee's job during the application season is not to complete the full inspection that mechanics conducted during the offseason but to watch out for immediate needs as the equipment experiences wear and tear.

you take any nurse tank on the highway. If a grower has an incident while hooking up to a set of tanks, any problem from the maintenance side of things might entangle the retailer in a lawsuit.

A number of considerations affect the transportation of nurse tanks:

Make sure the equipment is roadworthy. A nurse tank is only as safe as the condition of the structure (for example, running gear) it is attached to. A poorly maintained nurse tank structure compromises the integrity of the nurse tank itself. It is just as important to inspect the nurse tank's structure as it is to inspect the nurse tank itself.

Make sure the SMV emblem can be clearly seen and is not faded.

Make sure the 5-gallon water tank is full, and the hose is not brittle or clogged.

Make sure that the farmer's valve is completely closed and the dust cap is secured.

Make certain that the hitch pin, safety clip, and safety chains are securely hooked. One safety tactic is to cross the chains to form an X pattern under the tongue. In the event the nurse tank becomes detached, the chains will serve to cradle the tongue above the roadway (see volume 2 for more on hitches). Don't forget that safety chains are not hooked up only between the truck and the nurse tank. You should also hook up safety chains between nurse tanks if you have a trailing nurse tank.

NOTE TO INDIVIDUALS WHO WON'T USE CHAINS

Some farmers will not use safety chains even after you have asked them to. Individuals will let you know that it is their decision to use them or to leave them off. They might even say, "Mind your own business. I'm not going to use safety chains. If it rolls over, I don't want the nurse tank to take my truck with it." Imagine if, down the road, the tank rolls and hits an oncoming car. For that reason, every time a farmer refuses to put on chains, employees should note that on the order ticket in case this is called into question in a deposition or court.

Plan the travel route before you leave the facility. Do not forget: You are transporting a hazardous material. Your goal should be to get the tank to the field of application and the empty tank back to be refilled as directly as possible. When pulling a nurse tank from a retailer location to a field of application and back, do not make any unnecessary stops. Don't be tempted to stop at a drive-through, do business at the bank, or pick up supplies at the store.

Never transport anhydrous ammonia through town. It might be the quickest route, but it won't be the safest. There is no law against going through town with a nurse tank (unless an alternative route has been designated for moving hazardous materials), but you have to think of the consequences of having a roadside incident with a nurse tank in a populated area. It's worth an extra 5–10 minutes to take a route that does not go through town.

Figure 8.32. Taking ammonia tanks through town is not a good idea. Spending a few extra minutes to take an alternative route might be in your best interest.

Observe the 25 mph speed limit. The speed limit for a slow-moving vehicle is 25 mph. There are good reasons for this.

First, speeds greater than 25 mph increase stopping distances, whether you are pulling one tank or two, filled tanks or empty tanks, or two tanks side by side on a single running gear. The more weight a vehicle pulls and the faster the speed, the greater the distance it will take to stop.

Second, as speeds increase, nurse tanks begin swaying back and forth, which can cause the driver to lose control. At faster speeds, hitting a rough spot in the road will start the nurse tank weaving back and forth.

Third, driving too fast on tires that are rated to go only 25 or 30 mph can lead to blowouts and overturned nurse tanks.

Fourth, and last, an SMV sign tells the public and law enforcement that you are moving at 25 mph or less.

Figure 8.33. Nurse tank tires are often rated for only 25 mph.

Figures 8.34–8.36. Imagine how this would have ended if the train had hit the anhydrous ammonia tank.

Stop at all railroad crossings. Federal law requires equipment transporting placarded hazardous materials to stop at all railroad tracks.

Be prepared to deal with incidents that involve an anhydrous ammonia release. This means that you should always carry gloves and goggles for anhydrous ammonia when transporting it in case you can help mitigate the problem. No one plans for an incident on the road, but you should always plan to respond to one should the need arise.

APPLYING ANHYDROUS AMMONIA

Farmers also are involved in transporting nurse tanks from ag retailer locations to their fields or between fields. Farmers who take possession of nurse tanks are involved in the additional step of applying the ammonia from nurse tanks into the soil. In other words, those who make applications are actually doing a controlled release of anhydrous ammonia for plant nutrition.

There are a number of steps applicators must follow to safely apply the fertilizer as a nutrient:

Always wear the right PPE. Wear the right kinds of gloves and goggles when working on any aspect of anhydrous ammonia equipment. You should always put on the gloves and goggles before walking up to the nurse tank or applicator.

Figure 8.37. Anticipating what problems could occur helps to develop plans for dealing with a release.

Assume "live" equipment. Always assume that anhydrous ammonia is in the hoses and that the nurse tanks and application equipment are pressurized.

Install the coupler correctly. The quick connector/breakaway coupler needs to be correctly installed if it is to pull apart and stop the flow of ammonia. The purpose of the L-brackets is to allow a straight pull. Without the straight pull, the coupler will bind, and then the hose will either get pulled apart or the valve will be ripped from the tank.

The breakaway coupler is mounted on double L-brackets and secured to a mounting bar. The mounting bar has a double-bolt system attached to the double L-brackets along with a bushing for the breakaway coupler. The two L-bars swivel up and down and right to left, which keeps the hose in a straight line between the applicator and the nurse tank. If the tank separates from the applicator (for example, when a tongue or hitch pin breaks), the coupler should pull apart, causing the check valve to stop the flow.

It is critically important to follow manufacturer instructions for properly installing all breakaway couplers and to follow the recommended service periods. Most breakaway coupler failures are related to improper installation or lack of maintenance. This becomes an issue not only for farmers but for retailers too—some outlets will apply ammonia for 20–30 percent of their customers. The number of farms where the retailers apply the anhydrous ammonia is increasing each year.

Inspect and service the coupler on both ends. You need to inspect and service (that is, grease) the breakaway coupler on both ends. You also need to "exercise" the coupler in order to separate in the field of application. Inspection and service should be part of the preseason check before going to the field. When the breakaway coupler doesn't disconnect easily and smoothly, the hose can be pulled apart, or the farmer's valve (liquid withdrawal valve) on the nurse tank can shear.

Figures 8.38–8.39. Can you tell which breakaway system is correctly mounted? Actually, both are installed incorrectly. The top photo has two L-brackets that allow vertical and horizontal movement but is missing the mounting bar. The bottom photo has only one L-bracket and is missing the mounting bar.

Always use bleeder valves when disconnecting hoses.

Take care standing on tires. Be careful standing on tires, especially when it is wet. When you connect the hose from the applicator to the nurse tank, that usually means climbing on the front tire. Falling from the tire can cause more than personal injury—the hose may be damaged or broken, or the valve may be inadvertently opened.

Secure the safety chains. Ensure that the safety chains are secured with a good pin and keeper. The chains, clips, and hitch are not only for the nurse tank but also for the applicator to the tractor. These connection points are critical for keeping individual units together.

Never transport with the valve open. Never transport a nurse tank on a public road with the farmer valve open and the hose connected to the applicator. If you are driving a tractor with the application tool bar plus nurse tank, disconnect the transfer hose and make sure all tank valves are tightly closed.

Figures 8.40–8.41.
Primary valves on nurse tanks.

Shut off or close all valves at the tank source, which is especially important with double tanks on a single running gear. There are valves where the two hoses meet at the bulkhead along with primary valves on each tank (see arrows in figures 8.40–8.41). Shutting off the valves at the bulkhead does not shut off the primary valves at the tank. Failing to close the valves creates a live system. Whenever you tow or transport a nurse tank on the road (even behind an applicator), state and federal regulations require you to turn the valve off and disconnect the hose.

Mount safety water on applicators. Farmers who own their application equipment should mount safety water containers on their applicators to make sure water is available in case of a release. Never assume there is water in the tank. It's more common than you expect for the clear hose to pop out of the container, allowing the water to drain out.

Figure 8.42. Water is the only cure for dealing with anhydrous ammonia on the skin and reducing the severity of the burns.

The plastic of the clear hoses will degrade in the sun, causing brittleness and breakage. For these reasons we also suggest that anyone who makes applications keep a small squeeze bottle of water in their pocket and a gallon container in the cab. There may not be time in an emergency to access the 5 gallons of water on the nurse tank.

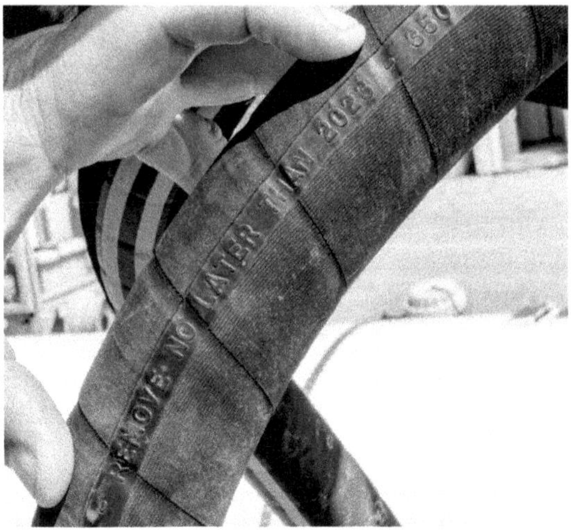

Figure 8.43. Hoses are labeled with expiration dates.

Inspect hoses. Check ammonia hoses for nicks, cuts, or soft spots before using them. Replace any damaged hoses before putting the unit back in service.

Check hose expiration dates. Make sure all hoses are within their expiration dates. Replace any hoses that are beyond those dates.

Check the O-ring gasket. Make sure the O-ring gasket in the withdrawal valve is present and functional. While wearing gloves, rub your fingers in the connector to make sure the gasket is there. Hand-tighten the hose end valve from the applicator to the nurse tank. Open the hose end valve all the way, then proceed by slowly opening the withdrawal valve on the nurse tank. This slow approach allows you to notice any leaks while the valve is slowly releasing product into the transfer hose.

Fully open the withdrawal valve. Make sure to open the withdrawal valve and hose end valve of the transfer hose to their fullest open position. If those valves are not opened all the way, it will impede the activation of the excess flow valve on the nurse tank.

HUMAN ERROR IS BEHIND MOST ANHYDROUS AMMONIA RELEASES

Retail employees and farmers who routinely handle anhydrous ammonia sometimes forget this fertilizer is an extremely hazardous material. During busy times of the year, retailers are handling several other fertilizer products, making sure that the bulk ammonia storage and nurse tanks are kept filled, and getting nurse tanks to the field safely. Some get lax in their anhydrous ammonia handling, because they have not experienced a release.

This brings up the question: "Does familiarity with anhydrous ammonia and fatigue from working long hours breed complacency and increase the problems we see when filling nurse tanks, transporting them on the road, and making the fertilizer application?" This question is important to ask, because accidental releases occur every year when handling anhydrous ammonia at the retail facility, transporting nurse tanks on the highway, and making field applications. In the midst of the demands of the season, it is easy to forget about the dangers of anhydrous ammonia when the job becomes monotonous, and the pace quickens as more growers place their orders for this product.

We know from experience that fatigue can lead to complacency when users bypass standard operating procedures in order to get nurse tanks filled quickly and transported out to customers. For example, a grower brings in a nurse tank to be refilled, and they are in a hurry to get back because heavy rains are predicted. The retail employee bypasses chocking the nurse tank tires and disconnecting it from the towing vehicle, trying to get the farmer back to his fields. The good news: Nothing happened that day, and the farmer was satisfied to get in and out quickly.

But what could have happened, or what might happen in the future given the same set of circumstances? Figures 8.44–8.45 show what can happen when a driver fails to chock the tires of his cargo tank delivery truck. A simple procedure was not followed, and the resulting release caused an early morning nightmare for a community.

Figures 8.44–8.45. A transport was beginning to off-load into a facility storage tank, but the operator failed to chock the wheels and set the parking brake. During the off-load, the transport rolled backward, which broke the piping on the transport and released more than 7,000 gallons of anhydrous ammonia. More than half the town had to be evacuated, and more than 25 residents were treated at the local hospital.

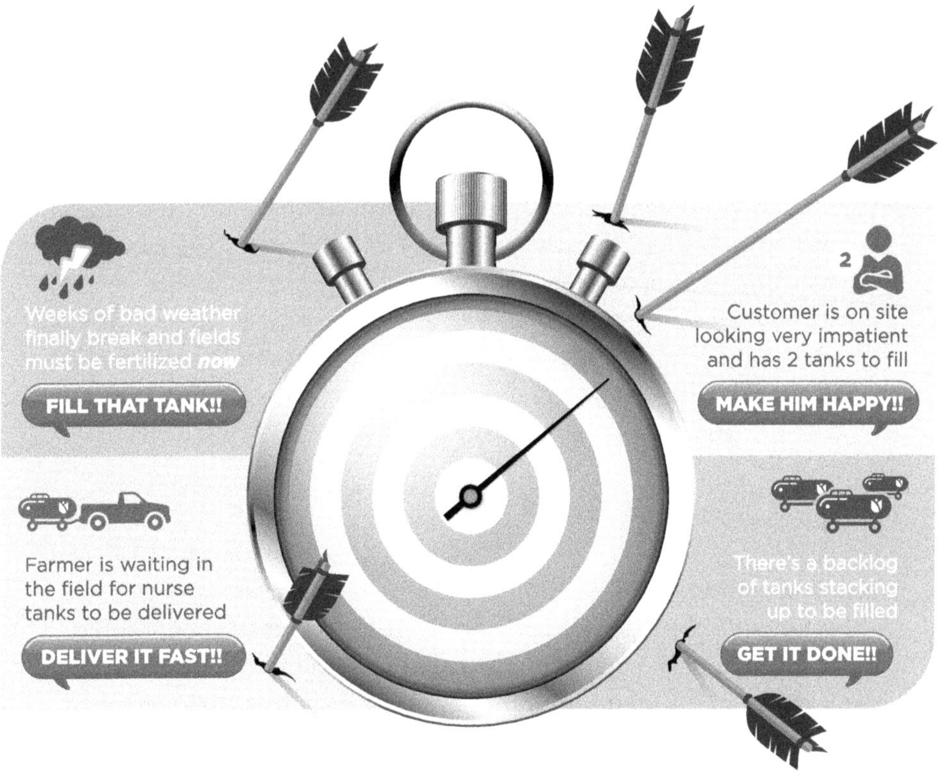

SPEED IS THE ENEMY OF PRECISION

It is impossible to work with precision when you are in a hurry — eventually, hurrying will lead to catastrophe. Slow down and stay safe!

The bottom line is that it comes down to the ag retail employees who are dealing with customers who need the product quickly. Those employees need to follow the safety procedures. Taking shortcuts during a short season can lead to errors when filling and transporting nurse tanks during spring and fall applications.

Consider these common factors at ag retailers:

- Retailers often limit which employees work full-time with anhydrous ammonia. However, during the spring and fall application seasons, inexperienced employees are often asked to assist.
- As the fertilizer season unfolds, employees work long hours each day and, at times, seven days a week. When there are rain delays, retailers face extra burdens when the weather clears enough for field applications.
- Filling nurse tanks and moving them is repetitious work. Employees on risers may be filling nurse tanks on one side, while disconnecting the hoses from the filled tanks on the other side. Meanwhile, people on the ground are connecting and disconnecting nurse tanks as one tank is filled and another is placed in line to be filled.

- The repetition of filling nurse tanks, transporting them to farms, making applications, and retrieving empty nurse tanks for refilling can become tedious. It is easy to become complacent and less focused when you work under these conditions. This can lead to accidents, such as driving away from the filling area with the hoses still connected, which could cause a release.
- Farmers recruit family and friends (who may not have any experience pulling or towing nurse tanks) to transport nurse tanks. The recruits sometimes do not understand how much distance they need to stop with the additional weight.
- Or an untrained employee may hustle to get tanks to a farmer who has run out of anhydrous ammonia, driving too fast or going through town to get it to the farmer who is waiting idly in the field.

HUMAN ERROR: AN AVOIDABLE CATASTROPHE

The driver of a tractor towing an applicator with two side-by-side ammonia tanks on the highway between fields found himself in a terrible predicament. One minute, everything was fine, but the next second he was involved in the release of 750 gallons of anhydrous ammonia in a populated area.

Eventually, the gas sent dozens of people to local hospitals. Among the injured were those working for fire departments and law enforcement. Authorities issued a one-mile shelter-in-place order. After attending to their patients, emergency room staff began complaining of ammonia exposure symptoms. The hospitals placed decontamination units outdoors to take care of the injured.

The tractor operator did not call 911—another motorist placed the call. That was the biggest problem. Had the tractor operator called in the release, he would have stated it was anhydrous ammonia. The motorist who called in the release stated it was a car fire. Emergency responders were confused about the source of the "smoke"—many thought it was a car fire. Many of the early responders were seriously injured when they came into contact with the ammonia cloud.

Initial media reports blamed the applicator who had been towing two side-by-side tanks with a tractor on a public road. That in itself was not the problem; the issue was pulling or towing ammonia tanks on a public road with the valves open and the hoses fully charged. The applicator was unable to turn off the valves due to the cloud. If he had just followed the regulations and used common sense, instead of traveling with the system charged, this incident would not have happened. This release was attributed to human error.

- Highway incidents are likely when drivers exceed the 25 mph speed restrictions for nurse tanks. A truck that weighs 8,000 to 11,000 pounds may transport two nurse tanks that weigh a total of 15,000 to 22,000 pounds. One problem with nurse tanks is that a high percentage are not equipped with brakes, because they are implements of husbandry. The truck must provide all of the braking power for a heavy load. Such a heavy weight pushing behind a truck means it will take longer than anticipated to come to a complete stop.
- When a nurse tank veers off the road and onto a shoulder at higher speeds, bad things happen. Most shoulders on rural roads drop quickly, resulting in the tank becoming unstable. Pulling back on the roadway often leads to losing control of the nurse tank, which could cause a rollover.
- Performing preventative maintenance during the offseason is critical, because equipment on nurse tanks can fail if defective, worn, or incorrectly installed. For example, using a high-pressure hose beyond its expiration date is trouble waiting to happen. Furthermore, bolts can break, or welds can fail on nurse tank carriages (also known as the running gear).

THE GOAL: REDUCE HUMAN ERROR

Most anhydrous ammonia releases are directly linked to human error. Reducing these mistakes is at the center of all training programs.

That's not to say that all anhydrous ammonia releases are a user's fault. In many documented cases, incidents are caused by those sharing the highway. Another driver may run a stop sign and crash into the nurse tank. This may rupture the nurse tank shell or shear off valves on top of the tank. What was supposed to be a normal day turns into a crisis with a major release. Emergency responders try to stop the situation, ambulances evacuate the injured, and nearby residents must remain in their closed-up homes.

Figure 8.46. A vapor release requires emergency responders to act quickly by shutting down the road and dealing with the plume of gas.

PREPARING BEFORE AN EMERGENCY OCCURS

Employee Training

Can you remember the initial anhydrous ammonia training you went through when you started working? Often new employees will watch a video or two about anhydrous ammonia. This is often followed by introducing employees to hands-on experience by teaming them with people who have worked with the product for years. Working at the riser allows new employees to actually connect and disconnect hoses and to learn the purposes and functions of the many valves. This one-two approach to safety training for anhydrous ammonia handling helps put the classroom information into real-world practice.

Can you remember the first time you were working on the riser or moving nurse tanks around? During the first week, you were probably very cautious about how you worked around ammonia and followed the policies of the facility to the letter. Did you notice how comfortable you became with anhydrous ammonia after about one month? Gaining confidence from experience is a good thing, but getting cocky will lead to problems. Management wants employees to stick to safe practices and use common sense.

Most ag retailers offer basic anhydrous ammonia training to the entire staff—applicators, field scouts, office personnel, salespeople, and agronomists. There are five questions every facility employee must be able to answer related to anhydrous ammonia:

1. Where do you go when a major release occurs?
2. Who do you call to get outside assistance?
3. Why is water so important when someone is exposed, and where is the water tank?
4. How do I operate the emergency shutdown system?
5. Who do I need to call at the facility when a release occurs?

FIVE THINGS EMPLOYEES MUST KNOW

When training employees, be sure they know these essentials

If employees can answer these five questions, then the training is deemed successful.

Why would employees who do not handle anhydrous ammonia need to know how to answer these five questions? Your life and theirs may depend on them doing the right thing during an ammonia release. In addition, shutting down the system may avert a catastrophic release, keeping a minor release from becoming larger.

This basic training is part of management's accountability to its employees and response teams. When emergency responders and the facility's risk manager arrive at the scene, a quick head count is the first step in ensuring that all employees are present and accounted for. When someone is missing, rescue teams will risk their lives to find and rescue that missing person.

Safety coordinators also can direct individuals at the meeting location to move to another location to allow emergency responders in and keep others out. In other cases, the employees can run errands or make additional phone calls when management asks them to. For instance, a manager may ask an employee to take a safety data sheet (SDS) to the emergency room if it was not provided when the ambulance left the facility with an injured person.

RUN UPWIND IF YOU CAN'T GET TO AN EVACUATION POINT

In the event of a catastrophic anhydrous ammonia release, you may not always be able to reach the designated evacuation site. When that happens, your goal is to save yourself by quickly running upwind from the release. You are in survival mode at this point. When it is safe to do so, you should make it to the evacuation point or at least call management to let them know you are safe.

Fire Department Training

Training fire department personnel (whether volunteer or paid) is important to ag retail facilities for two reasons:

1. Fire departments need the training, because anhydrous ammonia releases are something they seldom encounter.
2. Not all fire department volunteers come from agricultural backgrounds, so many are unfamiliar with anhydrous ammonia.

Training fire department personnel is critical to the well-being of everyone involved. This is especially true for urban fire departments that cover rural areas. As towns and cities encroach on farm ground, the resulting traffic exponentially increases the chance of a roadway accident involving nurse tanks.

Fire department training varies depending on the need. The training can be as simple as the fire department touring the facility each year. Those on the tour can ask questions and familiarize themselves with the facility's layout and operation. The walk-through also helps the fire department better understand what challenges they need to plan for in the event they are asked to provide assistance. In some cases, ag retailer trainers conduct in-house training programs or conduct hands-on exercises to train local fire departments.

Providing training to fire departments also builds understanding among ag retailers, farmers, and emergency responders. That becomes crucial when everyone must make important decisions in the event of a release.

REACTING DURING A MEDICAL CRISIS

Accidents happen, so what is your role if you are unlucky enough to be present at the time of an ammonia release? The truth is there is no playbook that lists the steps one should follow to deal with an anhydrous ammonia release or even injured people. It becomes especially complicated to determine what to do when injuries and ongoing releases occur simultaneously!

Steps for Dealing with Injuries

While no publication, video, or even experience can tell you exactly what to do when handling medical emergencies that involve a person exposed to anhydrous ammonia, there are eight steps you should take.

Step 1: Remain Calm

The situation is serious, and you have to convince injured people to follow your instructions. That will only happen if you remain calm.

KEEP FROZEN CLOTHING FROM RIPPING SKIN

People who have been injured may tell you their chests are on fire from the anhydrous ammonia on their shirts. Logically, the injured people want to pull their shirts off. You must convince them not to remove their clothes until they get into the water.

The anhydrous ammonia might have frozen their clothing to their skin. Only take off clothing after they are in the water and the clothing is thoroughly saturated and has thawed.

The good news is that the water will give exposed individuals almost instant relief and take their minds off removing their clothing. You just have to firmly say, "Hang on, let's get you to water as quickly as we can. Once you are in the water, the pain will be relieved, and then we can pull your clothing off."

Figures 8.47–8.48. Jumping into contaminated water can further aggravate skin injuries from ammonia exposure. Make sure your emergency water is always clean.

Step 2: Move Injured People Upwind

As soon as you're able, move injured people away from the ammonia release by going upwind from the release. A wind sock at the filling station can be helpful in indicating which direction the wind is coming from.

Step 3: Provide Water, Water, and More Water

In just about every scenario, take injured people to the facility's water tank, located upwind from the riser. Keep injured people in the tank until further help arrives. A nearby ditch or creek may provide alternative sources of water.

Step 4: Activate the Emergency Shutdown System

With injured people in the water tank, it is time to quickly activate the emergency shutdown system—if you can do so safely. Even if you activate the shutdown system, you still must call 911 to get emergency responders to the facility.

Step 5: Provide Only Life-Saving First Aid

When someone is injured from ammonia exposure, you should only (1) perform CPR if the person is not breathing or (2) stop any bleeding. Never put any ointment or salve on burns, because that can cause the ammonia to travel deeper into the skin. When that happens, the ammonia will be trapped in the body, and the water will not be as effective. Medical personnel will decide on the best treatment options for any injured person. We are not doctors, so don't pretend to be one.

Step 6: Call 911

Employees who call 911 should never be faulted for doing so if, in their best judgment, help was needed. Inform 911 dispatchers that you are reporting an anhydrous ammonia release. Anhydrous ammonia is a placarded hazardous material, an inhalation hazard, and has a 1005 decal.

Figure 8.49. Firefighters spray water to mitigate an ammonia release.

Dispatchers will also need to know if the release is ongoing, how many people are injured, and how severe those injuries are. The 911 dispatchers will ask additional questions that you should answer to the best of your abilities.

Step 7: Answer Questions from Emergency Responders

Be ready to provide a safety data sheet (SDS) to any responders. Responders will debrief you about what you know and what actions you have taken. Later on, your safety manager will ask similar questions when documenting the facts while they are still fresh.

Step 8: Send the SDS to the Emergency Room

Once emergency responders arrive, make sure to give them the SDS for anhydrous ammonia. The information on the SDS is important for emergency responders transferring injured people to emergency rooms. Just as important, the SDS also may help guide the responders on how to deal with the escaped vapor.

Four Scenarios

Now that you know these steps, consider these scenarios. How would you react in each one, using the eight steps as your guide?

Scenario 1

A person on the riser was exposed to anhydrous ammonia but remains calm. What do you do?

Scenario 2

The employee on the riser tries to disconnect a hose by cracking it without bleeding it off first. One of his coworkers on the riser is not wearing goggles. They scream in pain as the vapor finds its way into their eyes. What do you do?

Scenario 3

The person working the nurse tanks sees his coworker on the riser fall on the deck, seemingly disorientated and confused. The wind is blowing the release from the ruptured hose away from the riser. What do you do?

Scenario 4

As you deliver full tanks to a field, you find a farmer lying near an empty tank blinded by anhydrous ammonia with burns on his skin. The release is ongoing, but the wind is carrying the ammonia cloud away from the scene. What do you do?

The two main differences among the four scenarios are how serious the injuries are and whether the releases have stopped or are ongoing. The first question we should ask ourselves: Can the injured person be helped without much risk to the person offering assistance? Even so, it is important to remember that while we often categorize anhydrous ammonia injuries as minor or serious, they are all serious injuries. In spite of what injured people say or think, you should treat any exposure to anhydrous ammonia as an emergency.

ALWAYS SEEK MEDICAL ATTENTION

Every injured person, no matter how minor the injury appears, should seek immediate medical attention. Most injured people will say that everything is all right, so it's unnecessary to call 911 or go to a doctor. It is important to remember that ammonia can produce both acute (right now) and chronic (later on) effects. Anhydrous ammonia is always going to seek moisture, and in doing so, it will move deeper into the skin unnoticed.

An exposed person could spend the rest of the day feeling fine, but after taking a shower in the evening and lying down, that person may begin to experience symptoms. The exposed person may feel almost like they are having pneumonia-like symptoms—wheezing and gurgling in the lungs every time they breathe in and exhale out. These symptoms are caused by the ammonia reacting with water in the lungs.

This is why every injured person must be seen by medical personnel, no matter the perceived severity.

Play It Safe

You will have to make quick decisions based on the circumstances you find yourself in. You are now dealing with an injured person and an ongoing release that could bring you in contact with the deadly gas. If you are uncomfortable dealing with the injured person within a cloud, focus your attention on calling 911. The fire department will have self-contained breathing apparatuses that will enable them to deal with the release safely in the event they need to enter the vapor.

It is also true that you will be severely limited in what you can do if you turn over a nurse tank. If you don't have gloves and goggles with you, then you will not be able to do anything even if a person is down in the cloud. With appropriate PPE and the wind blowing away from you, then you might be able to shut off the release valve.

Our first reaction is to help injured coworkers by shutting the valve off and stopping the release. It seems unacceptable to be told you cannot help. But by trying to save them, you could become another victim. Think about the stories you've heard about attempted rescues in a manure pit, in a grain bin, or from a downed power line that ended up with others losing their lives while trying to save a family member or coworker.

You don't want to end up being one of the victims, and if you get hurt, it will only delay emergency aid to them. A situation that was bad has now gotten worse. We all have to rely on emergency responders to perform the rescue when we are unable to stop the flow of anhydrous ammonia. We hope that you never find yourself having to make this difficult decision.

But don't just take our word for it. The following true account could have had a much different ending, with the hero becoming the victim:

A hose ruptured on the riser while the pump was running. The employee on the riser was able to escape. He pulled on the emergency cable, but the system failed, and the entire facility was engulfed in ammonia vapor.

All but one person evacuated the facility as they were supposed to. The only person who didn't get out was the driver of a transport cargo tanker that was unloading at the bulk storage tank. When the wind shifted, the driver ran in and shut off the valve. He was not injured and was able to keep the release from getting worse.

While the outcome was good, it was probably not what should have happened. While the act was noble, did the end justify the means? Luck was on the driver's side that day. It goes without saying that calling 911 should be your first move. Do not die becoming a hero.

RESPONDING TO THE VAPOR CLOUD

Major releases are those that leave the boundaries of ag retailer facilities or result from ruptured nurse tanks on the roadway. People in surrounding communities may need to take protective measures to stay safe until the release is controlled and mitigated.

When you call 911, the dispatcher relies on your description of the incident when they notify fire departments, law enforcement, and emergency responders closest to the release site. It is typically up to the fire departments to decide on the best course of action to mitigate a release. While the fire department is heading out, they are formulating a plan of attack based on the information you have initially provided to the 911 dispatcher.

The questions the firefighters are trying to answer on their way to the anhydrous ammonia release site include:

- Are there injuries?
- How much ammonia was released?
- What is the source of the release?
- What direction is the wind blowing?
- Is the release moving upward, or is it hanging low to the ground?
- What is downwind of the release?
- How will I notify people downwind?
- Will the release require evacuation or sheltering in place?
- Which roads will need to be closed?
- From what direction should we approach the scene?
- Can the release be controlled?
- Are we equipped to safely shut down the release?
- Is it a “runaway” tank with no way of shutting or slowing down the release?
- Will I need assistance from other fire departments?
- Will we need to call in a hazardous material team?

As the fire department is traveling to the site, they are receiving updated information. While they establish their plan, they fully understand that conditions on the ground could change, which could alter what they do next.

Anhydrous ammonia releases grow more complex as the gas mixes in the atmosphere and moves with the prevailing wind. Incident commanders at the scene will have to decide whether to put water over the cloud or let the cloud dissipate into the atmosphere.

Incident commanders must determine the path of the moving cloud and how much ammonia is yet to be released. Gathering all of the facts dictates what actions they take to mitigate the release. It goes without saying that taking care of injured people is their top priority.

Emergency actions will vary depending on the specific circumstances of the incident. For instance, the fire department may arrive at the scene to find a ruptured nurse tank in the middle of 100 acres of farmland. The vapor cloud is going straight up into the atmosphere. They will likely decide to let the cloud dissipate into the atmosphere, because it poses no threat to those living in the area. From the fire department's perspective, this is a safer approach that protects the public and avoids exposing members of the fire department to the ammonia.

A FALSE SENSE OF SAFETY

Laws require every ag retail facility that distributes anhydrous ammonia to have two full-face gas masks with ammonia cartridges on standby. They also must have rubber rain suits, rubber boots, goggles, and gloves stored with the gas masks.

WARNING! If you have ever worn a respirator, you know they need to be individually fitted. How can you guarantee that these masks will fit tight around your face? You can't. These gas masks are for escape purposes only. They are not for performing rescues.

Running into a cloud wearing these emergency masks can have serious repercussions. Running into an ammonia cloud is analogous to a person jumping into a manure pit or a grain bin to help someone overcome with gas or engulfed in grain. Often people who offer assistance become victims themselves. Depending on the gas mask, it will be a matter of seconds (not minutes) before you can be overcome by ammonia vapor.

The air-supplied respirators that firefighters wear are designed to allow them to go into the release and attempt to deal with the problem head-on.

THE DECISION TO DO NOTHING

Sometimes emergency responders determine that the safest course of action is to do nothing: "We had an ammonia release from a nurse tank that happened on a rural road between two towns. The emergency responders shut the road down and let the cloud dissipate into the atmosphere. The wind was keeping it away from any potential threat to people."

ON-THE-SPOT DECISIONS ARE PART OF THE JOB

Emergency responders have to quickly assess the situation and make critical life-and-death decisions. Here is one account by an emergency responder:

> **I had a release right in front of the schoolhouse at eight in the morning. A valve had been sheared from the top of the tank. Some children were already at the school, and more children were arriving. That was a tough day for me.**
>
> **You don't want your actions and decisions to hurt people. Fortunately, the ammonia cloud went up into the atmosphere and went over the school. We had the children shelter in place, closed the school, and basically locked them in. We turned back all of the cars bringing children to the school. We couldn't take the chance of the ammonia cloud coming into contact with human life.**

Questions

Should the applicator have applied anhydrous ammonia in such close proximity to the school? Could that field have waited until a late afternoon or weekend?

If a house is downwind of an ammonia release or the release is moving toward a town, the fire department has little choice but to put water on the cloud. Furthermore, they must immediately inform those living and working in the cloud's projected path to shelter in place or, in extreme cases, evacuate.

Putting water over the cloud has a negative consequence: Anhydrous ammonia and water form aqua ammonia. Aqua ammonia that enters waterways can be toxic to aquatic life. Most fire departments understand this dilemma, but human life takes precedence over adverse environmental impacts. It is up to facilities to activate their emergency response plan to get bulldozers and backhoes working to prevent the aqua ammonia from entering waterways.

DO VALVES FUNCTION DIFFERENTLY ON OVERTURNED TANKS?

Look at the photo. If the nurse tank is upside down, does that change how the valves operate? Everything we communicate with our emergency signs, the colors of the valves, and all of the training indicates that orange is for liquid and yellow is vapor. But what happens when the tank is overturned?

The farmer's valve is now in the vapor space, and the vapor valves are in the liquid space. You should mention this to emergency responders who may be unaware of this fact.

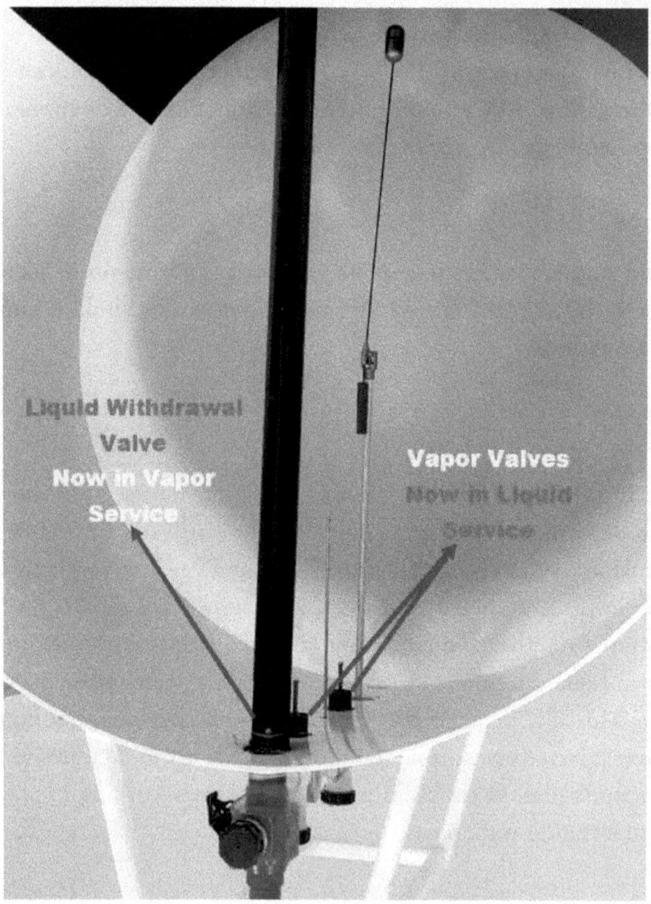

Figure 8.50. Taking an ammonia tank through a drive-through suggests complacency about the dangers.

You might ask, What does this have to do with you? If you are the one who called 911, you will be expected to provide what you know of the circumstances surrounding the release. You will also be expected to repeat what you told the 911 dispatcher when the fire department shows up and to bring them up to speed about what has occurred since the call.

We hope that you now can appreciate what it means to stay out of populated areas. Do you really want to pull a placarded hazardous material through town? Pick routes that stay clear of populated areas, and don't go through the bank or fast-food drive-through. It may not always be possible to avoid certain areas, but always do so when alternative routes are available.

MANDATORY REPORTING OF ANHYDROUS AMMONIA RELEASES

Federal and state regulations require you to immediately report anhydrous ammonia releases. The reporting threshold for anhydrous ammonia is when more than 18 gallons or 100 pounds is released into the atmosphere.

In all, you must notify three agencies when an ammonia release meets that threshold: the National Response Center, your state's department of environmental management, and your Local Emergency Planning Committee. Calling 911 is not mandatory, but it is included in the list of first contacts to call.

You are required to call the National Response Center within 15 minutes of the release. Obviously, if people are hurt, you are not going to call before you treat the people and move them to a safe area. Taking care of the injured is a higher priority than making the required phone calls. It's understood that the information may change as more information is collected, but the important part is calling 911 first and then the other three agencies soon after.

Figure 8.51. Calling the proper authorities in the event of an ammonia release is more than a formality. It is an important step in getting the problem resolved and limiting injuries to others and to the environment.

Always make the phone calls even if you are not sure that the release is above the 100-pound reportable threshold. There is no penalty if you call and authorities later determine that the release was below the reporting threshold. *Penalties only come into play when calls are not made within 15 minutes.*

Calling the agencies assumes that the threshold was exceeded. You can easily quantify the amount released by comparing the tank's weight before and after the release. We know what the overturned tank initially weighed at the facility before being sent out. The damaged tank can be reweighed to determine how much anhydrous ammonia remains. Subtracting the tank's weight after the release from the tank's initial weight will give you the amount of anhydrous ammonia released.

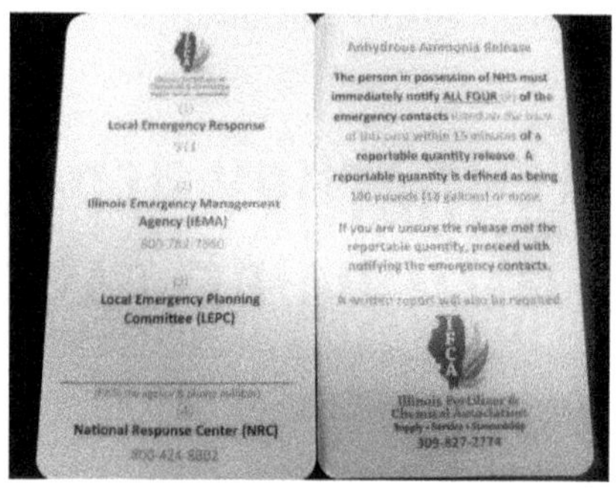

Who is responsible for reporting releases that exceed threshold amounts? Whoever is in possession of the tank is responsible for notifying the agencies. This means that a grower who unfortunately finds the retailer's nurse tank is leaking anhydrous ammonia on the roadway or in the field is required to call.

Figures 8.52–8.53. Carrying emergency contact information when using ammonia can speed up the process of getting the problem resolved. Time is of the essence once ammonia is released into the atmosphere.

Often farmers first call the ag retailer, believing the retailer is responsible for making the calls. However, because they were not in possession of the nurse tank, the retailer cannot make the calls but can provide the phone numbers to call. If the grower is incapacitated, the retailer will more than likely make the necessary calls on their behalf.

CONCLUSION

Ag retailers carry gasoline, diesel, pesticides, and fertilizers that they use or sell to farmers. Any spill or release of these products is important to the retailers and local communities where the facilities are located. This is why so much effort is required to train employees on how to prevent accidents and to educate fire departments on how to properly respond to them.

Of all the products stored at a retailer's facility, anhydrous ammonia is of greatest concern. This hazardous material can seriously injure people or even kill them. It is imperative that everyone handling anhydrous ammonia follow the policies established by their facilities to the best of their abilities.

Much of the training for anhydrous ammonia aims to reduce human error and lower the likelihood that a release will occur. When we cut corners, don't focus on the task at hand, or make a mistake, the result can be devastating not only for the facility's employees but to the surrounding communities.

Yes, it's true that employees receive the same information each year at training. With anhydrous ammonia, nothing really changes, so the information doesn't need to change either. But we are trying to make ammonia safety procedures so ingrained that employees know exactly what they need to do when they are handling this important, but dangerous, fertilizer.

Unfortunately, what also doesn't change from year to year are the mistakes that result in anhydrous ammonia releases. Retailers spend time and money on employee training, plumbing maintenance, and equipment repairs. But the money they spend on training and maintenance is of little value when someone decides to circumvent what they have been taught.

The bottom line is to respect, but not fear, anhydrous ammonia. A brand-new employee who lives in fear of anhydrous ammonia is something we don't want. On the other hand, we don't want a seasoned employee to think, "I have handled it all of these years and never had a problem. I just don't need to worry about it." We don't want you to be afraid of it, but we don't want you to be complacent.

There are only a few simple steps and guidelines you need to follow when working with anhydrous ammonia. Handled correctly, anhydrous ammonia is an effective and valuable fertilizer. Be informed, be safe, and put into practice what you have been taught.

A PASSING THOUGHT

Training is important. Here are the thoughts of one safety trainer:

> A safety coordinator can put things in place, but it is the employees who are my eyes on the ground. It is unfortunate that these types of releases occur. We don't make them up. There are some who think that we stay awake at night trying to invent these stories as a way of screwing up their day. Events like this are real, they happen. And so an employee has to speak up when something is not right. Working together allows us to prevent the releases.

Figure 8.54. Safety is always at the forefront when transporting and applying anhydrous ammonia.

■ ■ ■

SECTION 2

Handling Pesticides

9

Is This Pesticide Still Good?

With contributions from Joe Becovitz, Douglas Linscott, John Obermeyer, Bob Hill, and Kevin Leigh Smith

You may have seen supermarket shoppers reaching to the back of the cooler for what they perceive to be the freshest milk. You are used to checking perishable foods (milk, eggs, meat, bread, juices, and more) for the expiration dates or "best by" dates stamped on them to make sure you are not buying items that will soon spoil.

Figure 9.1. It is common to find expiration dates on food items, but this is seldom done with pesticide products.

Canned food items have expiration dates too. But unlike perishable products, their expiration dates are years after the products were processed and preserved. The products inside cans often include preservatives to maintain freshness, and those canned products were pasteurized and sealed against air, moisture, and microbial contamination.

Which raises the question: "Why can I open these canned products after the expiration date and they still look and taste good?" A product's expiration date and its actual shelf life may be different. With fresh, perishable products, the expiration date and shelf life are usually the same; however, the shelf life of nonperishable products often extends well beyond their expiration dates.

Consider that durable goods often provide warranty periods, which begin on the date of purchase and extend for months or years. With durable goods, the warranty is analogous to the expiration date. But whether it's a watch, a water heater, or a vehicle, we all expect our durable goods to function well beyond their warranty periods. In other words, a durable good's longevity (or shelf life) will most likely exceed its warranty (or expiration date).

One of the challenges we face as consumers is extending the life of products we purchase. For example, the life of a truck can be increased by servicing it at regular intervals, keeping the exterior clean, practicing safe driving habits, and reducing the number of miles you drive each year.

Expiration date, *warranty*, and *shelf life* are all terms that apply to pesticides too. Farmers and commercial applicators want to know how long pesticide products can be stored without losing their efficacy. How long a pesticide will remain effective becomes more difficult to determine if, in addition to being past its expiration or warranty date, the product was opened or was stored without protection from moisture, extreme heat, and cold.

If you leave a gallon of milk open on the counter overnight, you know it will sour even though it still has a week left on its expiration date. If you open a can of green beans and store it in the refrigerator, the beans will most certainly go bad well before the expiration date listed on the can. Likewise, how long a pesticide can last in storage depends on many factors, including its formulation, active ingredient(s), age, and storage conditions, as well as whether it remained sealed or was opened.

If you use a pesticide product beyond its expiration date, it may or may not be less effective. Then again, there's no guarantee that it hasn't degraded. There are too many factors that can degrade the product (like storage conditions) that are beyond the manufacturer's control. That is why manufacturers make no assurances or warranties about the quality of their products after the expiration date.

This chapter provides an overview of how to determine a pesticide product's storage stability and explores options for what you can do when these products expire. Our overall goal is to encourage you to follow best practices when storing pesticides, so you never have to wonder, "Is this product on my shelf still good?"

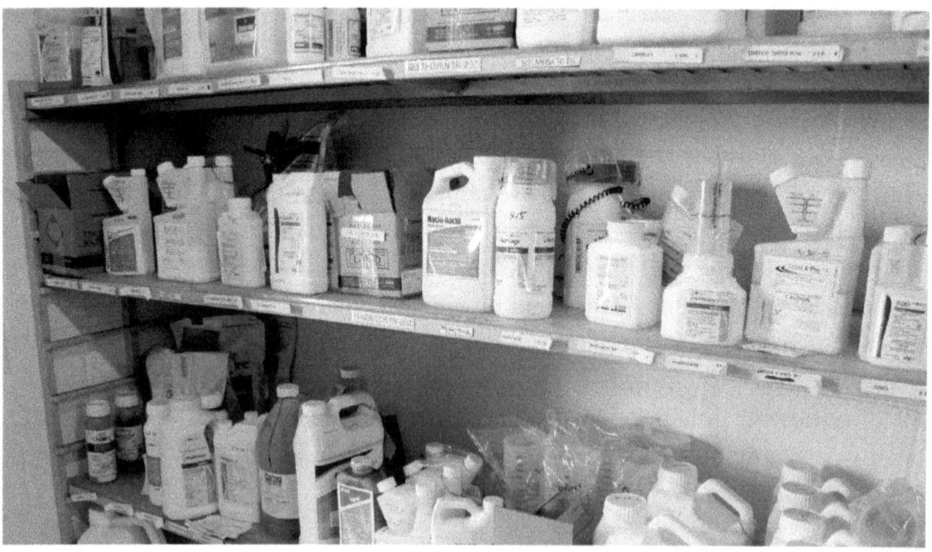

Figure 9.2. Companies and farmers invest in storing pesticides for future use. But if they fail to use pesticides before their expiration dates, it may cause poor performance, the loss of customers, and costly disposal.

CLARIFYING THE TERMS

Expiration date, *shelf life*, and *warranty* are often used interchangeably, but they have different meanings. Manufacturers set expiration dates. The expiration date tells you that a properly stored product will perform as expected if you use it before that date.

Warranties are normally provided for durable goods and are analogous to the expiration dates for perishable products. A pesticide's warranty assures you that there is a period of time that it is guaranteed to work; it will be replaced if it does not.

Shelf life is how long the product will remain effective after its expiration date and warranty have passed. Storage conditions greatly influence how fast a product degrades. That is why there is no set time for the long-term stability of pesticide formulations.

QUANTIFYING A FORMULATION'S STABILITY

The US Environmental Protection Agency (EPA), under the Federal Insecticide, Fungicide, and Rodenticide Act (FIFRA), requires all pesticide products to meet active ingredient stability and product quality specifications. No new pesticide product may be sold until the EPA assures that the product meets all these requirements.

- **Stable concentration:** The concentration of the active ingredient must remain stable when stored under conditions that simulate two years of real-time aging. While this is a general requirement, products will remain within their specification range much longer. This assures the active ingredient's potency meets its label claim while in storage and will reliably provide consistent field performance.
- **Stable physical structure:** The EPA requires the product to remain physically stable for two years (for example, it does not irreversibly clump or separate). To meet these requirements, manufacturers must develop and submit data from tests performed using good laboratory practices to demonstrate the product remains physically stable before it is released for sale. Manufacturers test products for an array of conditions (such as extreme temperatures) that products might encounter in storage.
- **Stable packaging:** The product formulation requires in-package testing to confirm that there is no risk of spills or leaks. Furthermore, manufacturers must ensure the packaging material itself does not create problems with the product formulation.

Testing the physical stability of a product focuses on how effectively its inert ingredients support the delivery of its active ingredient. In other words, does the product still mix, suspend, and disperse? Inert materials that degrade, or are used in insufficient amounts, may lead to premature product separation or even failures. Manufacturers need to optimize the inert ingredient levels to provide a product that is high enough in viscosity to remain stable and remixable but not so viscous that it becomes difficult to pump or pour.

Figure 9.3. Manufacturers are concerned about products that can no longer mix in water after storage.

When manufacturers evaluate a formula's physical properties, they also analyze the amount of active ingredient in the formulation. As a general rule, active ingredients are quite stable in the concentrated formula sealed in the container. But there are storage circumstances that would predictably lead to changes in the concentrated formulation.

For example, the product could thicken because the solvent evaporated, which could make the product difficult to pour, or cause the particles to clump and clog nozzles. In turn, these issues could cause poor mixing, foaming issues, and poor dispersion and suspension of oils and solid particles in the spray dilution.

In summary, pesticide manufacturers invest years of research and development and millions of dollars to develop new formulations and secure the necessary government registrations. However, even more important to a product's success is its ability to easily and reliably control pests. Manufacturers have a vested interest to serve their customers by providing products that meet the highest standards for quality, user safety, and consistency of performance. Products that are unstable or do not perform up to expectations will not attract repeat buyers over the long term.

STABILITY LINKED TO FORMULATION COMPONENTS

Pesticide formulations must allow for safe handling, dose control, storage, and field performance. Formulating a product involves adding various inert ingredients (nonactive materials), combined with some mechanical processing, to prepare an easy-to-use product. For this reason, product stability depends not only on the active ingredient but on the stability of the inert ingredients included in the formulation recipe.

Manufacturers must decide which formulation type they prefer for the active ingredient, then select appropriate government-allowed inert ingredients to prepare their formulations. The first requirement is to prepare a concentrated mixture that contains the active ingredient. If the active ingredient or any of the inert ingredients become unstable, then the product can fail to perform.

Each formulation type is unique in composition, and its responses to different storage conditions can vary widely. For example, liquid solutions like a soluble liquid

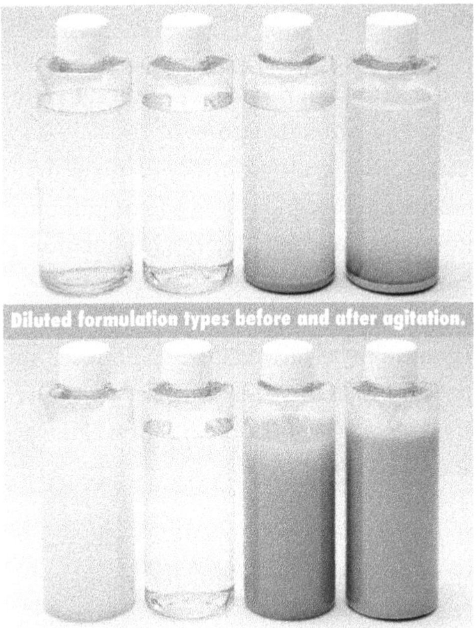

Figure 9.4. A formulation is a mixture that contains active ingredients and inert components that are effective and shelf-stable.

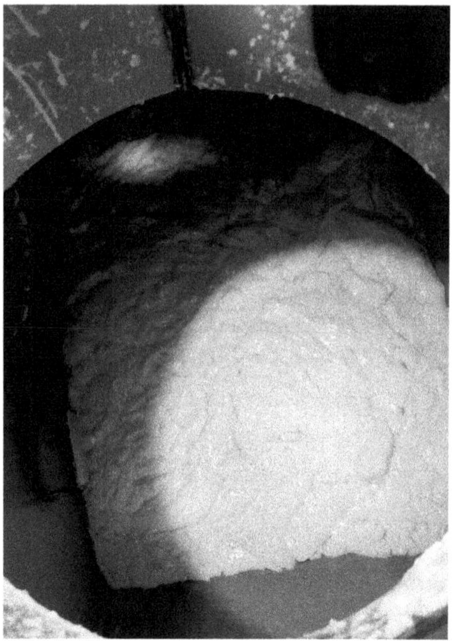

Figure 9.5. An unstable formulation concentrate can cause major problems in a spray tank.

(SL) or emulsifiable concentrate (EC) tend to be shelf-stable longer than a multiphase formulation like a suspension concentrate (SC) or concentrate emulsion in water (EW). A multiphase formulation is simply solid active ingredient particles that are suspended in water, or oil-phase active ingredient particles pre-emulsified in water.

Under harsh storage conditions, normally stable products can be rendered unsuitable for continued use. Likewise, when people use products beyond their shelf life, the products may suffer dispensing problems, poor dilution, loss of potency, or spray application difficulties.

Therefore, manufacturers examine how storage conditions affect different formulations, looking for changes in their chemical components (such as the effect on active ingredients) and for changes in the physical properties of the inert ingredients (such as clumping, failure to disperse, settling, and changes in pourability). The examples that follow summarize some of the stress tests manufacturers conduct on pesticide formulations to determine whether the active ingredients, inert ingredients, and formulation mixtures remain stable over a wide range of environmental conditions.

Testing the Effects of Heat

As they develop new pesticide products, manufacturers continually evaluate them using accelerated aging test methods. Although this may sound complicated, these tests simply assess a formulation's stability when stored at high temperatures. Because manufacturers can complete these studies in much less than two or three years, they are called *accelerated aging tests*. Such tests are typically most useful for predicting an active ingredient's stability.

Manufacturers conduct accelerated aging tests by subjecting their products to temperatures of 122°F (54°C) for at least two weeks and/or temperatures of 104°F (40°C) for eight weeks. The 122°F accelerated testing for two weeks normally predicts about two years of storage stability at ambient temperatures. In addition to predicting shelf stability, the results help manufacturers understand how sensitive the formulation is to much-higher-than-expected temperatures when stored.

These tests generally approximate a two-year shelf life under normal product use—they are more a rule-of-thumb indicator. In addition to accelerated aging tests, the EPA also requires real-time product stability testing before they finalize a product's registration. Real-time testing ensures that a product's formulation and packaging are compatible at ambient storage conditions of 68°F (20°C).

Testing the Effects of Cold

Freeze/thaw testing is perhaps the most intense physical stress test for liquid pesticide products. Users commonly store pesticides in unheated outdoor storage facilities where they can freeze. Cold temperature tests are meant to mimic those storage conditions. In contrast, storing pesticide formulations in temperature-controlled facilities that protect them from freezing can significantly extend their useful shelf life.

Manufacturers conduct cold storage tests (at less than 20°F, or −6.5°C) and repeated freeze/thaw cycle tests (from 20°F to 100°F, or −6.5°C to 40°C) for eight weeks to force the formulations to solidify. After thawing, the products are expected to recover their original forms.

These tests are especially challenging for water-based formulations such as suspension concentrates (SC) and suspoemulsions (SE), because solid water crystals form and then melt on warming. Cold storage testing for soluble liquid (SL) and emulsifiable concentrate (EC) products can cause the dissolved active ingredients to crystallize and precipitate out of solution. It is important to redissolve the crystals on warming prior to using the product. Undissolved crystals (even after agitation) can create problems with dose measurement for soluble liquids. In EC products, crystals can cause problems with spray mixing and can plug nozzles. To avoid crystallization problems, many EC formulation labels include cold temperature storage restrictions.

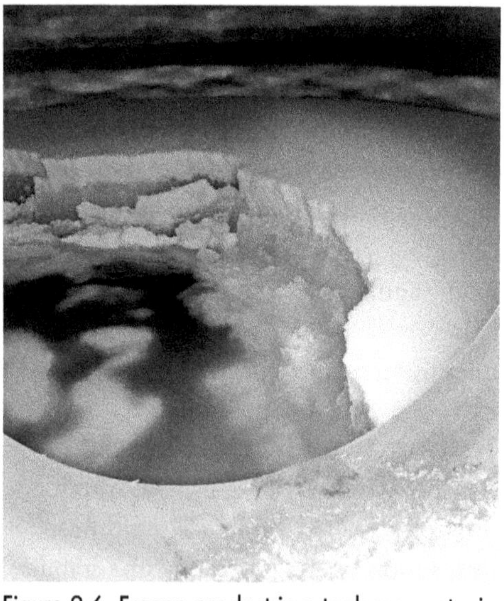

Figure 9.6. Frozen product in a tank may not mix back into solution when warmed.

Testing Viscosity

The viscosity of liquid pesticide formulations is another important property. For product development, the ideal viscosity offers easy handling and mixing but provides enough robustness to prevent the formulation from separating. Manufacturers prefer that a liquid product's viscosity remains consistent during storage testing.

Selecting the proper inert ingredients and the amount of active ingredient to combine in these formulations is critical. The viscosity of most formulations will normally change at different storage temperatures, but bigger changes in a formulation's viscosity are early predictors of longer-term product separation or undesired thickening issues. Generally, formulations kept at high temperatures have reduced viscosities (flow easier), while those kept at low temperatures tend to become more viscous (thicker).

Pesticides are often stored in unheated spaces and are used outdoors in spring, when temperatures are cold. So cold temperature storage and handling properties are critical for some products, especially bulk product formulations that require rapid pumping to efficiently transfer the product from a tanker truck to a bulk storage tank. Packaged products that become highly viscous are slow to pour, dilute, and mix—they also may hinder proper package rinsing.

If a viscous product tends to quickly settle to the bottom of the spray tank, that can cause more serious concerns. Settled products may require extended and more vigorous agitation to ensure uniform spray application mixtures. Settled products also may result in nonuniform product applications and make it more challenging to clean out storage and application tanks.

Testing Particle Size Stability

Formulations that contain suspended solids must meet particle size specifications. Particles must be capable of flowing through spray application equipment via a nozzle. Optimal particle sizes also ensure the most effective coverage of target pests. Product storage tests also validate the particle size—their goal is to determine that particle size does not change with extended storage times or normal variations in temperatures.

Figure 9.7. Products that fail to dissolve or suspend can plug filters and nozzles. This can affect spray coverage and create maintenance issues that lead to significant downtime.

Testing Particle Suspension in Spray Mixtures

For formulations that contain solid particles, another standard performance test is determining percent suspensibility. This test measures the physical stability of a spray dilution when it is left unmixed for 30 minutes, simulating poor tank agitation. Generally, it is desirable for at least 70 percent of the total particle mass to remain suspended over this test period. Suspensibility is a critically important test both for liquid products that contain suspended particles and for water-dispersible granule formulations and wettable powders.

As products age, or are subjected to varying storage conditions, their suspensibility can change. If the product's particle size changes during storage, that can result in diminished activity, poor control, and poor dilution stability. More severe changes in particle size can lead to buildup on filters and clogged nozzles. Suspensibility tests are also often repeated in water that contains varying levels of minerals as might be found in different water sources. Manufacturers often select the inert surfactant ingredients based on results from these evaluations (see chapters 12 and 13 for more).

Manufacturers also commonly run a second, related test, referred to as *resuspensibility*. This test lets the diluted particle suspension sit undisturbed overnight and then measures how much mixing is required to redisperse the particles that have settled out in the tank. As you might expect, manufacturers (and users) want such sediments to remix quickly once they start agitating the spray mix.

Testing Packaging Materials

EPA formulation registration requirements require manufacturers to test a formulation in its designated commercial package. In-depth testing examines whether the packaging material is compatible with the product formulation: that the package adequately protects the product and that the package does not break down or react with the formulation.

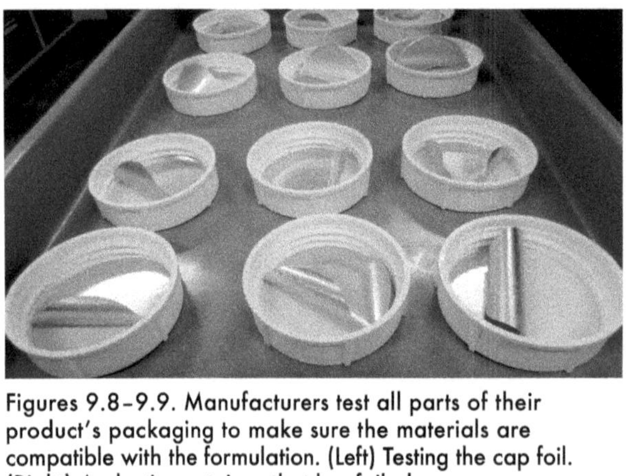

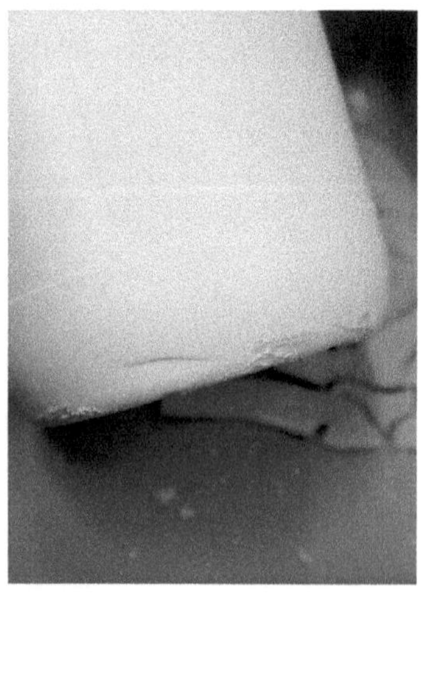

Figures 9.8–9.9. Manufacturers test all parts of their product's packaging to make sure the materials are compatible with the formulation. (Left) Testing the cap foil. (Right) A plastic container that has failed.

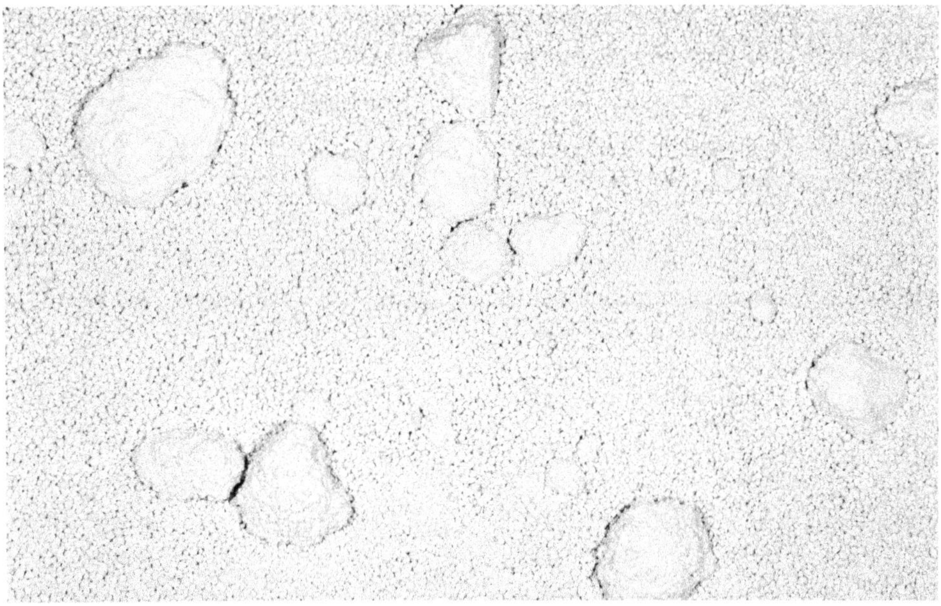

Figure 9.10. Dry materials will clump up when exposed to moisture.

Manufacturers run these compatibility tests at multiple temperatures and time points. They test materials and plastics for the bottles, caps, and seals. They test many materials, but not all of them suitably protect the product or are compatible with the pesticide product formulation. To test different packaging materials, a manufacturer will fill the proposed container with the formulated product, then inspect it for any splits, cracks, and softened areas over time, as well as any signs that the product was negatively affected by the packaging.

Dry formulations are often sensitive to moisture and may absorb water from humid air. For this reason, water-dispersible granules require moisture-proof, barrier-sealed packaging. Absorbing moisture may otherwise cause the granules to form lumps that will no longer disperse completely when added to the spray tank. This then causes mixing and spray-applicating problems for users.

It is important to keep dry products in sealed packages to preserve their shelf lives. In the case of dry fertilizer formulations that are coated with a pesticide, exposure to moisture can solidify the granules into a brick, rendering the product unusable.

DON'T CUT WATER-SOLUBLE PACKAGING

You should never store partially used, open bags of water-soluble products. Keep pouches in their sealed bags until you plan to use them to prevent them from absorbing moisture.

STORAGE RESTRICTIONS ON LABELS

In some cases, manufacturers are not able to build all the desired product attributes into a formulation. In such cases, product storage restrictions may offer acceptable solutions. Here are some storage restrictions taken from actual labels:

- Do not store above 100°F for extended periods of time. Storage below 20°F may result in formation of crystals. If product crystallizes, store at 50°F to 70°F and agitate to redissolve crystals.
- Store above 5°C (41°F). Keep unused water-soluble bags in resealed original container. Keep packages dry at all times. Keep packages out of rain. Keep from freezing.
- Keep from overheating and freezing. Store between 40°F and 77°F. Product should be used promptly after opening the package. Apply any unused product within 4 months of opening.
- Do not store for more than 30 consecutive days at an average daily temperature exceeding 100°F. If allowed to freeze, shake well to ensure the product is homogenous before use.
- Store above 10°F (−12°C) to keep product from crystallizing. Crystals will settle to the bottom. If allowed to crystallize, place in a warm room 68°F (20°C) for several days to redissolve and roll or shake container or recirculate in minibulk containers to mix well before using.

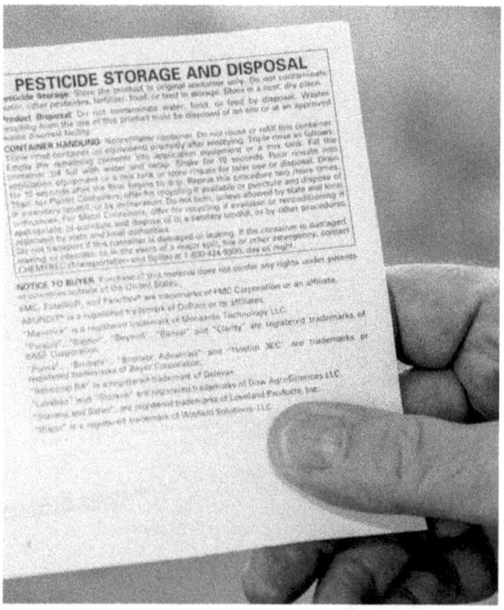

Figure 9.11. If a manufacturer's testing reveals known storage limits in stability, then the Pesticide Storage and Disposal section of the product's label directs users to follow specific storage restrictions. If users do not follow the label's storage restrictions, it is possible (and even probable) that the product will not handle or perform acceptably.

These examples demonstrate that products can have very specific storage restrictions and instructions.

ANSWERING THE QUESTION "IS THE PRODUCT STILL GOOD?"

What are the differences between the expiration date and the shelf life? Generally, US pesticide manufacturers do not print expiration dates on their containers. There are exceptions, such as for disinfectants (including germicides). Furthermore, limited warranties do not prescribe expiration dates.

Figure 9.12. Most pesticide products do not list specific expiration dates. There are exceptions, like disinfectants, which are mandated by the EPA to carry expiration dates.

Pesticide product manufacturers generally design their formulations to remain both chemically and physically stable for at least two to three years if sealed in their original packages. This ensures sufficient time for products to move through nationwide distribution channels to end users. This also provides farmers and commercial pesticide application industries some added time to use the products after they are sold without any concern about product integrity and effectiveness.

Storing Products Within the First Three Years

A manufacturer's goal is to ensure that customers who use their products during the first two to three years see no changes in the physical condition, handling, or performance of the products' formulation or packaging. In general, the two-to-three-year span becomes the unofficial expiration and warranty period for most pesticides.

If you don't see storage restrictions on the label, you can assume that the formulation is rather stable. But to be on the safe side, you should not keep products in inventory for more than three sales seasons. Consider the following storage recommendations to ensure that products remain effective for use:

- Read storage directions before purchasing pesticide products to make sure you can comply with them.
- Follow pesticide label storage restrictions to ensure product stability.
- Store pesticides in controlled environments to prevent temperature extremes.

Figure 9.13. Controlling extremes in temperature will extend the effective life of most pesticides.

- Use all products within the season of purchase if possible.
- Don't buy more than a two-year supply.
- Note the year of purchase on containers to identify the age of stored products.
- Rotate the stock on the shelf, oldest to front.
- When switching to new formulations, first use up the product you are replacing.
- Keep products sealed and lids tight to prevent evaporation and/or contact with moisture.
- Shake all containers before use to ensure active ingredients and coformulants (inerts) are uniformly mixed.
- Use opened containers and bags the same season.

Figure 9.14. The date of purchase has been written on this jug. It is a clear reminder of how long the product has been sitting on the shelves.

Figure 9.15. Dry products left unsealed are subject to degradation by moisture.

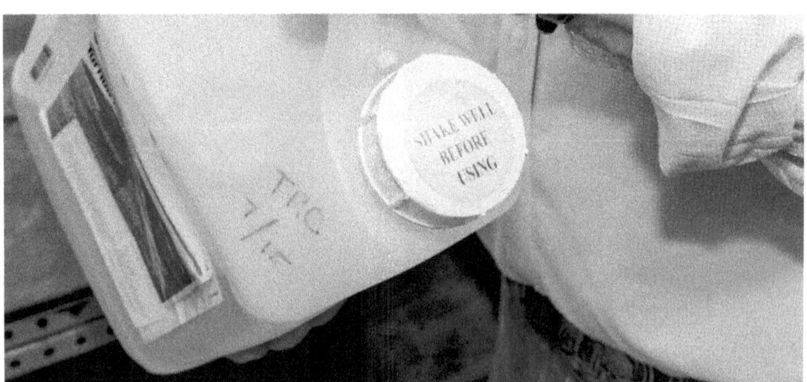

Figure 9.16. Remember to shake containers, especially when the product container instructs you to do so.

Testing Viscosity

The viscosity of liquid pesticide formulations is another important property. For product development, the ideal viscosity offers easy handling and mixing but provides enough robustness to prevent the formulation from separating. Manufacturers prefer that a liquid product's viscosity remains consistent during storage testing.

Selecting the proper inert ingredients and the amount of active ingredient to combine in these formulations is critical. The viscosity of most formulations will normally change at different storage temperatures, but bigger changes in a formulation's viscosity are early predictors of longer-term product separation or undesired thickening issues. Generally, formulations kept at high temperatures have reduced viscosities (flow easier), while those kept at low temperatures tend to become more viscous (thicker).

Pesticides are often stored in unheated spaces and are used outdoors in spring, when temperatures are cold. So cold temperature storage and handling properties are critical for some products, especially bulk product formulations that require rapid pumping to efficiently transfer the product from a tanker truck to a bulk storage tank. Packaged products that become highly viscous are slow to pour, dilute, and mix—they also may hinder proper package rinsing.

If a viscous product tends to quickly settle to the bottom of the spray tank, that can cause more serious concerns. Settled products may require extended and more vigorous agitation to ensure uniform spray application mixtures. Settled products also may result in nonuniform product applications and make it more challenging to clean out storage and application tanks.

Testing Particle Size Stability

Formulations that contain suspended solids must meet particle size specifications. Particles must be capable of flowing through spray application equipment via a nozzle. Optimal particle sizes also ensure the most effective coverage of target pests. Product storage tests also validate the particle size—their goal is to determine that particle size does not change with extended storage times or normal variations in temperatures.

Figure 9.7. Products that fail to dissolve or suspend can plug filters and nozzles. This can affect spray coverage and create maintenance issues that lead to significant downtime.

Testing Particle Suspension in Spray Mixtures

For formulations that contain solid particles, another standard performance test is determining percent suspensibility. This test measures the physical stability of a spray dilution when it is left unmixed for 30 minutes, simulating poor tank agitation. Generally, it is desirable for at least 70 percent of the total particle mass to remain suspended over this test period. Suspensibility is a critically important test both for liquid products that contain suspended particles and for water-dispersible granule formulations and wettable powders.

As products age, or are subjected to varying storage conditions, their suspensibility can change. If the product's particle size changes during storage, that can result in diminished activity, poor control, and poor dilution stability. More severe changes in particle size can lead to buildup on filters and clogged nozzles. Suspensibility tests are also often repeated in water that contains varying levels of minerals as might be found in different water sources. Manufacturers often select the inert surfactant ingredients based on results from these evaluations (see chapters 12 and 13 for more).

Manufacturers also commonly run a second, related test, referred to as *resuspensibility*. This test lets the diluted particle suspension sit undisturbed overnight and then measures how much mixing is required to redisperse the particles that have settled out in the tank. As you might expect, manufacturers (and users) want such sediments to remix quickly once they start agitating the spray mix.

Testing Packaging Materials

EPA formulation registration requirements require manufacturers to test a formulation in its designated commercial package. In-depth testing examines whether the packaging material is compatible with the product formulation: that the package adequately protects the product and that the package does not break down or react with the formulation.

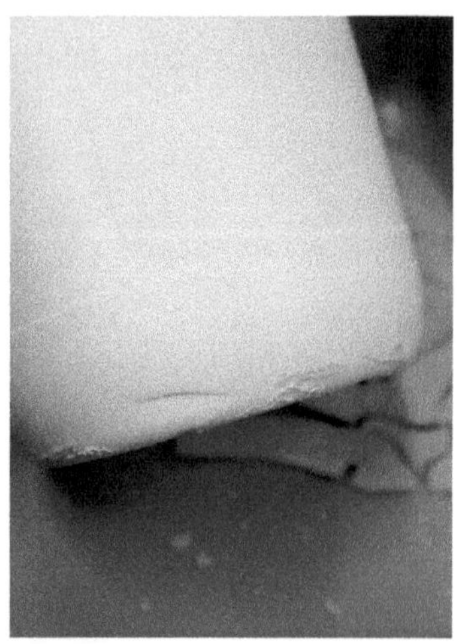

Figures 9.8–9.9. Manufacturers test all parts of their product's packaging to make sure the materials are compatible with the formulation. (Left) Testing the cap foil. (Right) A plastic container that has failed.

Evaluating Products of Questionable Age

There are many reasons why products may remain in storage longer than three years:

- Managing product inventory is not a priority.
- The user bought too much product.
- The user switched to a replacement product without using up current inventory.
- The product has limited use as part of a tank mix.
- The user is dissatisfied with a product's performance.
- Product labels are partially torn or missing.
- The EPA canceled the product registration, and the time to use up the existing stock has passed.
- The product was misplaced and found after a cleanup.

Many formulations are designed to be stable and effective for as long as possible, but none is permanently stable. Still we ask, Is this product on my shelf still good? If the product is still sealed in its original container and it's within two or three years of purchase, the answer is an easy yes. Just the opposite is true for opened products beyond three years.

The question then should be, When does a product go from being effective to needing to be disposed of? While the two-to-three-year storage period may have long passed, that doesn't necessarily mean the product is no longer effective. The answer is nearly impossible to pinpoint because there are so many interacting variables, including how old the product is, whether the container was opened, and under what conditions the pesticide was stored. It can be quite challenging to determine if the product will deliver consistent results, which ultimately is the key question.

Let's say you have 10 unopened 2.5-gallon pesticide containers that have been stored for an unknown number of years. You remember buying the product, but you switched to a different and more effective product soon afterward. You simply want to know whether the product in those containers is still effective, so you can at least recoup your investment in the product and avoid disposal costs. When deciding, you can take the following steps:

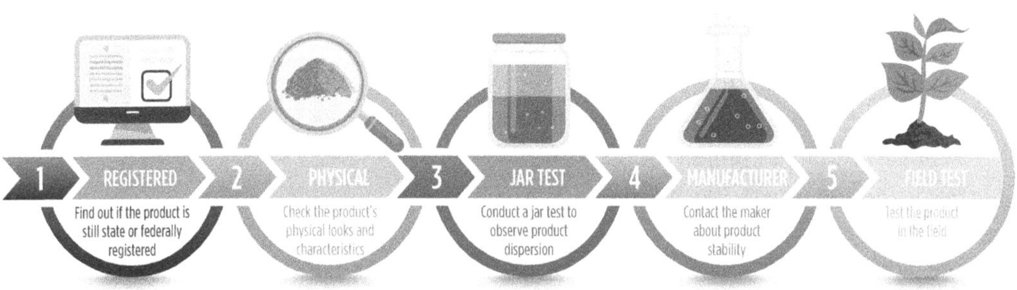

SHOULD IT STAY? OR SHOULD IT GO?

Five factors to consider when determining whether to keep your pesticide in inventory or dispose of it

Figure 9.17. Pesticide products stored for long periods often have unreadable or misplaced labels. Are you willing to take a chance that you know what is in the container? These products are now being stored for disposal, meaning the money spent was also wasted.

- Find out if the product is still registered (is it still legal to use?).
- Check the product's physical characteristics.
- Conduct a jar test.
- Contact the manufacturer or distributor representative.
- Test the product under field conditions.

Find Out If the Product Is Still Registered

If a product is no longer registered with your state or the federal government, then all of the other factors are moot. It does not matter if the product is viable or not—if it is no longer registered, you cannot use it.

You can check the product registration list from your state's department of agriculture to determine if the company has paid its annual fee. If the product remains registered, then you may use the product. If you have lost the product's label or the label is unreadable and you can't determine its identity, then it should be disposed of.

Figure 9.18. Old DDT may no longer be used as the EPA prohibits it. There are many products in storage that the EPA no longer allows to be used, such as this one.

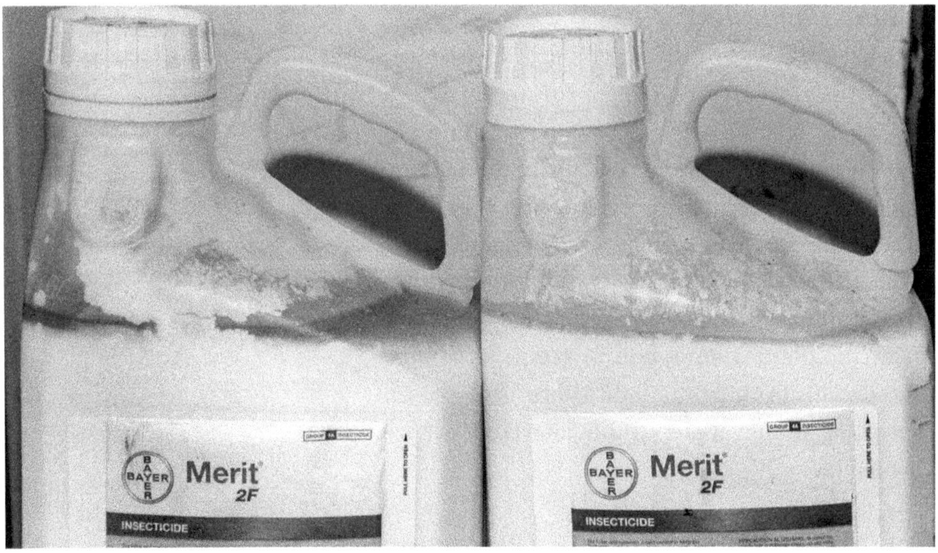

Figure 9.19. Separated product that cannot be remixed should not be used. It could be a sign that the products are no longer effective or could clog screens and nozzles.

Check the Pesticide's Physical Characteristics

Try to determine whether the product looks like it did when you first used it. Assess for changes in color, unusual odors, thickening, top-clearing, or noticeable sediment formation. If the ingredients have separated out, try gently inverting the closed container 5–10 times. If the product remixes and appears uniform, the product may still be acceptable to use.

However, you should consider disposing of the product if you observe any of the following:

- The product has clumped.
- There is accumulation on the bottom.
- The product does not resuspend.
- The product has an unusual odor or color.
- Separated layers remain after mixing.
- The package has been previously opened.
- A hard-pack sediment has formed on the bottom.

Figure 9.20. A useful test is to pour the product through an 80-mesh screen to see if it retains any particles. If it does, the product should probably be disposed of. Trying to use these products can lead to plugged screens and nozzles, costing downtime.

Conduct a Jar Test

A jar test allows you to determine if a product's ingredients mix well and remain uniform. To conduct a jar test, begin by mixing a minimum amount of product (follow label directions and use a small, clear container). Let the material sit for at least 30 minutes, and observe what happens.

Does the product settle out of solution within 5 to 10 minutes? Repeat the test to see if the solids in the bottom resuspend or redissolve. If the product fails to suspend, it should be disposed of. (For more details about conducting jar tests, see chapter 11.)

Figure 9.21. The jar test on this old product shows that it should be discarded. Being unable to mix in water is a sign that the inerts may no longer be working.

Contact the Manufacturer or Distributor Representative

If you have questions about the viability of a product, contact the manufacturer, using the customer service information found on the label. When you ask about the product's storage stability, you may be connected to a specialist who can better address your question.

Before you seek advice from the manufacturer or distributor, you should be prepared to provide the name of the product, its EPA registration and establishment numbers, and the barcode stenciled directly on the container. The manufacturer can use the barcode to determine the year and month the product was made, the location

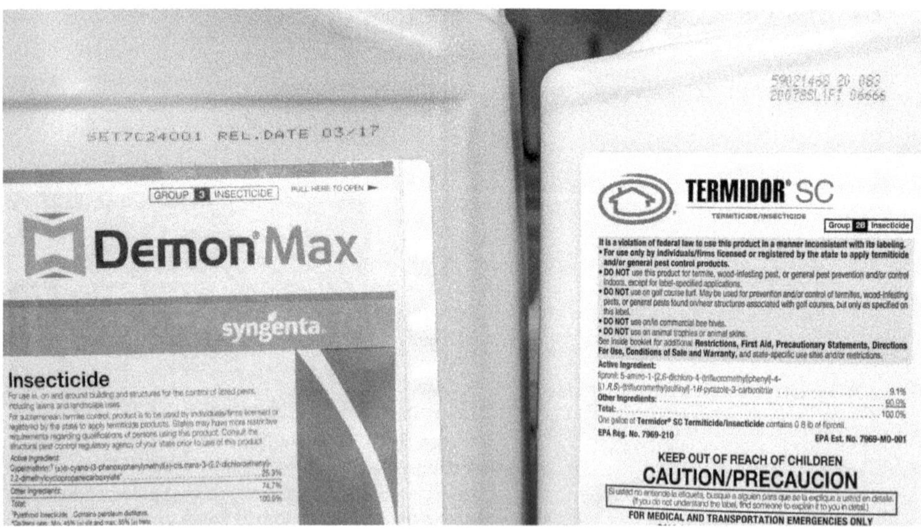

Figure 9.22. Manufacturers differ in how they code their products. The product on the left has a release date along with a barcode. The product on the right has only a barcode.

where it was packaged, and the batch and lot numbers. Often, a dribble down the pesticide container will smudge a portion of this code. If that code becomes indiscernible, then the product should be disposed of.

Once you provide all the relevant information to the manufacturer, they may be able to determine if the product is within its window of usability. The longer a product has been in the marketplace, the more likely the manufacturer has historical shelf-life information to help answer your questions about stability. The manufacturer may also ask you about your jar test results. If you observed that the product did not mix well, the manufacturer may instruct you to properly dispose of the product.

YOU NEED A DECODER RING TO DETERMINE A PRODUCT'S AGE

Finding the expiration date on a product container is impossible for end users. That's because manufacturer codes include dates that only they can decipher. A few companies will list the manufactured date on the container. While that will not indicate the expiration date, it will give you the product's age. This is an important first step in knowing if the product is still viable.

Test the Product Under Field Conditions

The only failsafe way to determine whether the product is still effective is to apply it in the field. Applying the pesticide on a small portion of a site will allow you to determine if the product still works without causing any unforeseen effects such as phytotoxicity. If the product does not provide control or causes negative side effects, then it should be disposed of.

However, do not conduct such field tests with fungicides. If the fungicide is no longer effective, the plants you wanted to protect may become infected with the disease. In many cases, there are no rescue treatments once a disease has infected the plant. If there is any doubt with a fungicide, it should be disposed of.

When Disposal Is the Only Alternative

If your product has failed the question, "Is this pesticide on my shelf still good?" then your only viable alternative is to properly dispose of it. Choosing to apply an old product may open you to liability concerns caused by uncontrolled weeds, continued damage from termites, or even destruction of natural habitat if the product does not control invasive species.

An EPA-canceled active ingredient will normally have a drop-dead end-sale date. Before that date, the product can be used according to the label, but after that date has passed, the product can only be disposed of by a company that has obtained the necessary hazardous waste permits.

Some states hold summer pesticide disposal days, during which people can bring canceled and unwanted pesticide products to a site. These products will be properly disposed of by a licensed contractor at no cost, or at a nominal fee for large quantities.

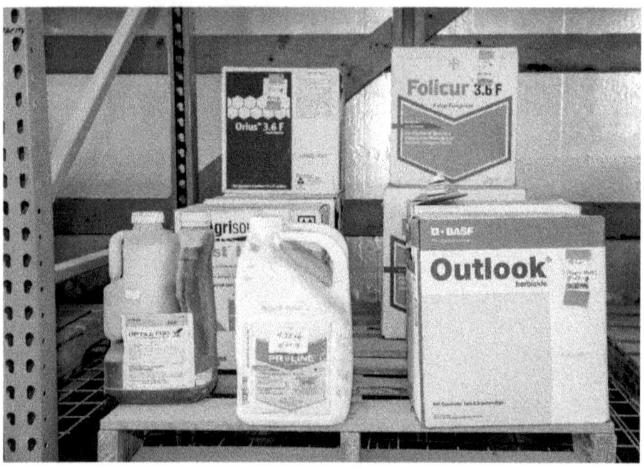

Figure 9.23. These products were identified as needing disposal, so they were segregated and are awaiting removal from the storage facility.

IT IS YOUR DECISION TO USE OR DISPOSE OF PRODUCT

The caveat to answering whether a product is still effective is that no one can guarantee how well a product will work that's been left in storage for any length of time. The most you can expect is a "best guess" judgment based on the information available. The responsibility for deciding whether to use the product lies with the user.

Manufacturers will seldom take and analyze samples involving a few containers. Instead, they will spend their resources when there are pallets of stored materials or hundreds of gallons of product involved. Compare the value of replacing the product with the cost if the product does not perform or if it creates application problems.

WARNING! WARNING! WARNING!

Be sure to monitor the condition of the packaging materials for products that are slated for disposal. We highly recommend using secondary containers to contain spills that might happen if the product's metal, plastic, or paper containers release their contents.

CONCLUSION

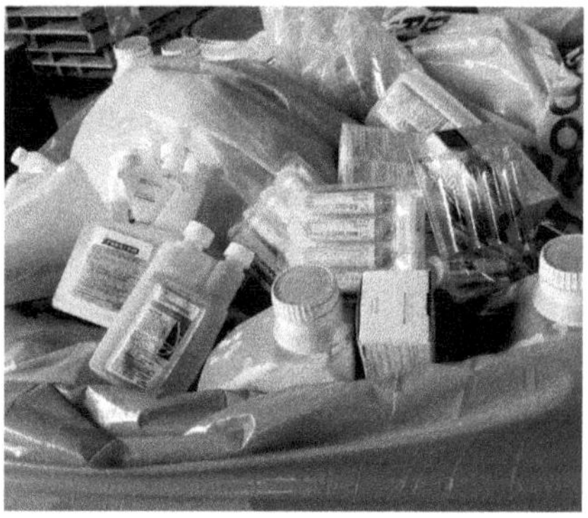

Figure 9.24. Expired products (many unopened) in the process of being properly disposed. What a horrible waste of money!

As an end user, you can control some of your chemical costs by purchasing and using what you need for a given season and then storing the product according to label directions. This is easier said than done because pesticide formulations often change, causing the product to be shelved in preference for what's new.

Over time, pesticide storage facilities often become a mishmash of older, forgotten, and partially used products that do not come with expiration dates stenciled on the containers. Managing this old inventory may lead you to ask some time-consuming and expensive questions. Is this product still legal to use? Does this product still control pests? Does this old product cause off-target injury? How much will it cost to legally dispose of the product?

This chapter provided commonsense guidelines that can help you better manage your pesticide product inventory. Whether it is putting a purchase date on the containers or rotating stock, these simple-to-implement procedures can help keep your inventory fresh. Keeping your formulations in tip-top condition is just as important as calibrating your equipment, training your employees, and serving your customers. By better managing your product inventory, you will no longer have to ask, "Is this pesticide on my shelf still good?"

Figure 9.25. Managing inventory takes effort, but doing so allows you to use the products efficiently and not make them a disposal liability in the future.

10

Measuring Pesticides

With contributions from Pat Hipkins, Dennis Nowaskie, Doug Linscott, Andrew Martin, John Obermeyer, Aaron Patton, Scott Gabbard, George Beestman, and Kevin Leigh Smith

Safe and effective pesticide applications require six equally important steps:

- Identify the pest correctly.
- Select the appropriate product.
- Calibrate equipment properly.
- Measure the pesticide accurately.
- Review the treatment site before and during application.
- Monitor the results.

Recertification workshops frequently cover pest identification, product selection, equipment calibration, site awareness, and monitoring. However, workshops rarely cover properly measuring pesticides. After all, how hard can it be? You fill a measuring device with product and add it to the tank, right? How much more is there to know?

It sounds simple, but it can be as complicated as it is critical. This chapter describes steps that pesticide applicators can take to correctly measure and mix products to improve results, save money, and enhance safety. Specifically, it explains some common myths about measuring pesticides, distinguishes between liquid ounces and dry ounces, and offers solutions to common problems associated with measuring devices for liquid and dry pesticide formulations.

THE OVERLOOKED STEP OF MEASURING AND MIXING

Measuring and mixing problems can undermine an effective application. If you do not accurately measure product—and use too little or too much—you can negate the reasons you purchased the product, offset your calibration efforts, and create problems at the job site. It can cause customers to complain, tarnish your reputation, lose business opportunities, affect your bottom line, and even violate the label or the law.

Imagine the problems that occur when an herbicide treatment fails to control miles of brush under a power line. In such a case, the company may be expected to redo the treatment, incurring additional expenses for manpower, fuel, wear and tear of equipment, and more herbicide just to do what should have been done right the first time.

Applying a product at lower than labeled rates can fail to control a pest. This, in turn, can lead to costly insurance claims, protracted lawsuits, and customer dissatisfaction. For instance, applying an insecticide at lower than labeled rates to control termites can lead to extensive structural repairs. If the application method, placement, and timing are correct, underapplying a fungicide or insecticide may cause fruit or vegetable producers to lose all or portions of their crops if the fungicide fails to control diseases or the insecticide fails to control insects. Algae and plant growth can become so thick on a pond or lake that it can make it unusable for boating, swimming, or fishing. These are just some examples of what can happen when products are not applied at the proper concentration.

Comparable problems can occur when products are applied at rates greater than stated on the label. Any application that exceeds the labeled rate violates state and federal laws and regulations, which can lead to significant fines being assessed against the individuals and companies involved in such applications. What's more, high rates can

Figures 10.1 - 10.2. The importance of having the correct amount of product in the tank cannot be overstated when you consider what could go wrong if you use the wrong rate on an apple orchard, ditch bank, pond, utility line, golf course, or cornfield.

injure crops or leave residues that exceed federal tolerance standards. In some cases, overapplication can lead to carryover problems, which means a new crop planted the next season could be injured. High rates are also more likely to move off-site, potentially harming nontarget organisms and the environment.

All these problems are detrimental to your professionalism, credibility, safety, and profitability. And yet you can avoid these problems simply by following product labels and applying the proper amount of product each time.

READ PRODUCT LABELS FOR CORRECT RATES

Pesticide labels provide guidance for both practical and legal (approved by the Environmental Protection Agency [EPA]) product use. Product labels state use rates as well as giving directions for diluting the product to make the finished spray mixture.

Putting the right amount of product in a tank requires following the instructions on the label. While that sounds simple enough, some labels can be difficult to read. Sometimes that's because small package sizes require ultrafine print. If that's the case, consider viewing or printing the product labels online; however, remember that the label on the container is the official (legal) label.

It may take more than just a few minutes to read through the label to find the correct rate for a specific pest and location. The first step is to locate the section of the label that describes the site or type of application you want to make. Rates depend on the crop, environment, target, or type of application. For example, herbicide rates to control brush depend on whether the method is a foliar spray or a basal bark, dormant, or cut stump application.

Labels may provide a single rate for specific uses. However, labels can also state a rate range. In such cases, the actual rate you choose may depend on factors such as the predominant pest, pest density, soil type, temperature, crop stage, carrier, and application method. For example, if the pest density is low, you might apply at the lower end of the range; if the soil is high in organic matter, you might apply at the higher end of the range.

Rates may also vary by state or region. And different crops sometimes have different use rates for the same product. You must select the most appropriate legal rate within a given label range and determine the best precautions for a site and situation.

MEASUREMENT OF LIQUID PRODUCTS

Pouring the right amount of a liquid into a measuring container is straightforward. All you have to do is calculate how much product you need (based on the label rate and your equipment output), fill to the appropriate line in the measuring container, and pour the contents into the tank. However, not all measuring devices are accurate.

It's easy to check the accuracy of any measuring container. For a few dollars, you can buy an accurate graduated cylinder from a science supply store. Because research is usually conducted in metric units, the measurement lines on graduated cylinders are usually in metric units (such as milliliters).

You're not actually going to use the metric cylinder to measure the product. Instead, you use it to check the accuracy of your measuring device. For that, it's important to know the following equivalents:

8 fluid ounces = 1 cup = 236.5882 milliliters

Collect 237 milliliters of water in the graduated cylinder, and pour that into the measuring device you want to evaluate. Read the measurement at the water line. If the water level is at 8 ounces or the 1-cup line, then you know your container is accurate.

Here are several suggestions to make it easier to measure liquid pesticides accurately:

- **Don't answer phones or talk to people when you're mixing products.** Keep distractions to a minimum until you have finished the job.
- **Try not to use measuring containers that include both metric and English units.**
- **Look for liquid measuring containers with pour spouts and handles.** This minimizes the risk that you'll spill any of the product you carefully measured.
- **Buy measuring containers with graduations that are easy to read.** Select ones with large, well-spaced lines and numbers that are etched (not painted) and highlighted with color. This will make it easier to read the units on the container. Painted lines on measuring containers can wear off or be dissolved by some products such as surfactants and organic solvent-based formulations.
- **Replace measurement containers when numbers and markings become difficult to read.**

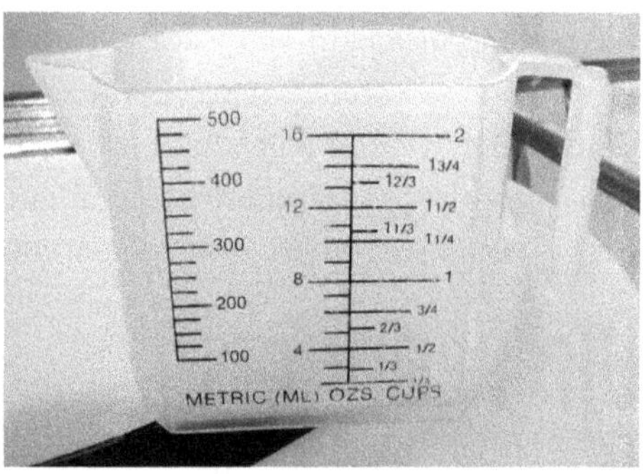

Figure 10.3. A cup with both metric and English units can be confusing and lead to measuring errors.

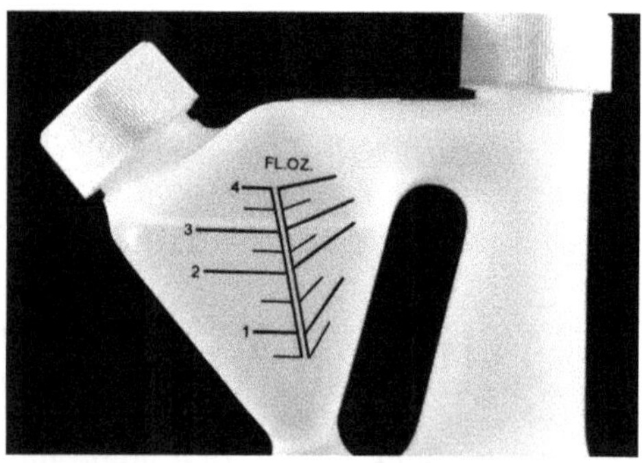

Figure 10.4. Many pesticide products come with their own measuring devices. "Tip and pour" products are easy to measure and can be safer to use.

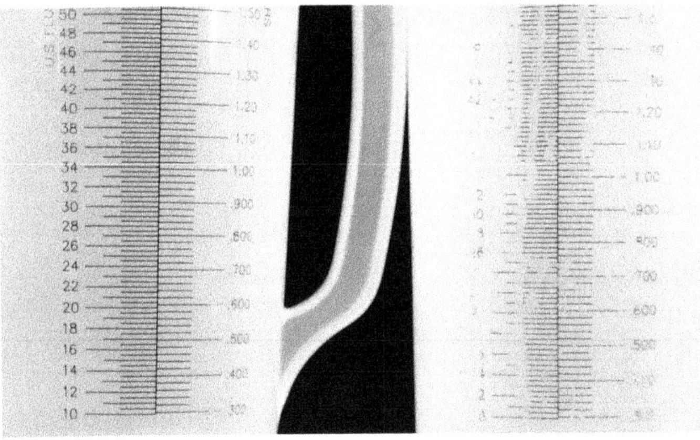

Figure 10.5. Worn cups like the right-hand one here should be replaced. Also, these cups start at 10 ounces and should not be used to measure smaller amounts of product.

- **Use taller and narrower measuring containers.** Taller, narrower measuring containers typically have more graduation marks and more space between lines on the container compared to shorter, wider ones. Your measurements will be more precise.
- **Read measuring containers only when placed on flat, level surfaces and at eye level in a well-lighted place.**
- **Never write on or add marks to your measuring device.** Always use a graduated device that measures accurately with manufacturer-applied measurement marks instead of adding your own.
- **Always rinse out measuring devices in between measuring different kinds of pesticides.** Residues from one kind of pesticide (like an herbicide) may contaminate a different kind of pesticide (like an insecticide or fungicide). These interactions could potentially injure crops, people, or the environment.
- **Always shake the product container before measuring the liquid.** This will ensure that the inert and active ingredients are well distributed prior to measuring.

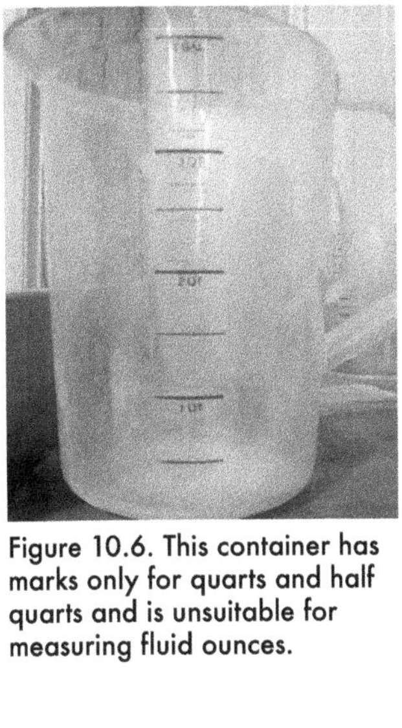

Figure 10.6. This container has marks only for quarts and half quarts and is unsuitable for measuring fluid ounces.

Figure 10.7. Agitate containers to ensure product is well mixed.

Figure 10.8. Some products require users to shake the container to mix up the formulation. Note the product separation and the "shake well" instruction. We suggest shaking all products before use to ensure a well-mixed product.

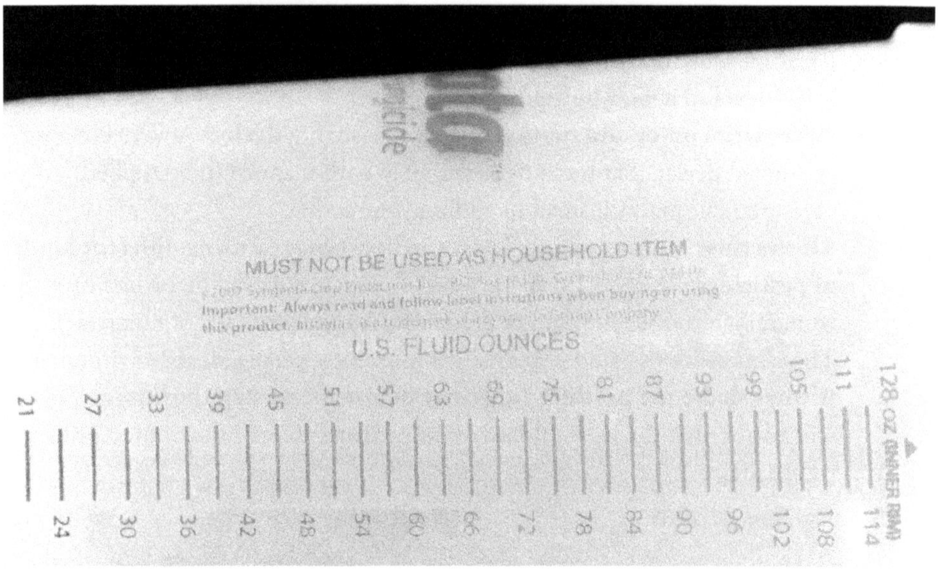

Figure 10.9. Never use pesticide measuring devices for any other purpose.

LIQUID AND DRY OUNCES ARE NOT CREATED EQUAL

We're accustomed to measuring liquid formulations in ounces. An important fact you may have forgotten is that *a liquid ounce is a volume measurement, while a dry ounce is a measure of weight.* This distinction directly influences the accuracy of a measurement.

If you measure 3 ounces of salt by volume, you have more material than if you measure 3 ounces of salt by dry weight (see figure 10.10). If you use a device designed to measure liquid volume when you should be using a device for dry formulations, you change the intended application rate drastically.

When you pour 1 liquid ounce into a measuring container, it always represents 1/128 of a gallon. This is called a *volumetric measurement* because 1 ounce of any liquid always fills the same volume—it takes up the same amount of space.

Figure 10.10. Fluid ounces and dry ounces are different measures.

A *dry ounce* is equal to 1/16 of a pound. The amount of a material it takes to get 1 dry ounce depends on the density of the material. For instance, 4 dry ounces of coffee creamer *weigh* the same as 4 dry ounces of salt. However, because salt is denser than coffee creamer, the latter takes up more space. The volume occupied by a given weight of dry material depends on the density and particle size of the product.

To make measuring their products easier for users, manufacturers have designed volumetric tubes specifically for their products so that you don't have to weigh dry products using a scale. This is the norm for dry flowable (DF) and water-dispersible granule (WDG) formulations. WDGs and DFs are all granulated products that are designed to be diluted and dispersed in water.

The manufacturer weighs 1 ounce of the product and then places it in the measuring tube and marks the level on the side of the tube. This process is then repeated to produce a series of graduations. The marks and lines represent the number of dry ounces—*not* fluid ounces.

Figure 10.11. All of these containers have 4 dry ounces (by weight) of material, but the amount in each jar is different based on the product's density.

There are some things you need to know about retail-supplied measuring devices:

- **Before you measure, check whether a measuring device is meant for fluid ounces or dry ounces.** If it is for liquid ounces, the measuring device should clearly say liquid ounces, fluid ounces, fl. oz., or liq. oz. Many measuring tubes look similar, so be careful.
- **Be cautious of measuring devices for closely related products that are formulated differently.** Only use a fluid ounce device for measuring liquid products and a dry measuring device for dry products.
- **Be cautious of containers that have different sets of graduation marks indicating both liquid and dry ounces.** When hurried or interrupted, the person measuring the product might use the wrong set of measurement lines.

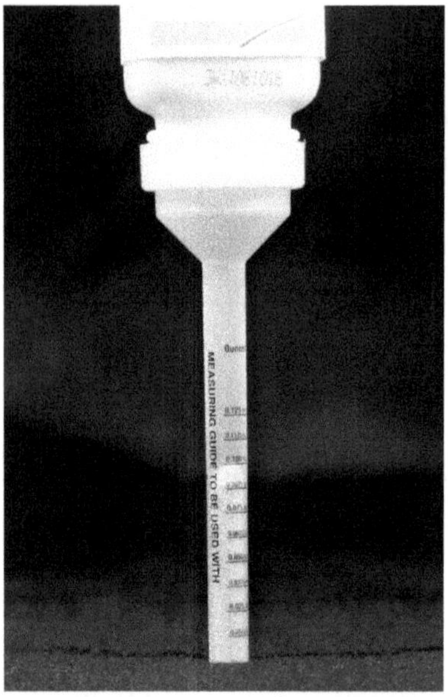

Figure 10.12. A product comes with its own measuring device to measure dry ounces.

Figure 10.13. The specific volume of a water-dispersible granule (WDG) pesticide product depends on its density. The volumes of WDG products vary by their ingredients and the size and shape of their granules.

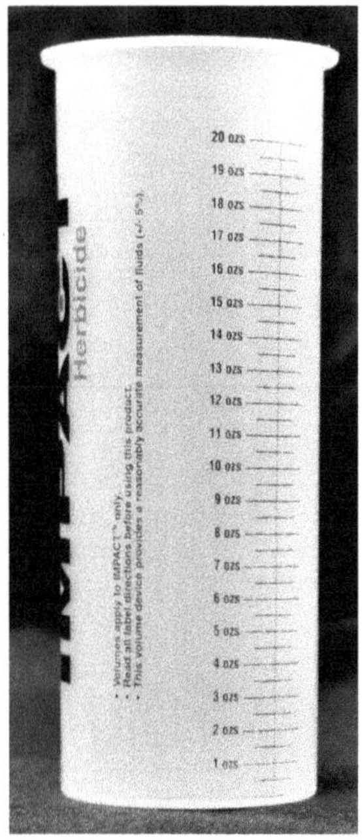

Figures 10.14–10.15. These containers look similar, but one measures fluid ounces, and the other measures dry ounces of a specific product.

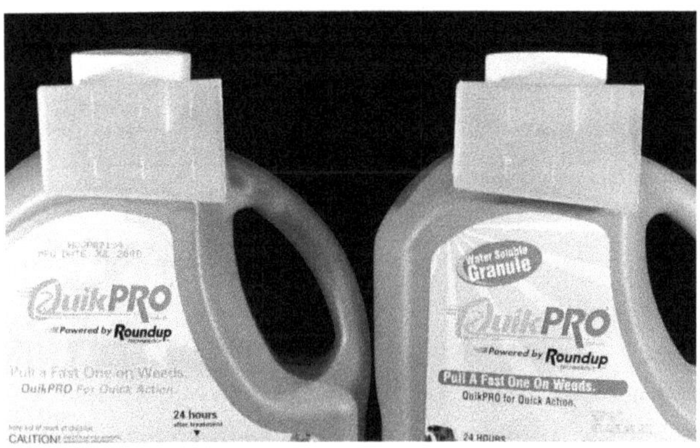

Figure 10.16. Quick Pro® is a soluble granule. The measuring cups attached to these containers are different colors but measure the exact same amount of dry ounces. A similar liquid formulation of this product requires a green measuring cup instead. Throw in other cups for products with similar-sounding names, and it is no wonder that measuring errors are made. Always keep measuring cups with the container to avoid inaccurate application rates.

Figures 10.17–10.18. This Sahara® pitcher displays fluid ounces on one side but has dry ounces and instructions to use only for Sahara® on the other.

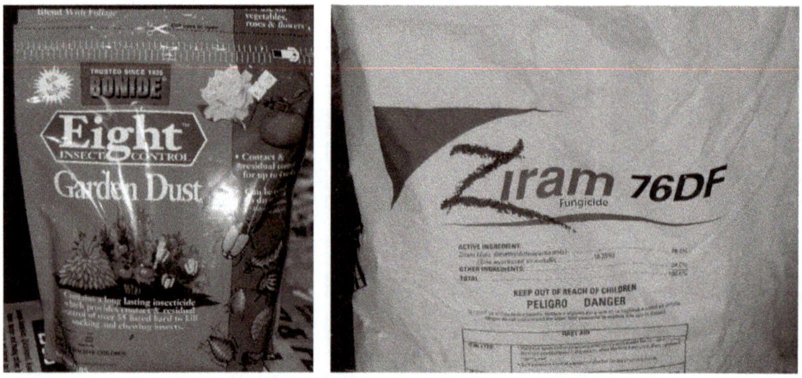

Figures 10.19–10.20. Some dry products are intended to be applied as is, while others will be mixed in water. The Garden Dust here is applied dry, while the Ziram® DF (dry flowable) should be mixed in water.

MEASUREMENT OF DRY FORMULATIONS

Products that are meant to be diluted in water (WDG, DF) generally include the term *dispersible* to acknowledge that the product is intended for spray applications. Be aware that some dry granular products are designed to be applied dry and are not intended for spray applications. Such products generally are designated as *granulars* (GR or G).

Wettable powders include the *wettable* descriptor to denote the products are intended for dilution in water (unlike dusts). Consult the product labels or consider diluting a small sample of the product in water to observe its dispersion (or lack of dispersion) if you are uncertain.

Because many WDG products are mixed in water and applied using ounces per acre, being off by even just an ounce or less will affect the results. The following suggestions will help you measure dry pesticide formulations accurately:

- **Remember that measuring devices (scoops or tubes) are product-specific, and read and follow the measuring device instructions.** Many clearly state that they must be used only for the products they came with and then discarded. If you substitute a measuring tube meant for another product, you almost certainly will mix the pesticide incorrectly.
- **Do not accept tubes other than those shipped with the product at the time of purchase.** It's common for manufacturers to ship fewer measuring tubes than the number of pesticide containers in a package. When you purchase individual packages from a case at the retailer, make sure to ask for the tube that came out of the box in which the product was shipped.
- **Use the measuring scoop that comes with dry materials.** The scoops, like the tubes, are product-specific. Sometimes they are built as part of the cap, or they are provided separately.
- **Keep WDG measuring devices with the product.** Always keep these measuring tubes with the corresponding pesticide container.

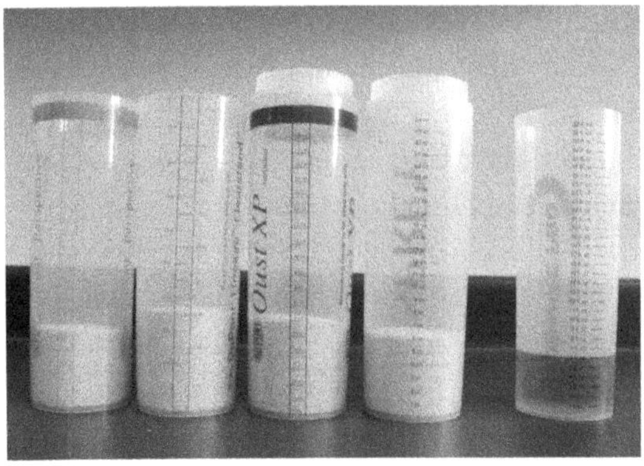

Figure 10.21. These WDG tubes clearly show how the graduations for measuring the weight of dry materials are product-specific. Each tube contains 6 dry ounces of product. For comparison, see 6 fluid ounces in the tube at far right.

Figure 10.22. There are only four measuring tubes but eight containers. You should have a measuring tube for each lot of material you purchase.

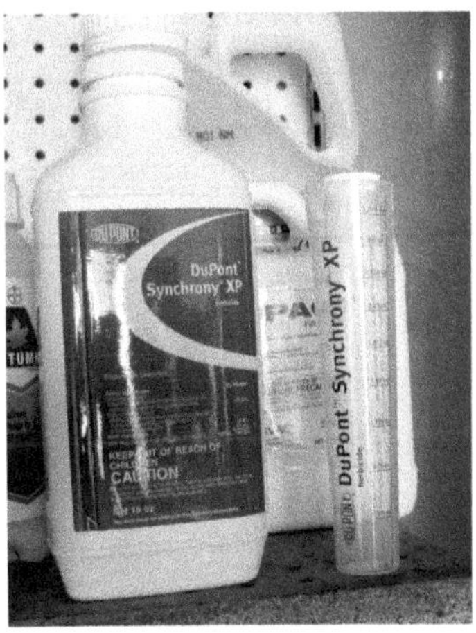

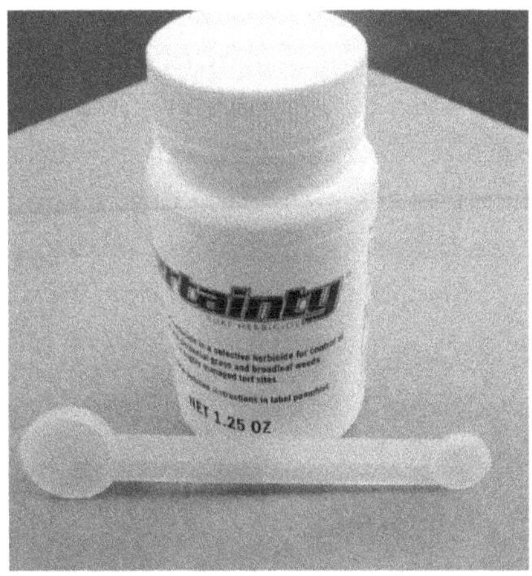

Figures 10.23–10.24. It is important to use the measuring device that comes with a specific product. The dry ounce units on a measuring device are specific to that product.

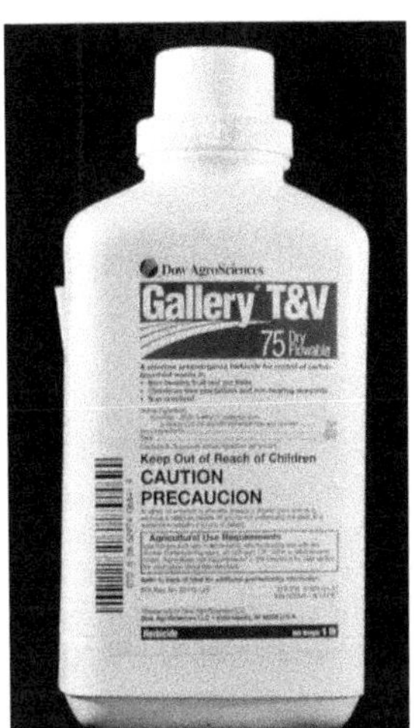

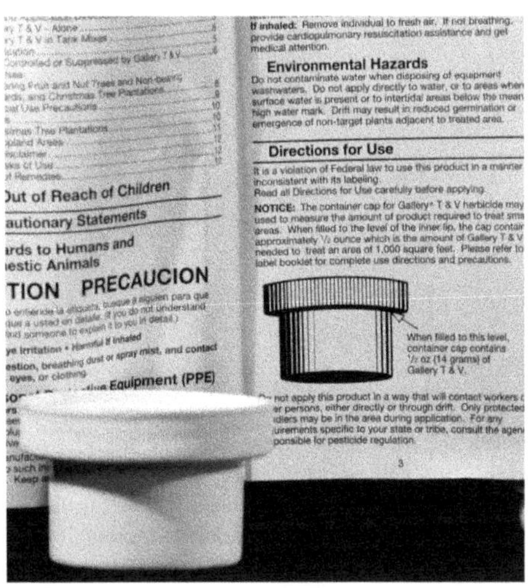

Figures 10.25–10.26. Some product caps serve as the measuring device. In such cases, product labels provide specific instructions using the cap.

- **Do not reuse leftover measuring tubes.** If the manufacturer changes the formulation, the formulation's density may also change. Also, if a new batch of product has a different particle size, the way the WDG particles settle into a tube will change. In such cases, the manufacturer will calibrate new measuring devices to most accurately dispense the product. For example, notice the slight variations in the 1-ounce markings between different batches of Oust XP® and Escort XP® (see figure 10.27). The tubes' designs and colors differ, indicating a change in the formulated product. These examples show why it's important to follow the manufacturer's instructions to "dispose of after use." In many applications, these slight differences can make a significant difference in product performance.
- **Turn the container over several times before pouring the material into the measuring tube.** This ensures a more uniform mixture of the product.
- **Do not tap down the material in the measuring container.** Package engineers calibrate product measuring devices based on the practices expected from typical customers. These calibrations match the loose bulk density of a product. If you tap the volumetric measuring device to minimize a product's volume, you may exceed your desired application rate.
- **Seal the top of the measuring device and then gently invert it, then return the sample to the upright position with minimal disruption.** If you do this and assess the resulting WDG product volume, you can improve the consistency of your volume measurements.

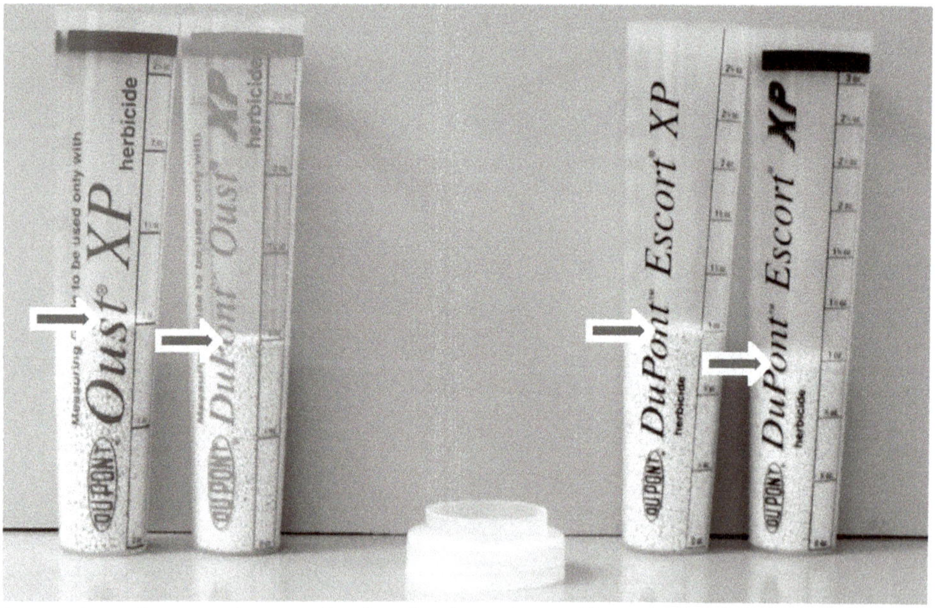

Figure 10.27. This example shows that the tubes have changed from one batch to another. The manufacturer indicated this by changing the color for the two Oust XP® tubes on the left and by using bold lettering and a dark line for the two Escort XP® tubes on the right. Even small changes can make a difference in how well a product works.

- **For the greatest accuracy, weigh dry flowables on a scale.** The variance in normal product density, user technique, and general measurement accuracy combined could in the worst cases approach 15 to 20 percent. Manufacturers often suggest weighing out products when more precise measurements are required.

  Select a scale that allows you to tare (zero) out the weight of the container (a plastic cup in figure 10.29) in which the dry materials will be placed. In this example, the plastic container was weighed. With the push of the tare or zero button, the weight reads zero. Now, when you weigh the material, you will get a reading that does not include the container's weight.

  Follow the scale manufacturer's instructions to check and adjust the balance calibrations, as needed. It is also a good practice to check the scale's accuracy using a known weight standard.
- **Reduce the amount of measuring.** You can reduce measuring error by using the whole container of product at once, since manufacturers fill their containers by weight, not volume. This also saves time.

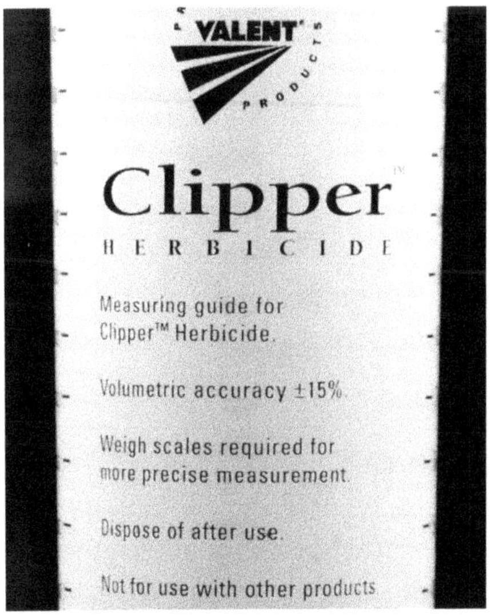

Figure 10.28. A measuring tube can only be used for that specific product and only for that batch of product purchased. Tubes for the same product can change from one purchase to another, which is why it is best to discard them after use.

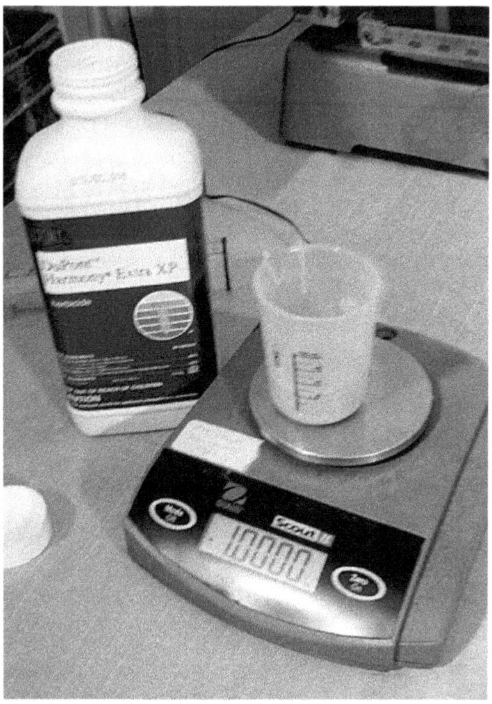

Figure 10.29. For dry products a scale is the most accurate way to measure.

YOU DO THE MATH

Let's say you have an 800-gallon tank.

If you apply 10 gallons of water per acre, how many acres will the tank cover?

If you apply 20 gallons of water per acre, how many acres will it cover?

If you apply 15 gallons of water per acre, how many acres will it cover?

The answers normally given are 80, 40, and 60 acres.

However, the 60 acres answer is wrong—the correct answer is 53 acres. Many people assume that you can split the difference between 80 and 40 acres and come up with 60, just as 15 is halfway between 10 and 20 gallons. But the math doesn't work that way: 800 ÷ 15 = 53.3. Always write down your math to double-check your calculations.

TANK SIZE AND SPRAY MIX AGITATION

There are several other factors to consider if you want to get the correct rate applied to the treatment site. Two of these are tank size and spray mix agitation.

Tank Size

You probably assume that the markings on your tanks are accurate, whether they hold 5, 100, 250, 500, or 1,000 gallons. Unfortunately, this is not always the case. It is important to know how much is actually in your tank, since the amount of the product's active ingredient that you need depends on the volume.

You can use a flowmeter to check the markings on a tank. Fill the tank and stop the flowmeter at the increments marked on the tank. The flowmeter and the tank marking should match up. As with the scale, make sure the flowmeter is calibrated according to the manufacturer's instructions.

Another way to measure tank size is to weigh the truck or trailer carrying an empty tank on a truck scale. For example, weigh a 500-gallon tank on a truck or trailer with the tank empty. Then fill the tank with water to the 500-gallon mark, and reweigh the truck. Subtract the original truck weight with the empty tank from the second weight, with the tank filled. Divide the resulting number by 8.3 pounds (1 gallon of water weighs approximately 8.3 pounds) to calculate the number of gallons in the tank. It should be close to 500 gallons. If not, remark the 500-gallon graduation mark on the tank.

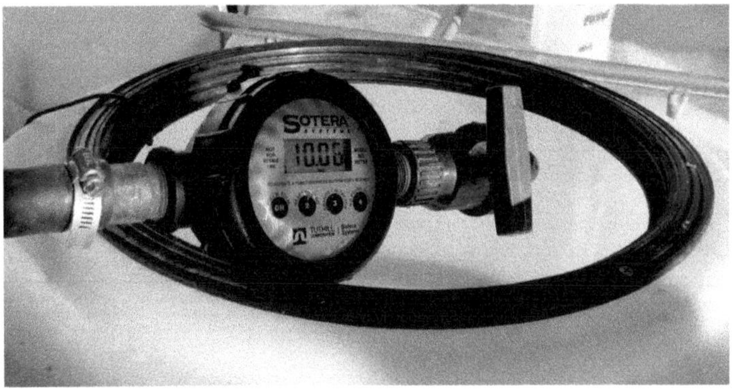

Figure 10.30. Many applicators ignore the marks on their application tanks and use flowmeters to measure water. This can be particularly helpful if you are mixing on less-than-level ground.

Figures 10.31 – 10.32. With today's sophisticated equipment, electronic flowmeters can be a great asset when filling sprayers. In this case, the digital readout is mounted in a position that makes it easy to read during the filling process.

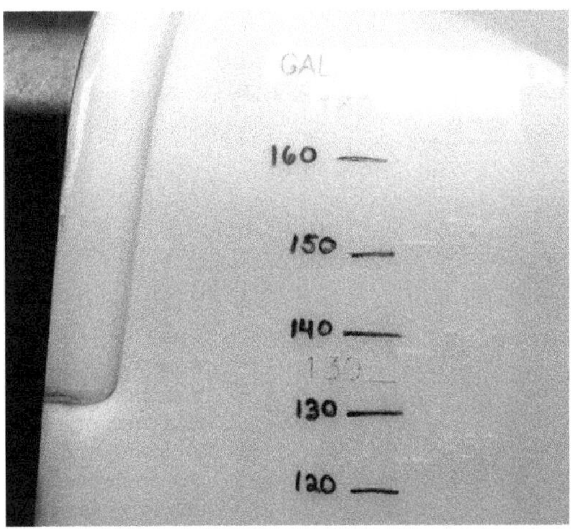

Figure 10.33. Measuring the tank to ensure that its markings are correct is an easy step that can lead to more precise applications. Notice the difference in where the 130-gallon mark is and the actual measurement taken with a flowmeter. Make sure the tank is level when remarking.

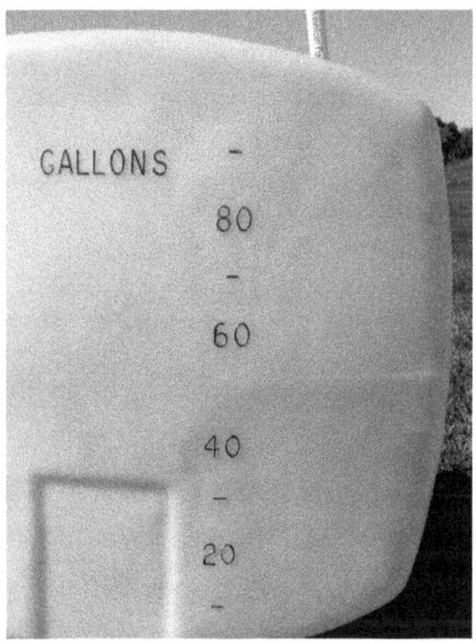

Figure 10.34. Where is the 100-gallon mark on this 100-gallon tank? Tanks should be clearly marked for capacity.

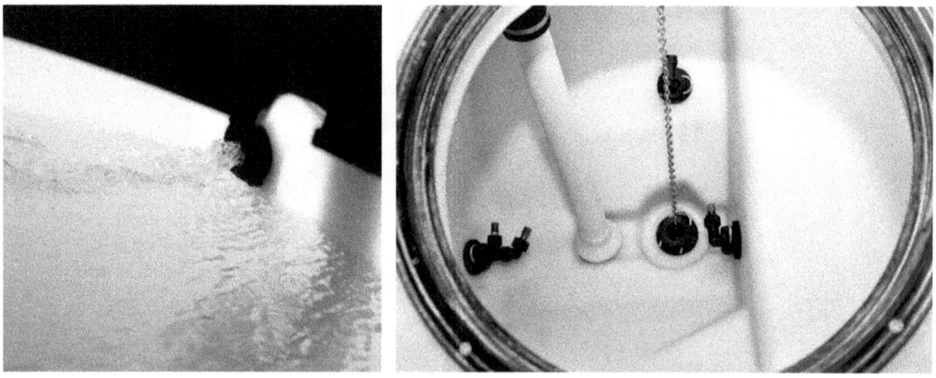

Figures 10.35–10.36. Polyethylene (poly) tanks may employ pressurized bypass agitation to maintain a uniform mixture. In some cases, valves can be installed to better direct the incoming water at the bottom of the tank.

Spray Mix Agitation

It is also important to assure that you apply crop protection products consistently across the entire treated area. WDG products are formulated to disperse into very fine particles (about 1–15 microns in diameter) when diluted in water. The preparation of these products includes particle size reduction. WDG products are then granulated (agglomerated) to minimize their dustiness for user safety. The granulation process also significantly increases the product density for easier handling for end users.

Spray tank agitation is a basic requirement for WDG products. Generally, users fill their spray tanks to half to two-thirds of their full volume, begin spray tank agitation, and then add the WDG product to the spray tank. Then they finish by filling the spray tank to its full volume for spray application. This mixing method allows sufficient time for the WDG product to wet and disperse into the water fully.

It is important to constantly agitate these spray mixtures to prevent solids from settling out of suspension over time. Fiberglass and steel tanks may be equipped with paddle-type mechanical agitators.

Figure 10.37. Make sure to recirculate the hose reels at the beginning of the day. Materials may have settled in the bottom of the tank overnight.

Figure 10.38. Without agitation, product will settle at the bottom of the tank. As a result, solution taken from the bottom of the tank will have a much higher concentration of the product than solution taken from the top of the tank. This can lead to inconsistent results and unwanted damage.

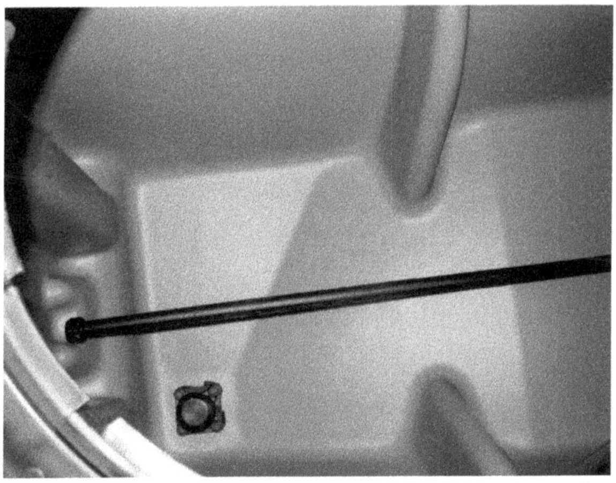

Figures 10.39–10.40. There are many ways to agitate tanks and help to keep product suspended in water. A sparger tube is a tube with holes mounted on the sidewall of the tank. It is normally in the lower third of a tank with the spray directed downward, which causes swirling up and over the top of tank solution.

If you will apply a WDG formulation with a backpack sprayer, be sure you can agitate the spray suspension. Use a backpack with paddles or some other type of agitator in the tank. When using a hand sprayer, simply shaking or inverting the closed applicator intermittently is generally adequate. The frequency for agitation may vary by product but is best determined by visually looking for obvious stratification from top to bottom.

A best practice to assess remixing intervals is to dilute some product in a glass jar (jar test; see chapter 11) and monitor how long it takes for sediment to form on the jar bottom and for stratification to appear in the upper liquid.

Preparing a slurry of WDGs that is then added to the tank is generally not required but may be practical in some cases, such as when adding WDG formulations to concentrated liquid fertilizer mixes or micronutrient spray mixes. A slurry also may help avoid other spray mix antagonisms or to speed up dispersion when the spray water is extremely cold (see chapter 11).

Figures 10.41–10.44. Making a slurry helps to solve many problems when mixing dry materials.

CONCLUSION

Figures 10.45–10.46. Applicators used to measure products by the pound and gallon, as shown in these photos. Today the trend is to measure by the ounce. This requires a greater level of precision in measurement.

Precision in pesticide application depends on the equipment and applicators working in unison to properly deliver an appropriate dose of a labeled product safely to a given site and pest. Today's products are generally more pest-specific and use lower amounts of active ingredient per unit of area. Therefore, there is less room for error before and during an application. Over- or underapplying pesticides can be expensive for you and your customers. You also could harm the environment. What's more, overapplying a pesticide is illegal.

The precision instruments available in today's spray technology are only as good as what you put into them. Seemingly minor errors in measuring pesticides into the spray tank may have a large impact on the end results. If measuring devices or tank marks are inaccurate, that can multiply errors.

Know and understand the differences between liquid and dry ounces, both common units of measure for pesticide products. If you use the measuring devices enclosed with pesticide packaging, discard them when you have used up the product. Keep in mind that manufacturers often change their product formulations, with only subtle differences in the label and package.

The common instruction to "read and follow label instructions" has new meaning. When you look at the big picture in spray technology, realize that the subtleties of measuring the product or carrier are often neglected and may be the difference between application success and failure. Using an inaccurate measuring container, following the imprecise measurement marks on a spray tank, confusing dry with liquid ounces, or using a WDG volumetric measuring device intended for another product will limit application accuracy. These issues show just how important it is to correctly measure pesticides for the best performance. Investigate and, if necessary, correct measurement errors in the pesticide application process. Don't let hidden pitfalls derail your pesticide application.

11

Avoid Tank Mixing Errors

With contributions from Melissa Olds, Raymond Cloyd, Bryan Young, Doug Linscott, Jason Deveau, Jim Reiss, Aaron Patton, Bill Johnson, Tim Overley, and Kevin Leigh Smith

Once a company formulates a new product, they test it with other products. When testing possible tank mixes, a company selects products that growers or commercial applicators might apply together during the growing season. However, manufacturers cannot test all the possible tank mixes applicators can conceive.

The more tank-mix partners you add, the greater the potential for incompatibility. A tank-mix incompatibility can be costly. There is the additional cost and difficulty of cleaning out the sprayer, plus you may have to make repeated applications to control the original pest problem.

Certain processes for mixing products together help ensure that they work effectively. This chapter focuses on factors influencing product compatibility. Understanding these factors (including formulation, mixing order, and carrier type) will help you effectively combine crop protection products. If you don't mix products together the right way, you will encounter costly and time-consuming problems with your mix.

Figure 11.1. When you follow the correct mixing order and other mixing principles, you can avoid creating problems like this in the spray tank.

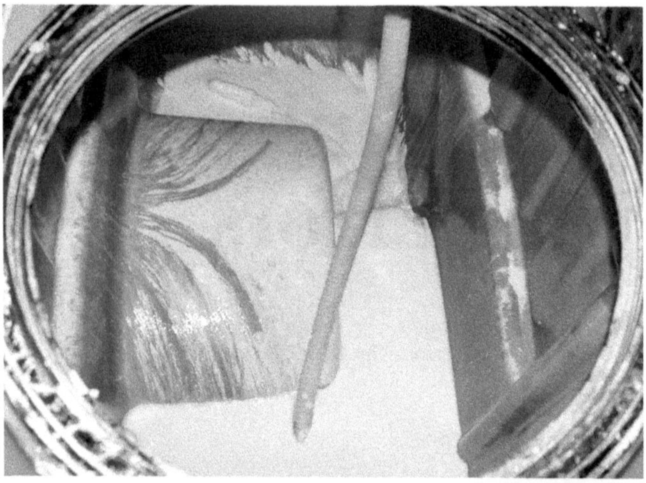

Figures 11.2–11.3. Instead of making a faster and more efficient application, this applicator has a mess to clean up. Always wear personal protective equipment when dealing with physical incompatibility problems just as you would when mixing and spraying the products.

TANK MIXING BASICS

It is common practice to combine multiple crop protection products into a single spray application. This is called *tank mixing*. The primary benefits of tank mixing are efficiency and convenience. Tank mixing may also:

- Permit applicators to make more timely applications based on weather conditions and crop or pest life stages
- Improve the control of certain pests
- Broaden the spectrum of the target pests controlled
- Allow products to be applied together with fertilizer, biostimulants, and plant growth regulators
- Reduce expenses by decreasing input costs—including labor, fuel, and equipment wear—resulting from multiple passes across a targeted area
- Reduce crop damage and soil compaction associated with making multiple applications

- Decrease the likelihood of pesticide resistance by applying products with multiple modes of action together
- Enhance the efficacy of active ingredient(s) by adding spray adjuvants

Tank mixtures can include:

- Pesticides—herbicides, fungicides, insecticides, miticides
- Fertilizers—such as urea ammonium nitrate
- Adjuvants—such as surfactants, crop oils, and water conditioners
- Spray dyes
- Micronutrients—such as iron, zinc, boron, and manganese
- Macronutrients—such as sulfur
- Biostimulants
- Plant growth regulators

Most spray applications use water as the liquid carrier, but water quality can vary greatly. Specifically, water's pH, mineral content, and temperature can influence how chemicals dissolve or suspend, interact with one another, and generally perform. Therefore, it is important to consider the water source when tank mixing (see chapter 12).

Each formulated product has a set of complex chemical structures. As a result, each tank-mix partner you add increases the potential for incompatibilities (negative interactions). Incompatibilities arise when one product influences how other products in the mixture disperse, mix, and deliver. There are two basic kinds of tank-mix incompatibility: physical and chemical.

TWO FAILED TANK MIXTURES COMPARED

Physical Incompatibility

Tank mixes with physical incompatibility problems may separate into layers (that is, oil and water), and solids may settle faster than normal. In severe cases, physical incompatibility may cause the solution to form gels or cause solids to clump. When these things happen, operators may have to drain and flush the tanks and then remove and clean all filters, screens, and nozzles. Examples of physical incompatibility include:

- Dry products fail to disperse or suspend properly in the solution. When this happens, sediment can form a cake-like layer on the bottom of the tank, or undissolved particles can clog screens and nozzles.
- Liquid solutions curdle and thicken into a paste or gel, which can't be sprayed and is difficult to clean out of the tank.
- Oil residue coatings or films collect on tank walls and rubber hoses.
- Active ingredients separate into distinct layers in the spray tank.
- There is excess foaming from trapping air in the tank mixture.

Physical incompatibility can cost you time and effort while you clean out gels, solids, and hard-packed sediments. Additionally, it can be expensive to remediate or safely dispose of the residues and any rinsates you generate while cleaning the sprayer. In severe cases, if you don't clean out the equipment completely, you can damage the crop.

Figures 11.4–11.7. Chemicals that do not mix well can plug filters and screens, gel up in the tank, or turn semisolid.

Chemical Incompatibility

Chemical incompatibility can negatively affect spray quality, product uptake, and plant surface retention of the application. The effect of a chemically incompatible tank mixture may not always be obvious. Visible crop injury (phytotoxicity) may occur a few hours to several days following an application. Sometimes reduced efficacy is the only observable effect of chemical incompatibility. This may require additional applications and can decrease yield, harm crop quality, or both.

Most chemical incompatibilities arise from surfactants in the tank mix that are not compatible in mixtures. For example, an oil adjuvant added to an herbicide can increase cuticle penetration, but it also can result in scorching or bleaching on crops that might not otherwise be affected by the active ingredient.

COMPOSITION OF INGREDIENTS

When discussing pesticide products, it is common to focus on active ingredients; however, they are only one component of a formulated product. Physical compatibility problems may not directly involve an active ingredient. Rather, inert ingredients (coformulants) in the product may cause incompatibility.

In water-dispersible granule (WDG) and soluble granule (SG) formulations, inert ingredients primarily help dissolve or disperse the active ingredients in a solution. In liquid formulations, inert ingredients may include emulsifying surfactants, oils, solvents, preservatives, and thickening agents that prevent particles from settling. See the sections "Solubilized Active Ingredients" and "Solid (Dispersed) Active Ingredients" later in this chapter.

FACTORS THAT MAY LEAD TO INCOMPATIBILITY

Common tank mixing mistakes include:

- Combining formulations that have not been tested together (see "Jar Test")
- Applying new tank mixtures using multiple products (especially mixes that contain oil adjuvants)
- Not following label instructions
- Mixing products in the wrong sequence
- Adding products when there is insufficient water in the spray
- Adding products to the tank mix before previous products have had time to dissolve and disperse
- Mixing with too much agitation (which can cause foaming with liquid flowables) or too little agitation (which can cause dry formulations to not disperse well)
- Not considering compatibility when substituting water with fertilizer
- Not considering slower mixing times due to cold water

The key to creating a stable tank mix is understanding the factors that influence compatibility. Next, we describe eight actions that will help you avoid tank-mix incompatibilities.

1. Check Containers First

Formulations such as suspension concentrates (SCs), oil dispersions (ODs), suspo-emulsions (SEs), and capsule suspensions (CSs) contain active ingredients that can separate into layers. The layering effect is similar to what you might see in a bottle of Italian salad dressing or a can of oil-based paint if you leave them undisturbed for an extended time. Before you add these formulations (or any product) to your tank, check the container or storage tank.

These pesticide products require thorough mixing before you use them. When you ensure that the content inside the containers is homogeneous (uniform), you ensure that the active ingredient(s) and inert materials are appropriately distributed throughout the container before you add them to the spray tank.

Figure 11.8. Different formulation types.

2. Follow Label Instructions

Product manufacturers develop their mixing instructions based on:

- Laboratory tests and field tests of potential tank mixtures
- Mixing and handling feedback from customers, field researchers, and application specialists

The manufacturer checks and validates this information and communicates it to customers through product labels, websites, fact sheets, technical bulletins, and user group presentations. To benefit from all this information, carefully read product labels before you start mixing. When it comes to mixing instructions, a product's label specifies:

- Products that have been tested for compatibility with it
- The order in which you should add products
- The recommended carrier volume
- The need for a tank-mix adjuvant
- General recommendations for tank-mix agitation
- Product sensitivities to extreme temperatures
- Warnings about spray water carrier pH or mineral content
- Products that should not be tank mixed with it

Be careful when mixing products that have radically different water requirements. The label is the only source of this information. One pesticide product might need a pH above 7 to maximize its solubility and performance, yet another might require a more acidic environment (pH less than 7). Conditioning the water for one diminishes the performance of the other, and adjusting the water to accommodate both products reduces the performance of both.

As a general rule, do not combine products that have contradictory pH requirements. A second, less preferable option for mixing opposing products is to use water suited for the product that you are most dependent on. Then, if the label allows, use a higher rate of the second product to overcome poor water quality.

It is critical to understand label directions. Furthermore, since labels change, applicators should read the label each time they purchase a product. Manufacturer websites and customer service links are useful resources you can check when dealing with compatibility.

3. Mix in the Proper Order

The order in which you add tank-mix partners can reduce physical incompatibilities. Moreover, knowing how the active ingredient will disperse or dissolve will ensure a compatible tank mix.

Solubilized Active Ingredients

One type of soluble active ingredient formulation is a soluble liquid (SL). Generally, manufacturers prepare these formulations as salts of the active ingredients. This formulation type is water based, so when you add it to a spray tank, you dilute the SL even further. A solubilized active ingredient remains dissolved in solution when it is concentrated in the jug or after you mix it with spray water. You need only mild agitation.

Other examples of soluble active ingredients include oil- or solvent-based formulations—such as emulsifiable concentrates (ECs) and microemulsions (MEs)—where the active ingredient remains dissolved in solvent. These solvent-based formulations require you to add emulsifying surfactants to the formulation concentrate to prevent rapid oil/water phase separation.

With phase separation, the solvent-dissolved active ingredient oil layer floats to the top of the tank and forms a distinct and separate layer from the water. To avoid this mixing issue, emulsifying agents are added to the formulation to reduce separation of the oil-and-water emulsion into layers. These emulsifiers create a milk-like emulsion, keeping the oil droplets small and numerous, to create a uniform diluted spray mixture.

A SL formulation that contains predissolved active ingredients means there will be less settling in an unagitated spray tank. However, there are limits to solubility. Once water reaches its saturation concentration, any remaining product will fall out of solution—just like if you add too much sugar to iced tea.

Solid (Dispersed) Active Ingredients

A dispersed solid active ingredient has been reduced to a very small particle size, and these particles are then either suspended in a water-based liquid (suspension concentrate) with a thickener or clumped into a WDG formulation. In both cases, the active ingredient remains a particle the entire time, from the container through application.

In general, solid active ingredient formulations are not very soluble in water or fertilizer. These formulations rely on physical agitation to keep small, solid particles suspended in solution. Imagine shaking a snow globe to make it snow. Unless you keep shaking, the snow rapidly settles. Solid active ingredient formulations are similar.

Without agitation, the active ingredients settle to the bottom of the tank over time. Some formulations will settle to the bottom of the tank without forming a hard layer. Agitating the mix will easily resuspend them. Other formulations may form a hard pack, which makes it difficult to resuspend the particles. In any case, the longer the suspension remains without remixing, the more difficult it becomes to resuspend into a uniform, sprayable mixture. Obstructed hoses or piping may also prevent the product from being remixed.

The active ingredients in liquid suspension formulations are suspended in a liquid such as water or oil. They include suspension concentrates, oil dispersions, or capsule suspensions. Suspoemulsion (SE) formulations contain suspended particles and an oil active ingredient (or adjuvant) that has been premixed to ensure they are physically compatible. In essence, SE products form an emulsion in the container, while EC products form an emulsion when mixed in water.

Dry solid formulations, like WDGs and wettable powders (WPs), need to be dispersed in water. Each granule needs to break into smaller primary particles (2 to 10 microns in size) that can pass through screens and spray nozzles. You should mix these formulations first in water using moderate agitation. This allows the water-soluble surfactants that hold the small particles together to dissolve and release them into the suspension.

One of the more common tank-mix problems with WDG and WP formulations occurs when you add oil-based products or adjuvants to the spray mix before adding the solid formulation products. Since crop oil is less dense than water, the oil will stay on the liquid's surface. Consequently, when you place the dry formulation in the spray tank, it first encounters a layer of oil, which coats the dry formulation. And this oil coating the dry formulation prevents water from reaching the pores of the dry formulation, which limits its ability to suspend or disperse.

The end result is that the dry formulation cannot break down into primary particles, which leads to larger solid particles settling or clumping in the bottom of the tank and hoses. These large particles can clog nozzles and filters. And even if the incompletely dissolved WDG particles do pass through screens and nozzles, they may not effectively control the target pest—the particles can be too large to cover the area or for the plant to take up.

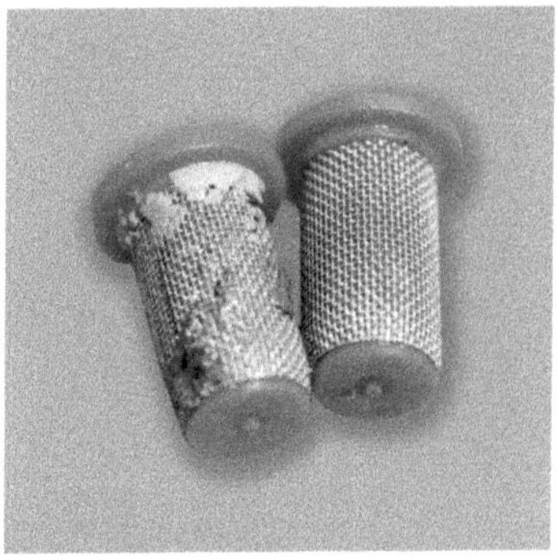

Figure 11.9. Products that only partially break down in water can plug up screens and nozzles.

4. Use the Right Water Volume

The amount of water you use when preparing a tank mix can be critical. You need to have a minimum volume of water in the tank before you add the first product. Think of sugar dissolving in a glass of water. Once the water is saturated, the sugar will no longer dissolve and will settle to the bottom of the glass. However, if you increase the volume of water in the glass, the extra sugar will dissolve into the water.

Problems can occur if you have only 5 to 20 percent of the water volume in the spray tank, vat, or inductor cone when you add the first products. This can create a very concentrated mixture that completely saturates the water. This is a common issue for most products.

Furthermore, compatibility problems may occur sooner when you use low initial water volumes. SLs are essentially salt solutions. When you use low initial water volumes to combine SLs, solubility becomes a factor—just like for the sugar in the tea. Active ingredients precipitate out of the spray solution when the active ingredients can no longer dissolve in the carrier due to changes in water properties or to interactions with other products.

It is important to follow label recommendations. Typically, directions state that the tank must be half full of water before you add the first product. In addition, using less carrier volume may cause dry materials to clump or form gels. The more carrier that is available to disperse and/or dissolve products, the less likely the product will settle to the bottom of the tank or clog sprayer lines.

Figure 11.10. "Here [is] . . . what happens when you mix glyphosate and 2,4-D without having at least 50 percent water in the tank. The sediment had to be removed from the injection system on our spray truck. The tank was easy to clean but not the injection lines. Some had to be replaced."

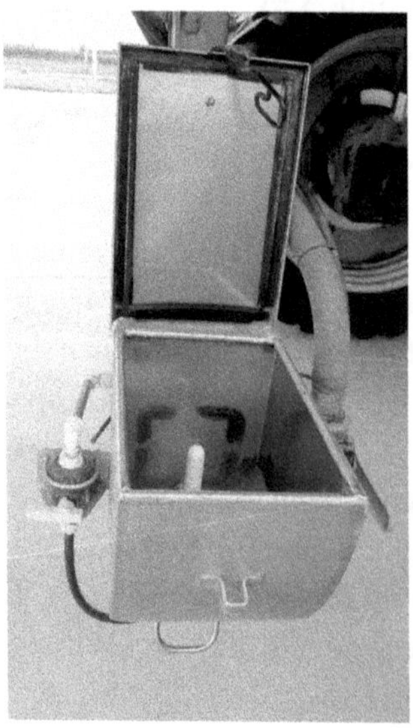

Figures 11.11 – 11.13. These inductors and vats are commonly used for mixing pesticides. Do not put more than one product in these mixing containers at a time. The risk of incompatibility is greater when you use low water volumes and high rates of crop protection products. Having more water in the tank decreases the chance of problems.

5. Be Patient When Adding Products

You should always agitate products in the tank for a while before you add another product. Adding products in rapid succession may result in physical incompatibility. Patience is especially critical when the active ingredients do not immediately dissolve in water. Remember, it takes time for some products to completely disperse or suspend.

If you add products too quickly (that is, one right after the other), you can interfere with the dispersion and mixing process. Just because a product appears to have dispersed, that does not mean the product has completely suspended or dissolved. Products do not mix instantly (that is, in 2 minutes or less), especially when you mix in cold water (lower than 42°F).

6. Agitate Properly

You should agitate tank mixtures before and during spray applications. If you do not agitate tank mixtures, then products that have dispersed active ingredients (including WDGs, WPs, suspension concentrates, oil dispersions, capsule suspensions, ECs, and suspoemulsions) will settle out over time.

That said, some sprayers may have agitation that is too aggressive. Mechanical agitation (such as paddles in airblast sprayers) and hydraulic agitation (associated with high-capacity pumps) can agitate tank mixes too much and make the tank water look like it was agitated by a high-speed food blender.

Too much agitation may result in problems like foaming. Foam is a problem when you add WDG or WP products to the spray tank. The dry formulations remain on top of the foam and will only partially hydrate. Since the products do not have

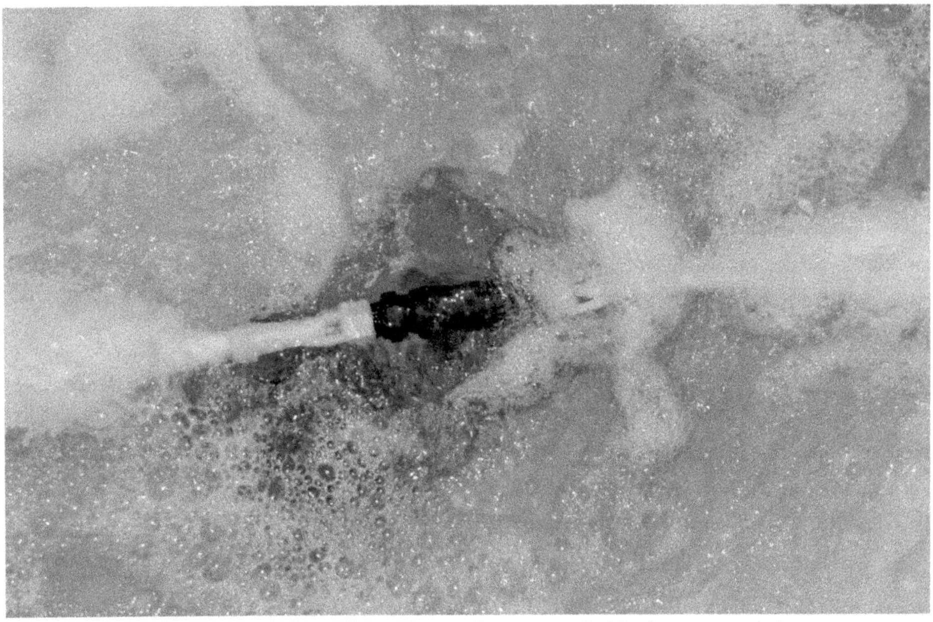

Figure 11.14. Tank agitation helps to keep the product suspended in the spray solution or to resuspend products following spraying delays.

adequate water available to properly disperse, the products tend to simply swell and then sink to the bottom of the tank. This prevents uniform dispersion, may clog the pump intake completely, or may cause the pump to lose suction and stop pumping.

Overagitation is also a concern when using dispersed active ingredients—such as emulsions in water (EWs) or ECs that form emulsions. Overagitation may affect the surfactants attached to the surface of a solid active ingredient and cause the surfactants to separate from the active ingredients. This destabilizes the tank mixture and causes clumping. In this case, too much agitation may create a spray solution that resembles cottage cheese curds.

Overagitation can cause certain polyacrylate or polyacrylamide thickener-type anti-drift agents to lose their effectiveness from mixing shear. Over the course of the application, the anti-drift agent's effectiveness will diminish as the remaining liquid continues to be recirculated through the mixing pumps. For this reason, anti-drift agents are often added last.

You need to use moderate agitation. Overall, you want the bypass agitation to be just enough to prevent particles from settling on the bottom of the tank. Providing moderate agitation and using an antifoaming agent early in the mixing process will ensure that all products disperse quickly without foaming. You have moderate agitation when you can observe the top layer of the spray tank mixture moving gently when you look inside the tank. If the spray mixture looks like boiling water, that is too much agitation.

7. Adjust for the Carrier

The two most common carriers used for applying pesticides are water and liquid fertilizers (that is, 32-0-0, 28-0-0). About the only similarity water and liquid fertilizer share is that they are both liquids.

Generally, water is a better carrier for tank mixtures, because most products are designed to be used with water. Liquid fertilizers are a concentrated salt solution (they contain charged ions such as sodium, potassium, chloride, calcium, phosphate, and/or ammonia), so there is less free water available to disperse dry formulations and other products in the tank mixture. You must be cautious when you mix multiple products in a liquid fertilizer solution, because compatibility problems are likely. Always conduct a jar test.

For example, when using 32 percent urea ammonium nitrate (UAN) liquid fertilizer, the high loading of fertilizer saturates the water. This makes it impractical for dissolving other products directly into the spray tank. If you were to add WDGs to 32 percent UAN, you could end up with partially dissolved granules that would block screens. This is why dry product formulations often recommend that you premix the product separately in water (sometimes called a *slurry* or *preslurry*) before you add it to a liquid fertilizer carrier.

Remember that the salts in liquid fertilizers can react with the inert ingredients in your product. Such reactions may cause incompatibility issues.

TWO MOST COMMON CARRIERS
Treat them differently

8. Account for Product and Water Temperatures

Water temperature can influence the rate at which products dissolve, disperse, emulsify, and flow. Think about adding sugar to your iced tea again. When you put sugar into cold tea, it settles directly to the bottom. However, when you add sugar to warm tea, the sugar dissolves quickly. In general, more material dissolves in warmer water than in colder water.

Dry formulations and liquid flowables take more time to disperse when the carrier temperature is cold. When mixing products into cold water, allow at least 3 to 5 minutes for each tank-mix partner to dissolve and disperse before adding the next product. Alternately, consider preparing a slurry in warmer water and then adding the slurry to the cold water. Visually inspect the spray mixture closely before adding other products, especially before tank mixing with other oil-based formulations.

Water-based formulations of pesticides, including suspension concentrates, concentrated oil-in-water emulsions, suspoemulsion premixtures (SEs), and even some soluble liquid (SL) products tend to thicken when cold. When cold, these products also pump and pour more slowly and often require more time to fully mix in cold water simply because they have a higher viscosity than when they are warm. When these products are warm and you add them to cold water, they generally dissolve quite well;

however, it may still help to slow the rate at which you add the product to the spray mixture and slightly increase the agitation.

Oil-based products—including ECs, oil dispersions, and crop and seed oils—are the most sensitive to water temperature. Very cold (or very warm) spray water tends to change the product's emulsification behavior. In cold water these formulations can form gels that are difficult to disperse. Also, temperature extremes tend to lead to less stable emulsions that separate more rapidly when left unagitated. In worst cases, this can also lead to product oiling in the spray mixture—oily product residues form on spray tank surfaces.

Generally, it is best to apply oil-based products shortly after you mix them. If you don't agitate such mixtures, it makes matters worse and can contribute to tank-mix compatibility problems with other pesticides. In addition, you may need to clean out your tank more rigorously before using it again.

And cold water can cause problems with water-soluble packets. These packets work a lot like the prepackaged dishwashing soaps that readily dissolve in warm water. The packaging dissolves more slowly in cold water. If the packaging does not completely dissolve, it can plug nozzle and pump intakes. Thus, use extra caution if you use very cold (<50°F) or very warm water (>80°F). If you must use cold or warm water, be prepared to add some mixing time.

GENERAL RECOMMENDATIONS FOR MIXING ORDER

In the past, pesticides and similar products were liquid formulations that contained a single active ingredient—they were SL, flowable (F), or EC formulations. New formulation types can contain multiple active ingredients and be applied at lower rates. The technical complexity of some of these newer formulations can create new and unexpected tank-mix compatibility concerns.

Formulations like ECs can be used at lower rates (ounces per acre rather than quarts) because they use solvents to dissolve the active ingredients. However, as described earlier in this chapter, if oil-based products (or ECs) are added before other products, interactions may result, especially with WDGs.

Therefore, it is important to understand mixing order. As we've already noted, formulations change, so you must constantly examine key tank mixes before applications. Effectively mixing multiple products is not that difficult as long as you follow the label information and understand a few basic mixing principles. And if you haven't made a particular tank mix before, you can test the mixture in a jar before messing up hundreds of gallons (see "Jar Test"). In the following, we present some general recommendations for tank mixing order when using water as a carrier and when using liquid fertilizer as a carrier. Remember: Always wear personal protective equipment when handling pesticide products.

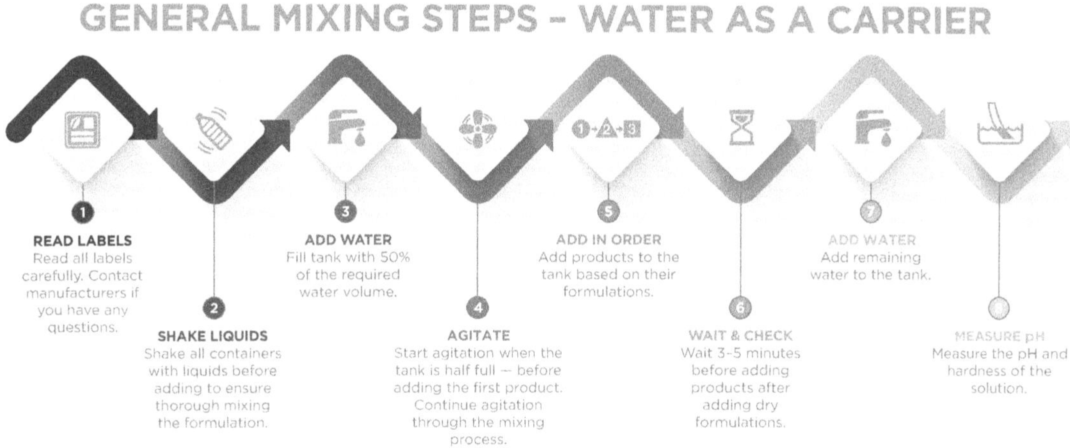

When Using Water as the Carrier

When you use water as a tank-mix carrier, follow this general mixing order (also see the "General Mixing Steps—Water as a Carrier" graphic).

1. **Read all product labels.** First, know what kind of formulation the product is. Some products will list their formulation types after the brand names, but many do not. If necessary, contact the manufacturer to determine the pesticide formulation type. Second, look for label information about optimal water pH, the influence of hard water, and the need to add adjuvants like surfactants or oils. Finally, you should always defer to product labels—some may call for a different order than that described here.
2. **Shake all liquid product containers.** Before you dispense product into the tank, shake the container to thoroughly mix the active ingredient and inert ingredients together. Bulk tanks require recirculation, and totes need to be recirculated, or preferably remixed with a tote mixer.
3. **Fill the spray tank with 50 percent of the required water volume.**
4. **Start the agitation and continue through the mixing process.** Never add multiple products to the inductor at the same time. Before adding the next product, flush the inductor with clean water.
5. **Add products based on formulation type in the order shown on p. 262.**
6. **Wait 3 to 5 minutes before adding products after dry formulations.** Dry formulations include WDGs, soluble granules (SGs), and dry flowables (DFs). If you use water-soluble packets, visually ensure that the bags have fully dissolved and the product has dispersed before you add more products to the spray solution.
7. **Add remaining water.** You should continue agitation until the tank mixture appears uniform.
8. **Inspect solution for signs of incompatibility such as separation.**

ADD PRODUCTS IN THIS ORDER

WHEN USING WATER AS A CARRIER

01 WATER SOLUBLE PACKETS

Make sure that water soluble packets fully dissolve and disperse in the water before adding other products to the mixture.

02 DRY FORMULATIONS

Give more time for dry formulations — such as copper, water dispersible granules (WDG, WG), wettable powders (WP), and soluble granules (SG) — to absorb water. This ensures that the product dissolves and disperses properly in the mixture. SG products will dissolve upon dilution.

03 AMMONIUM SULFATE

Add dry or liquid ammonium sulfate (AMS) only.

04 DRY OR SOLID ANTI-DRIFT AGENTS

Add dry or solid anti-drift agents.

05 COMPATIBILITY AGENTS & ANTI-FOAMERS

After adding these products, wait 2–3 minutes before adding other products.

06 DISPERSED LIQUID FORMULATIONS

Add liquid formulations that have a dispersed active ingredient:

- Suspension concentrates (SC)
- Flowables (F or FL)
- Suspo-emulsions (SE)
- Emulsions in water (EW)
- Micro-capsule (CS)

07 LIQUID DRIFT RETARDANTS

Add polymer-based drift retardants before EC formulations. This allows the polymer to absorb more available water and avoid clumping.

08 REMAINING LIQUID FORMULATIONS

Add the remaining liquid formulations:

- Emulsifiable concentrates (EC) — these oil- or solvent-based products require more time to disperse throughout the solution
- Oil dispersions (OD)
- Solutions and soluble liquids (S or SL)

09 ADJUVANTS

Add adjuvants:

- Crop oil concentrates (COC)
- High-surfactant oil concentrates
- Methylated seed oils
- Nonionic surfactants
- Spreader-stickers
- Water conditioning agents (with or without NIS or oil)

10 MICRONUTRIENTS & LIQUID FERTILIZERS

Add micronutrients and liquid fertilizers:

- Zinc
- Boron
- Manganese
- Fertilizer
- Copper
- Acidifers

9. **Measure the pH of your spray solution.** Measuring the pH lets you determine the proper levels of water conditioners or ammonium sulfate to add. In general, add pH adjusters at the end of the mixing process to ensure that the final spray solution is in the pH range specified by the product labels. In other words, determine the actual pH of the tank mixture and its desired pH before you add a pH adjuster. But use pH adjusters with care. Acidifiers can lower the pH of the tank mixture too much, causing some active ingredients to precipitate out of solution or to volatilize.

Keep in mind that defoamers, pH adjusters, and compatibility agents are not subject to typical mixing-order protocols. Product labels will indicate if such products are required. Sometimes a label will direct you to add these additional products to the tank just before you add the pesticide.

Antifoamers suppress foam during the mixing process. Do not add the entire amount of antifoamer to the tank mixture at the beginning; instead, divide into two or three parts. Add the remaining amounts at the end of the mixing process when the tank is nearly full.

When Using Liquid Fertilizer as the Carrier

When you use a liquid fertilizer (32-0-0, 28-0-0, 10-34-0) as a tank-mix carrier, follow this general mixing order (also see the "General Mixing Steps—Fertilizer as a Carrier" graphic).

1. **Read all product labels.** First, know what kind of formulation the product is. Some products will list their formulation types after the brand names, but many do not. If necessary, contact the manufacturer to determine the pesticide formulation type. Second, look for label information about optimal water pH, the influence of hard water, and the need to add adjuvants like surfactants or oils. Finally, you should always defer to product labels—some may call for a different order than that described here.

2. **Shake all liquid product containers.** Before you dispense product into the tank, shake the container to thoroughly mix the active ingredient and inert ingredients together. Bulk tanks require recirculation, and totes need to be recirculated, or preferably remixed with a tote mixer.
3. **Fill the spray tank with 50 to 75 percent of the required fertilizer volume.**
4. **Never premix products with fertilizer in an inductor.** Instead, premix them with clean water. Some products are not designed to be added directly to liquid fertilizer mixtures.
5. **Add products based on formulation type in the order shown on the next page.**
6. **Add remaining fertilizer to the sprayer.**
7. **Inspect solution for signs of incompatibility such as separation.**
8. **Continue to agitate mixture until it is applied.**

Figure 11.15. An example of an inductor.

Figures 11.16–11.17. (Left) When you add a dry herbicide directly to 32 percent UAN, it sits on the top. (Right) Even after agitation, the herbicide looks like it is not fully solubilized in the solution. This shows the need to preslurry the herbicide with water before adding it to the tank mix.

Figures 11.18–11.19. (Left) The herbicide was prediluted in water and then added to UAN fertilizer. (Right) After agitation, the diluted herbicide mixture combined with the fertilizer carrier to create a more uniform solution that could be sprayed without clogging nozzles.

ADD PRODUCTS IN THIS ORDER

WHEN USING FERTILIZER AS A CARRIER

01 PRESLURRIES OF WSP & DRY FORMULATIONS

Make a preslurry for water soluble pouch packages (WSP) and dry formulations in clean water. Dry formulations of copper, water soluble packets, water dispersible granules (WDG, WG), wettable powders (WP), and soluble granules (SG) must be dispersed in water before adding to the fertilizer carrier.

02 PRESLURRIES OF DRY ANTI-DRIFT AGENTS

Preslurry dry drift retardants separately in water before adding to the ferilizer carrier.

03 COMPATIBILITY AGENTS & ANTI-FOAMERS

After adding these products, wait 2-3 minutes before adding other products.

04 DISPERSED LIQUID FORMULATIONS

Add liquid formulations that have a dispersed active ingredient:

- Suspension concentrates (SC)
- Flowables (F or FL)
- Suspo-emulsions (SE)
- Emulsions in water (EW)
- Micro-capsule (CS)

05 LIQUID DRIFT RETARDANTS

Add polymer-based drift retardants before EC formulations. This allows the polymer to absorb more available water and avoid clumping.

06 REMAINING LIQUID FORMULATIONS

Add the remaining liquid formulations:

- Emulsifiable concentrates (EC) — these oil- or solvent-based products require more time to disperse throughout the solution
- Oil dispersions (OD)
- Solutions and soluble liquids (S or SL)

07 ADJUVANTS

Add adjuvants:

- Crop oil concentrates (COC)
- High-surfactant oil concentrates
- Methylated seed oils (MSO)
- Nonionic surfactants (NIS)
- Spreader-stickers

08 MICRONUTRIENTS

Add micronutrients:

- Zinc
- Boron
- Manganese
- Liquid copper

JAR TEST

Tank mixtures are susceptible to variations in chemistry, equipment, and water quality. To ensure that the products will mix together well, you should conduct a jar test before mixing in the tank. Instead of mixing (and potentially wasting) large volumes of products in the spray tank, the jar test allows you to test product combinations for compatibility first.

Jar tests are particularly important when you mix products in a liquid fertilizer carrier and when you mix more than two products (pesticides, adjuvants, fertilizers, micronutrients) in water—especially if you are mixing products you have not used before.

USE THE SAME WATER

Use the same water source and temperature for the jar test that you will use in your tank mix. You should actively practice this for all ingredients, since bulk tanks may be colder during mixing than a small sample from a 2.5-gallon jug mixed indoors.

A jar test simulates what occurs in a tank mixture and will provide evidence of physical incompatibilities such as separation, settling, inversions, and oil residue buildup. Always follow product label directions on how to conduct jar tests for specific products, because different products may require different water volumes.

Figure 11.20. When you use less water, many products will be incompatible (left bottle). When you start off with more water, products mix better (center and right bottle).

TEST FAMILIAR MIXES, TOO

You may use the same brand of products from one year to the next. However, manufacturers often change the inert ingredients in their products. These changes might alter how the product interacts during mixing. So even if you're making the "same" tank mix you've always prepared, you should still conduct a jar test to help identify potential compatibility issues resulting from even slightly different inerts.

How to Conduct a Jar Test

The labels for most products provide instructions for jar tests—they usually involve measuring products with cups, teaspoons, and tablespoons into glass or plastic containers. However, commercial compatibility test kits are also available. If you use a commercial test kit, always follow its instructions.

To perform a jar test without a kit:

1. Measure 17 ounces of water into a 1-quart glass jar. Use the same water you will fill your spray tank with.
2. Determine how much of each spray component to add to your jar using tables 11.1 and 11.2.
3. Add the spray components to the jar in the order provided in table 11.3. Stir the mixture thoroughly after adding each component.
4. After mixing all the components, let the solution stand in a ventilated area for 10 to 15 minutes and observe the results. Look for any signs of incompatibility. These signs include large flakes, gels, sludge, separation, and precipitation or signs of chemical reaction (such as heat or strong odors). If possible, let the mixture sit for longer, 30 minutes or overnight, and check again. If you do not see any physical signs of incompatibility, test the mixture with a spray bottle on a small area where you intend to apply it. Observe the results.

Remember, a compatibility test will only reveal the physical incompatibility of products in a tank mix. A compatibility test will not reveal any other form of antagonism, such as products that might inactivate one another or cause phytotoxicity in plants. The only way to know about potentially incompatible products for sure is to read and follow all product labels, ask the pesticide manufacturers, or contact university specialists.

Table 11.1. **Liquid Measures** to Add to a Jar Test

Emulsifiable concentrates (E, EC), suspension concentrates (SC), flowables (F), liquids, and solutions[1].						
AMOUNT TO ADD TO 1 PT OF LIQUID IN JAR	**APPLICATION RATE OF FINISHED SPRAY** (gallons/acre)					
	25	**20**	**15**	**12**	**10**	**5**
1/8 teaspoon	4 fl. oz. per acre	3.3 fl. oz. per acre	2.4 fl. oz. per acre	1.9 fl. oz. per acre	1.6 fl. oz. per acre	0.8 fl. oz. per acre
1/4 teaspoon	8 fl. oz. per acre	6.7 fl. oz. per acre	4.8 fl. oz. per acre	3.8 fl. oz. per acre	3.2 fl. oz. per acre	1.7 fl. oz. per acre
1/2 teaspoon	17 fl. oz. per acre	13 fl. oz. per acre	9.6 fl. oz. per acre	7.7 fl. oz. per acre	6.4 fl. oz. per acre	3.3 fl. oz. per acre
1 teaspoon	2 pts. per acre	27 fl. oz. per acre	19.2 fl. oz. per acre	15.4 fl. oz. per acre	12.8 fl. oz. per acre	6.7 fl. oz. per acre
1 ½ teaspoon	3 pts. per acre	38.4 fl. oz. per acre	28.8 fl. oz. per acre	22.4 fl. oz. per acre	19.2 fl. oz. per acre	9.6 fl. oz. per acre
2 teaspoon	4 pts. per acre	51.2 fl. oz. per acre	38.4 fl. oz. per acre	30.4 fl. oz. per acre	25.6 fl. oz. per acre	12.8 fl. oz. per acre
1 tablespoon	6 pts. per acre	76.8 fl. oz. per acre	57.6 fl. oz. per acre	46.4 fl. oz. per acre	38.4 fl. oz. per acre	19.2 fl. oz. per acre

[1] Table used with permission of Precision Laboratories.

Table 11.2. **Dry Measures** to Add to a Jar Test

Wettable powders (WP), dry flowables (DF), soluble powders (S), water dispersible granules (WDG)[1].						
AMOUNT TO ADD TO 1 PT OF LIQUID IN JAR (Level Spoons)[2]	**APPLICATION RATE OF FINISHED SPRAY** (gallons/acre)					
	25	**20**	**15**	**12**	**10**	**5**
1/8 teaspoon	4 fl. oz. per acre	3.2 fl. oz. per acre	2.4 fl. oz. per acre	1.9 fl. oz. per acre	1.6 fl. oz. per acre	0.8 fl. oz. per acre
1/4 teaspoon	8 fl. oz. per acre	6.4 fl. oz. per acre	4.8 fl. oz. per acre	3.8 fl. oz. per acre	3.2 fl. oz. per acre	1.6 fl. oz. per acre
1/2 teaspoon	1 lb. per acre	12.8 fl. oz. per acre	9.6 fl. oz. per acre	7.7 fl. oz. per acre	6.4 fl. oz. per acre	3.22 fl. oz. per acre
1 teaspoon	2 lb. per acre	1.6 lb. per acre	1.2 lb. per acre	15.4 fl. oz. per acre	12.8 fl. oz. per acre	6.4 fl. oz. per acre
1 ½ teaspoon	3 lb. per acre	2.4 lb. per acre	1.8 lb. per acre	1.4 lb. per acre	1.2 lb. per acre	9.6 fl. oz. per acre
2 teaspoon	4 lb. per acre	3.2 lb. per acre	2.4 lb. per acre	1.9 lb. per acre	1.6 lb. per acre	12.8 fl. oz. per acre
1 tablespoon	6 lb. per acre	4.8 lb. per acre	3.6 lb. per acre	2.9 lb. per acre	2.4 lb. per acre	19.2 fl. oz. per acre

[1] Table used with permission of Precision Laboratories.
[2] Assumes a product weight of about 0.5 oz. per tablespoon.

Table 11.3. General Mixing Order for a Jar Test[1]

ORDER	INGREDIENT
1	**Sulfonylurea herbicides**[2]
2	**Compatibility Agents** Buffers Acidifiers
3	**Dry Products** Ammonium sulfate (AMS) Water-soluble packets (WSP) Wettable powders (WP) Dry flowables (DF) Water dispersible granules (WDG) Soluble powders (S)
4	**Liquid Products** Suspension fertilizers/micronutrients Flowables (F) Suspension concentrates (SC) Drift retardants Micro-encapsulated (ME) Emulsifiable concentrates (E, EC) Solutions and soluble liquids Liquid fertilizers Chelated micronutrients Growth hormones
5	**Adjuvants** Crop oil concentrates (COC) High-surfactant oil concentrates (HSOC) Methylated seed oils (MSO) Nonionic surfactants (NIS) Water conditioning agents

[1] Table used with permission of Precision Laboratories.

[2] Add sulfonylurea herbicides before reducing pH, or create a slurry in 5 gallons of water at pH 8.5–9.5, and then add the slurry to the water once completely dissolved. Use 2–4 ounces of household ammonia in 5 gallons of water to raise pH to desired level.

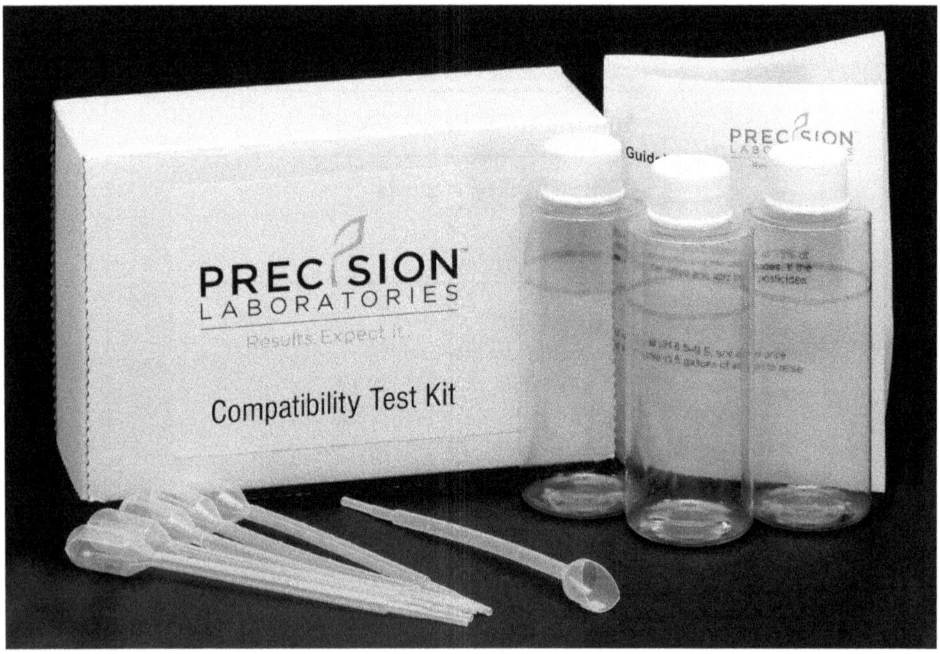

Figure 11.21. A commercial jar test kit.

One example of a test kit is the Precision Laboratories Compatibility Test Kit (https://www.precisionlab.com). The instructions for this kit are based on 1 milliliter (mL) representing 1 gallon of liquid. Thus, 100 mL is equal to 100 gallons of water. The clear plastic polyvinyl chloride container that comes with the kit holds 120 mL but is marked at the 100 mL level. The mixing container is calibrated based on the number of acres a 100-gallon tank will cover.

Physical incompatibility problems revealed by jar tests may include the following:

- The spray mixture is not uniform.
- Material settles to the bottom unusually fast.
- Particles float on the surface of the solution.
- Residues accumulate on the sides of the container.
- Large particles are present that will not pass through screens, filters, or spray nozzles.

We recommend that you keep detailed records for specific tank mixture preparations. Keep track of problem tank mixes (so you avoid repeating them) and successful mixtures for easy reference in the future. Be sure to note the following:

- Carrier (water or fertilizer)
- Water source (pond, city, well)
- Water temperature
- Water volume at the time you placed the first product into the tank (half or three-quarters of a tank)
- Fertilizer source (32% UAN, 28% UAN)

- Carrier volume applied per acre
- pH and hardness of water before and after tank mixing
- Product trade name(s)
- Formulation types
- Mixing order
- Rate of each product applied
- Agitation time between adding crop protection products

DEALING WITH INCOMPATIBILITY PROBLEMS

You cannot resolve many incompatibility problems by simply adding more water or adding a compatibility agent to the mixture after the fact. Therefore, contact the dealer or the product manufacturer(s) to assess available options. A manufacturer's representative might ask:

- What does the tank mixture look like?
- Which specific brand-name products did you use?
- What was the mixing order?

In some cases, fixing the problem may be as simple as adding more water, detergent soap, nonionic surfactant, compatibility agent, or pH adjuster. Collect some of the failed tank mixture and add a compatibility agent, pH adjuster, hot water, or detergent to see if that fixes the problems.

Figure 11.22. This photo shows the mess you can create in a tank if you mix multiple herbicides together with little to no water (left bottle). The center bottle confirms that diluting the incompatible mixture will not resolve the issue. Only mixing products in the correct order (right bottle) will yield a homogeneous and sprayable tank mix.

However, if a compatibility problem is not isolated to the spray tank (and is associated with nozzles, pumps, and hoses), then you may need to partially or completely dismantle the sprayer to ensure you have removed all of the residue from the equipment. Some tank-mix compatibility problems may be irreversible. Even if you manage to get the product back into what appears to be a sprayable form, you should be asking, "Will the mixture still work?" You should directly ask the manufacturers these questions.

HOW TO AVOID INCOMPATIBILITY PROBLEMS WHEN TANK MIXING

CONCLUSION

This chapter provided guidelines for proper mixing. The techniques we described will ensure that you successfully mix products without any problems. The time to experiment with an untested tank mix is not on the day of application. If you find yourself in a situation where you simply cannot get the products applied in time and attempt to put more products in the tank (something you've never tried before), you are taking a major risk. Once you mix products together, there's no recovery. You have essentially jumped from the plane without your parachute! Thus, applicators who prepare tank mixtures containing multiple products should plan for success each step of the way.

■ ■ ■

12

The Impact of Water Quality

With contributions from Donald Penner, Bill Johnson, Larry Bledsoe, Norm Wagoner, John Garr, Kiersten Wise, John Obermeyer, and Arlene Blessing

Effective pesticide applications require attention to factors that influence product performance: product selection, label instructions, equipment calibration, and application timing. Applicators also learn about product performance from trial and error and from industry and university recommendations. Experience and education allow applicators to see what happens to control programs when rates are changed, products are compared, and various application techniques are attempted.

However, one factor that doesn't get much attention is the quality of the water used to spray the product. Water often comprises 95 percent (or more) of the spray solution. What effect might it have on product performance? Research clearly shows that the quality of the water used for spraying can affect how pesticides perform.

Figure 12.1. Water may contain minerals that can impact water quality.

Figure 12.2. It is often difficult to visually observe pesticide reactions to water pH. But this pesticide mixture changed color when using water with a different pH, which may indicate a chemical compatibility problem. The manufacturer must determine whether the color change reduces the product's efficacy.

So why do we seldom notice something as obvious as the quality of water used in the spray tank? For the most part, water is viewed as a relatively clean input; if it runs clear, we don't give much thought to its purity. Concise, easy-to-read information on how water quality affects pesticide performance is scarce.

What kinds of problems can poor water quality cause? Water quality parameters such as acidity and dissolved minerals can interact with the active and/or additive ingredients of the pesticide product. Poor water quality can adversely influence the pesticide by reducing solubility and decreasing absorption by the target pest, resulting in inferior performance and the need for retreatment.

Reduced product performance may not be obvious. In some cases, water may reduce the pesticide's effectiveness only slightly, yet enough that tolerant or tough-to-control weeds, insects, or diseases are not well controlled. An applicator might blame the pesticide, add another product to the tank mix, blame other factors for lack of performance (e.g., weather, resistance), or increase the application rate, thereby masking water's effect on the product's performance.

Checking water quality is important. Time spent addressing the quality of water used in the spray tank can pay big dividends. This chapter provides an overview of water quality and related factors known to affect pesticide performance. Testing methods and options to improve the quality of the water used are discussed.

BASIC TERMS USED TO DESCRIBE WATER QUALITY

Water is a simple molecule composed of two hydrogen atoms attached to one oxygen atom. It's one of nature's most remarkable liquids, capable of dissolving or suspending minerals and organic matter.

The chemical characteristics of water may change slightly (or significantly in some cases) as it moves through the atmosphere, through the soil profile, or over the soil surface. For example, rainwater falling to earth turns slightly acidic as it interacts with atmospheric gases. The slightly acidic water may turn alkaline as the calcium in the limestone layers in the ground solubilizes in the water as it percolates downward. Understanding what is in the water can help determine its quality and its suitability for pesticide application.

Debris. This includes material floating or suspended in water, such as leaves, sticks, seeds, and other solid waste materials. Debris generally is found only in water obtained from lakes, ponds, rivers, or ditches. Appropriate filters can remove debris and prevent the plugging of spray tips.

Suspended Solids. Materials such as silt, clay, and organic matter can be suspended in water but will settle to the bottom of a tank if the water is left undisturbed. *Turbid* is the term for water in which suspended solids can be seen floating.

Dissolved Minerals. All water sources in nature contain dissolved minerals such as calcium, magnesium, and iron. Unlike suspended solids, these minerals do not settle. We all have heard, "My water tastes great, just like the water I grew up with." It is the minerals dissolved in water that give it a good taste. Distilled water, from which the minerals have been removed, tastes flat.

The concentration of minerals such as calcium, magnesium, and iron in water is important in describing water quality. Technically, water hardness is a measurement of the total amount of calcium and magnesium ions in water: The greater the concentration of these and similar minerals, the harder the water. Conversely, water becomes soft as dissolved calcium and magnesium ions are replaced with sodium or potassium ions.

Total hardness is measured in parts per million (ppm) or in grains of calcium and magnesium per gallon of water (see table 12.1). One grain (65 milligrams) corresponds to approximately 17.1 ppm. Example: A water sample could contain 20 grains per gallon. A common aspirin is about 5 grains in weight, so 20 grains would be equivalent to four aspirin tablets dissolved in each gallon of water, or 342 ppm.

Table 12.1. Categorization of Water Hardness Concentrations

Parts per Million (ppm)	World Health Organization Water Classification
0–114	Soft
114–342	Moderately hard
342–800	Hard
>800	Extremely hard

Sodium and calcium chlorides also affect some pesticides but often are overlooked; their presence results in water described as *saline*. Saline water is common in arid regions, especially where crops are irrigated; in areas of salt-rich soils; and along seacoasts. Saline water is not detected using standard tests for hard water.

Figures 12.3–12.5. (Top left) Two bottles of distilled water with the minerals removed. (Top right) Adding calcium carbonate to the bottle on the right. (Right) Calcium carbonate dissolved in the water.

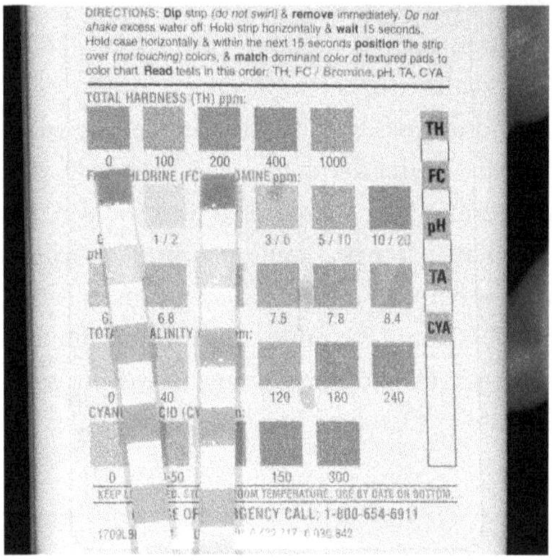

Figure 12.6. Using a test strip to measure hardness. The hardness of the distilled water (left strip) is 0 ppm, while that of the water with the dissolved calcium (right strip) is between 200 and 400 ppm. The pH did not change.

Water pH: Acidic or Alkaline. The pH value describes the acidity (concentration of hydrogen ions) or alkalinity of any solution. The scale ranges from 0 to 14:

pH < 7 ... considered acidic (low pH)

pH = 7 ... considered neutral

pH > 7 ... considered alkaline (high pH)

Small changes on the pH scale represent large changes in acidity. For example, a pH of 5 is 10 times more acidic (i.e., contains 10 times more hydrogen ions) than pH 6 and 100 times more acidic than pH 7. This 10x relationship between the values is the same going both up and down the pH scale (e.g., a pH of 9 is 100 times more alkaline than a water solution measured at pH 7).

Table 12.2. Examples of pH Values for Common Substances

pH Value	Substance
14.0	sodium hydroxide
12.6	bleach
11.5	ammonia
10.2	milk of magnesia
9.3	borax
8.4	baking soda
8.0	seawater
7.4	human blood
7.0	distilled water
6.8	tea
6.7	milk
6.0	atmospheric water
5.0	pickle juice
4.5	tomatoes
4.2	orange juice
4.0	wine and beer
2.8	vinegar
2.2	lemon juice
2.0	stomach acid
1.0	battery acid
0.0	hydrochloric acid

Figure 12.7. Sensitive paper strips allow you to evaluate the water that will be used in your spray tanks.

WATER CHEMISTRY IMPACTS PESTICIDE PERFORMANCE

Since water normally is used to deliver a chemical to the target pest, it should be considered the foundation for the application process, but pesticide applicators often ignore (or underappreciate) water quality. Whether the water is from a well or from a lake, stream, or pond, it may be the deciding factor between ineffective and optimum product performance.

Poor water can lessen the effectiveness of the pesticide in the spray mixture, influence spray quality, and affect how the pesticide interacts with the atmosphere and the leaf surface. Here are some examples of water quality's effect on pesticide performance:

- Certain pesticide molecules can be deactivated when they bind with cations (positively charged molecules) in the water.

- Spray water that is too acidic or too alkaline may degrade some pesticides or make them less soluble in the mixture.
- Smaller spray droplets may evaporate after exiting the sprayer nozzle before they even reach the target.
- The wind can blow smaller spray droplets away from the intended target (this is called *spray drift*).
- Spray droplets can bounce, shatter, or roll off the leaf surface (a lack of spray retention).
- Spray droplets that remain on the leaf surface may form a bead and fail to spread.
- The deposit formed after the spray evaporates may not have the right chemistry or enough active ingredient present.
- The active ingredient contained in the spray residue on the leaf surface may fail to penetrate the leaf.

All these potential failures link to the molecular, chemical, and physical properties of the water in the final spray mixture. Spray deposit retention on the leaf and the chemical composition of the spray droplet ultimately determine the effectiveness of the pesticide application. Here are some specific circumstances that hinder pesticide performance:

Turbidity. Suspended, positively charged organic pesticides are attracted to and bind with negatively charged particles found in water. Some products (e.g., glyphosate) bind to suspended sediments, rendering the product unavailable for plant uptake. One label states, "Product performance may be significantly reduced if water containing soil sediment is used as a carrier. Do not mix this product with water from ponds or ditches that is visibly muddy or murky."

Pesticides have indexes called the soil sorption coefficient (Kd) and the soil organic carbon sorption coefficient (Koc). Both coefficients reflect how strongly the pesticide

Figure 12.8. The bottle on the left contains distilled water with zero hardness; the bottle on the right contains hard water. A material that mimics glyphosate was added to both bottles of water. Notice that the water on the left remains clear, indicating the added product is in solution. The water on the right is cloudy, indicating that the calcium has bound to the glyphosate mimic. Chemical incompatibilities such as these are not always visible.

binds (adsorbs, or sticks) to soil particles and particles suspended in water; the process is called *adsorption*. Herbicides with high Kd or Koc values bind tightly to soil as well as sediment and organic matter in water. So, the more sediment and organic matter in the water, the less herbicide remains available to bind to the soil or be taken up by plant tissue. The Kd or Koc values can be obtained by asking the manufacturer for product-specific information about the chemical and physical properties of the pesticide.

Water Hardness. Water hardness can affect some pesticides negatively. As in magnets, opposite charges attract: Negatively charged pesticide molecules attach to the positively charged iron, calcium, and magnesium molecules (cations) in hard water. The binding of pesticides with these minerals creates molecules that cannot enter the target pest, that enter at a much slower rate, or that precipitate out of solution.

The following cations, if present in water, can cause problems and may contribute to water hardness. They are listed in the order of greatest potential to bind to pesticides:

- aluminum (Al^{+++})
- iron (Fe^{+++}, Fe^{++})
- magnesium (Mg^{++})
- calcium (Ca^{++})
- sodium (Na^{+})

The chemical characteristics of the pesticide change once the pesticide recombines with positively charged ions such as calcium or magnesium. The label for one herbicide indicates that a water conditioner "may increase the performance of this product on annual and perennial weeds, particularly under hard water conditions."

In one sense, the more the pesticide is bound to minerals, the more "diluted" the product becomes in the tank. In some cases, the chemically altered molecule may be unable to dissolve in water, penetrate the leaf tissue, attach to the site of activity in the pest to disrupt biological functions, or perform as a pesticide. These effects are not limited to the spray tank environment but extend to the spray solution on the leaf surface, which can affect product uptake.

Water pH. Pesticides normally are formulated as weak acids or as neutral to weakly alkaline products. As a general rule, herbicides, insecticides, and fungicides perform best in slightly acidic water, pH 4–6.5. Pesticides such as the sulfonylurea herbicides perform better in water that is slightly alkaline (above pH 7). When water pH falls outside of the preferred upper or lower boundaries, product performance can be compromised. In some cases, the pesticide can fall out of solution.

The pH of the solution also can influence how long a pesticide molecule remains intact. A pH that is higher or lower than optimal causes some pesticides to begin degrading or hydrolyzing. When a weakly acidic pesticide is placed in water that is slightly acidic, it stays largely intact. Certain insecticides and fungicides have been shown to break down in alkaline water, and the effect of pH usually proceeds faster as the temperature of the water increases.

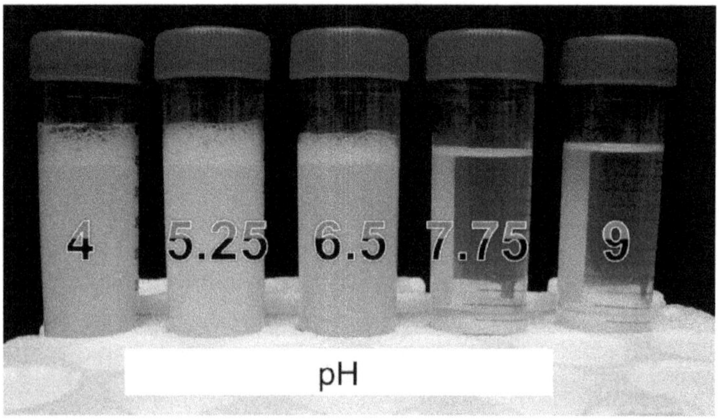

Figure 12.9. A solution's pH can dramatically affect the solubility of some products. Sulfonylurea herbicides are far more soluble in alkaline solutions (jars on the right) than acidic solutions.

Many products have a weak electrical charge. The pH also can change the chemical charge of a pesticide molecule, limiting its ability to penetrate the leaf cuticle and reach the site of action, thus reducing its efficacy.

The Properties of Water Molecules

Water is anything but simple. Water is a biologically active molecule with chemical and physical properties that make it an amazingly complex compound. Here are eight properties of water that you should know.

1. Water Is Not Pure

We assume water from a spigot is pure because it looks clean and tastes good. But that water typically contains varying amounts of dissolved minerals such as calcium and magnesium—they give water its taste. We further assume that if water is suitable for drinking, then it's adequate for mixing pesticides. This is not necessarily true.

Calcium, magnesium, and iron dissolve in water as it moves through the soil profile (in the case of water from wells) or moves across the surface (in the case of surface water). So most water isn't pure because it contains some amounts of these dissolved minerals. Water's interaction with the environment determines whether it is hard or soft, acidic or alkaline. All of these factors can affect pesticide performance.

2. Water Quality Can Vary Dramatically Within a Few Miles

Most people know water is significantly different across regions of the United States. The source of the water, limestone deposits, well depth, recharge area, and annual rainfall amounts all influence water quality. Even water from two sources that are close to one another can differ significantly in chemical composition. The properties of surface water may be completely different from those of the groundwater in the same location, even on the same farm. These differences are not obvious to the naked eye and can affect pesticide performance.

3. Water Is a Polar Compound

Water (H_2O) consists of two hydrogen atoms (H) and one oxygen atom (O). The oxygen atom in water has the effect of drawing the electrons it shares with hydrogen atoms. As a result, the electrons spend relatively more time in close proximity to oxygen.

Electrons are negatively charged, so they make the oxygen atom somewhat more negative. At the same time, each hydrogen atom becomes somewhat more positive. This unequal charge makes water a polar compound—one side is positive, the other negative, like the two sides of a magnet. Because of these polar properties, water is extremely attractive to other polar compounds, including itself. This property allows water to act as both a solvent and a carrier for pesticides. It allows water to form a cohesive lattice.

Water has strong *cohesive* properties (interconnectedness) rather than *adhesive* properties (like a paste). Think about water droplets running down a windshield. As a droplet moves down and hits another, both droplets merge into a single drop. This interaction is stronger than the spreading of the same droplet across the windshield and is responsible for the effect we call *beading*.

Because of this, there are even times when water can move up or against gravity. For example, when the upper layer of soil begins to dry, capillary action can wick water up into the soil profile, due to water's affinity for itself. So, as water evaporates at the surface, the strength of association between water molecules draws more water from below the surface up to replace what was lost.

4. Water Is an Excellent Solvent

Water's polarity allows some products to disperse and dissolve in it. Polar pesticides are attracted to water just as water is attracted to itself. When this attraction occurs between water and a polar organic solid, the solid material is said to go into solution. Dissolving sugar or salt in water illustrates water's effectiveness as a solvent. The molecule of the herbicide glyphosate is polar, so it dissolves in water like table salt would. Because it bonds with water, glyphosate will dissolve and remain in solution without further agitation. It is said to form a true solution in water.

Some pesticides are oil products that are insoluble in water. These products are significantly less polar than water molecules or completely nonpolar (that is, they are neutral or have no charge). Instead of dissolving in water, these products are dispersed in it and never form a stable mixture by themselves, even after thorough mixing. Without agitation, nonpolar pesticides will stratify into distinct layers on the top or bottom of a spray mixture, depending on their density relative to water. This is why manufacturers formulate some pesticides with emulsifiers to prevent separation.

5. Water Has High Surface Tension

Have you ever noticed that a water droplet sitting on a leaf is almost always spherical? That's because the properties of water and air create surface tension, the force that allows a liquid to resist an external force (like air). The surfaces of plants also play a role.

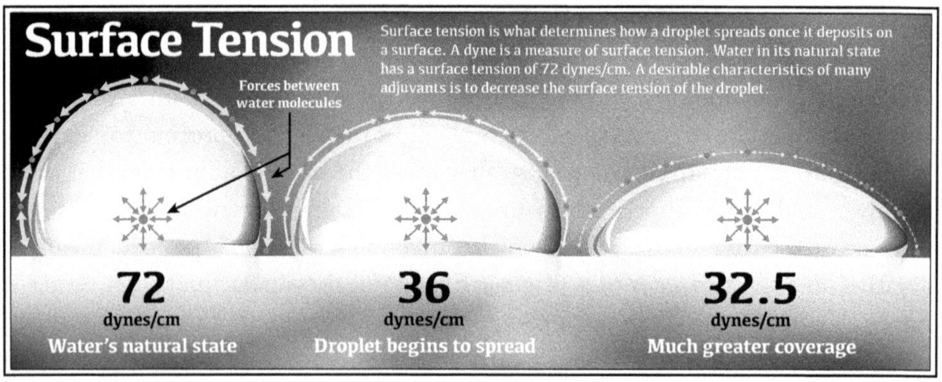

Figure 12.10. The left droplet has a high contact angle, which means it covers less of the leaf surface. The right droplet has a low contact angle, which means more coverage.

As long as a droplet has high surface tension, the water droplet will remain upright and not spread over the leaf. Such droplets have a high contact angle, which means the droplet covers less surface area. A lower contact angle means the droplet covers more surface area.

For pesticide applications, when a water droplet doesn't spread and has a high contact angle, it is generally perceived as a poor application. Too much spreading and too low a contact angle may also present problems. Finding the right balance of spread and contact angle is important for the specific pesticide.

6. Water Is a Volatile Molecule

Volatility is the tendency of a substance to vaporize. Water is a volatile ingredient in spray applications, especially when the relative humidity is low. Volatility affects the spray droplet in two ways. First, the spray can evaporate before it contacts the plant. Second, the droplet on the leaf can evaporate before the pesticide has a chance to enter into the plant. Neither is desirable.

7. Water Has No Inherent Buffering Capability

A solution that resists changes in its pH is a called a *buffer*. For example, a pH 5 solution is acidic. Let's say you add a pesticide into the mix that has a mildly alkaline pH of 7.5. If the solution's pH remains at or near 5, then the original solution is considered a buffer because it can resist the shift in its pH.

In contrast, water is not a buffer and will take on the characteristics (pH, buffering) of the products placed in it. As a solvent without buffering capabilities, water's pH can readily shift up or down if you add even a small amount of an acidic or alkaline material. The degree of pH change in a water-based pesticide spray mixture depends on how much material you add and the material's actual pH and solubility. Manufacturers often formulate a given solution's pH to optimize performance. For example, Monsanto optimizes the formulation of Roundup® so the solution has a pH of 4.5–5.5.

The Properties of Pesticide Molecules

Each pesticide has unique chemical and physical properties that can influence how it interacts with the leaf surface and carrier (water or otherwise). Many pesticides respond to differences in pH, dissolved minerals, and water temperatures. Here are four things to remember about the properties of pesticide molecules.

1. Pesticide Solubility and Degradation Rates Vary

A spray solution's pH plays an important role. The solution's pH determines how much of the pesticide will go into solution (solubility) and how quickly the pesticide will degrade or break down once dissolved.

Each pesticide molecule has a pH range that is optimal for its solubility and chemical stability in water. A pesticide's half-life describes how long it takes for half of a pesticide to be destroyed or degraded. Changes in pH can make a pesticide's half-life range from minutes to months. For example, some carbamate insecticides have a half-life of 10 minutes in alkaline solutions. And sulfonylurea herbicides are far more soluble in solutions that are more alkaline.

2. Opposites Attract

Negatively charged pesticides bind to positively charged molecules in water. For example, the active ingredient in formulated glyphosate has a negative charge, so it associates with minerals in soil or in water that have a positive charge. The associations can be strong or weak. The stronger the association, the more likely it is that pesticide performance will be poor.

When pesticides bind to certain minerals (particularly iron, calcium, and magnesium) plants do not absorb them as well. This is because the pesticide, after binding to the minerals, may have become too large to penetrate the plant.

Frequently, the combined molecule (pesticide and mineral) has a reduced solubility, which may result in solid particles precipitating out of the spray mixture. When this happens, less of the active form of the pesticide is available. This process is called *hard water antagonism*—the pesticide is still there but in a form not readily available to the plant. The more minerals the water contains (the harder it is), the more certain pesticides will be deactivated by binding to minerals like calcium, magnesium, and iron.

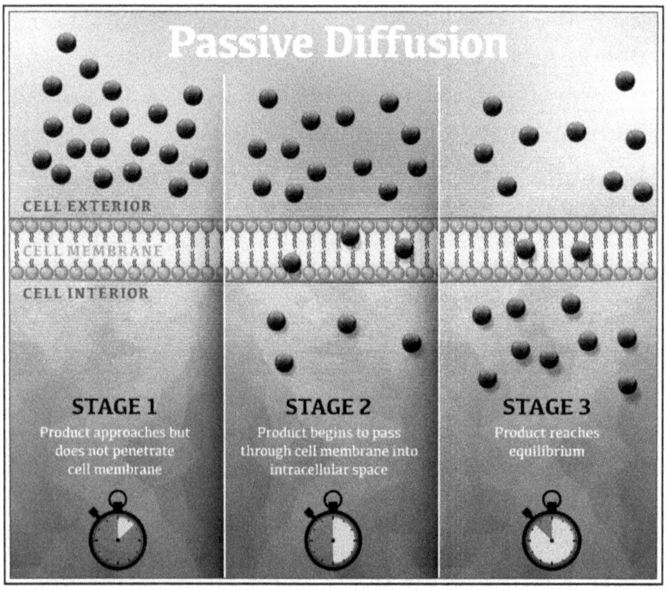

Figure 12.11. During uptake, pesticides move into the plant by passive diffusion out of the spray deposit, across the leaf cuticle or cellular membrane, and into the leaf.

3. Pesticides Enter Plants Through Passive Diffusion

Many contact insecticides and fungicides work at or near the leaf surface and have limited movement through plant tissue. All herbicides and locally systemic insecticides and fungicides must enter a leaf to be effective.

Systemic herbicides must, after entering living plant cells, move relatively long distances in the plant to reach the growing points in the roots and shoots. Successful pesticide movement in the plant depends on many forces, including pH, solubility, and whether it moves in the phloem or xylem. Successful absorption of systemic pesticides depends on:

- Getting the right size of spray droplets
- Delivering droplets uniformly across leaf surfaces
- Keeping droplets on leaves long enough

Systemic pesticides enter plants through a process called *passive diffusion*, which means that the pesticide moves from a region of higher concentration (typically the spray deposit) to a region of lower concentration (typically the inside of a plant). You can see passive diffusion at work if you place a drop of dye in a glass of water. It takes only a few minutes for the dye to disperse throughout the glass—it is moving from a higher to lower concentration. You've also experienced passive diffusion if you've ever caught the aroma of coffee or freshly baked bread drifting from the kitchen to another room.

Gradually, the pesticide molecules that have been dissolved in the spray solution migrate into the plant. The time it takes for an effective amount of the pesticide to move into the plant can vary greatly depending on various factors:

- The pesticide and its properties
- The plant species
- The plant's leaf structures (see chapter 13 for more on this)
- The thickness of wax layers on the leaf

4. Environmental Factors Play a Role

Environmental conditions at the time of a spray application affect pesticide performance—especially when they influence the rate of evaporation. When humidity is low, water droplets in a spray solution can evaporate quickly, leaving behind a dry pesticide residue on the leaf surface. In this dry, solid phase, pesticides with polar characteristics (most commonly water-soluble pesticides and their salts) either are unable to penetrate the leaf surface or penetrate it slowly. Pesticides with nonpolar characteristics may still penetrate the leaf, but movement is much slower than when the deposit is still in liquid form.

WHAT DO PESTICIDE LABELS INDICATE ABOUT SPRAY APPLICATION WATER?

The pesticide label may or may not specify the need for water conditioners, additives, or adjuvants. The following are some pesticide label statements addressing water quality.

Regarding suspended solids and turbidity:

- "Reduced results may occur if water containing soil is used, such as visibly muddy water or water from ponds and ditches that is not clear."

Regarding water hardness:

- "The addition of 1–2 percent dry ammonium sulfate, by weight, or 8.5–17 pounds per 100 gallons of water, may increase performance of product."
- "Recommended nitrogen-based fertilizers include liquid fertilizers (such as 28% N, 32% N or 10-34-0) and may be applied at the rate of 1.25–2.5 percent per 100 gallons of spray solution. Spray grade ammonium sulfate may be used at 12–15 pounds per 100 gallons of spray solution."

Regarding water pH:

- "Do not add pH-adjusting agents."
- "Do not use nonionic surfactants or other additives that alter the pH of the spray solution below pH 5; spray solutions of pH 6.0–8.0 are optimum."
- "Do not use spray additives that reduce pH of the spray solution to below 3."
- "Do not use with spray additives that alter the pH of the spray solution below pH 5 or above 9, as rapid degradation can occur."
- "The spray mixture pH should be 4 to 7 for good efficacy."
- "Best results obtained at spray solution pH of 6.0–8.0."
- "Product is unstable under highly alkaline conditions. Do not use this product in water with pH values above 8.0 unless a buffer is added. If necessary, water should be buffered to neutral (pH = 7.0) before adding this product to the spray tank." (In this example, *buffering* means lowering the pH and holding it there initially; it may return to the original pH over time because of the water chemistry.)

In some situations, a pesticide label may prohibit the use of water conditioners or other additives. One label states, "Do not apply tank mixes that contain ammonium sulfate (AMS) or crop oil concentrate to any food/feed crop listed on this label. For food/feed crop uses, do not use liquid fertilizers that contain ammonium sulfate (AMS) as a source of nitrogen, as commodities derived from the crop may contain residues that exceed established tolerances." Another example: "Do not add buffering agent or pH adjusting agent to the spray solution when this herbicide is the only pesticide used."

Applicators could assume that you don't have to worry about the mixing water and product performance if the manufacturer has not addressed this issue on the label. But look at the half-life (fate) of the product in the spray solution under different conditions. For instance, one product might have a half-life of 2 minutes at pH 9 but a half-life of 10 hours at an acidic pH of 5 (see table 12.3). Thus, water quality is important even when the label does not address the issue.

TESTING THE WATER

Determining water quality and choosing a specific water conditioner and amount to use requires knowledge of the water to be used for the specific application. Testing the water is the key to ensuring the best performance of the spray application.

Temperature. Thermometers can provide temperature information quickly. If formulation stability can be affected by temperature extremes, alternate water sources may be required. Another approach is to store the water in the sprayer or tank, indoors or outdoors, until the water reaches the desired temperature.

Figure 12.12. Herbicide in a pond. Notice the cloudy water, which is an indication of pesticides tied up by hard water or sediment. This ultimately reduces the effectiveness of the pesticide product on weeds in the pond. The product label states, "The harder the water or the greater the algae concentration, the higher the required dose of this product."

Suspended Solids. Whether the water is from a well or a surface source, solids can settle in the tank and cause equipment problems. Sedimentation and filtration can be used to remove suspended solids for some applications; otherwise, finding an alternative water source may be necessary. Glass jar tests of standing water can indicate the likelihood of a problem with suspended solids.

Dissolved Minerals. The composition of water (e.g., hardness, pH, and iron) can vary widely among wells in close proximity to each other. What is dissolved in water depends on the composition of the soil and the underlying bedrock. The depth of the well and type of aquifer also influence water quality. Each well is unique in terms of its water chemistry and turbidity.

Water in creeks, ponds, and reservoirs can differ greatly. The dissolved minerals and suspended sediments from all sources flowing into a reservoir or pond are mixed together, giving the water a unique chemistry profile. Filtration is recommended whenever surface water is used.

Water quality can be checked anytime. Addressing water quality a few weeks prior to the pesticide application season will give you time to purchase kits, ask questions, and evaluate whether the water needs to be improved for the pesticide chemistries you will be using. Water also may be tested quarterly. After a few years of testing, you will probably notice that the water quality doesn't change much. Surface water is much more variable than underground sources; thus, underground sources often are easier to manage.

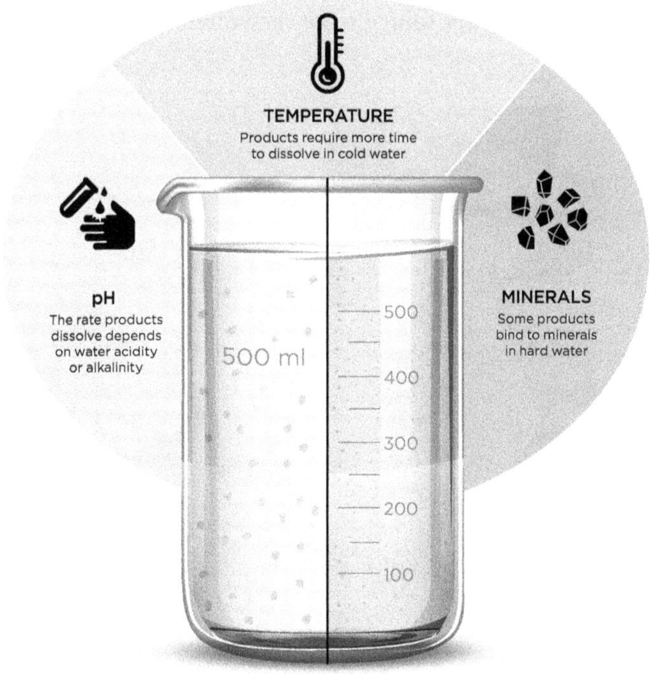

IMPACT OF WATER ON CROP PROTECTION PRODUCTS

Water-Testing Approaches

Two options for testing the water are hiring a commercial vendor and purchasing do-it-yourself water-testing kits.

Take a Water Sample to a Professional. Farmers and businesses such as nurseries, greenhouses, and lawn care and pest control companies often use the same water source throughout the season. If you use the same water source consistently, you can plan ahead by having the water tested by a professional in your community who manages water, water quality, and water testing. Examples include local companies whose primary business is to condition water for industrial, research, and residential customers (e.g., drinking water and swimming pools).

These companies routinely conduct tests for iron, pH, and water hardness; they can provide a complete analysis of your water. If you choose this route, be sure to ask the vendor the following:

- Can you test for iron, pH, and hardness?
- What will be the cost?
- How much water will you need to run the suite of tests?
- Do you have guidelines or special containers for collecting and transporting the water samples?

Use Do-It-Yourself Test Kits and Meters. Industries such as right-of-way, aerial, and forestry pest management use multiple water sources as they travel over wide geographic areas to apply pesticides. They generally don't have enough advance notice to arrange testing of each water source since they often don't know where their next source of water will be.

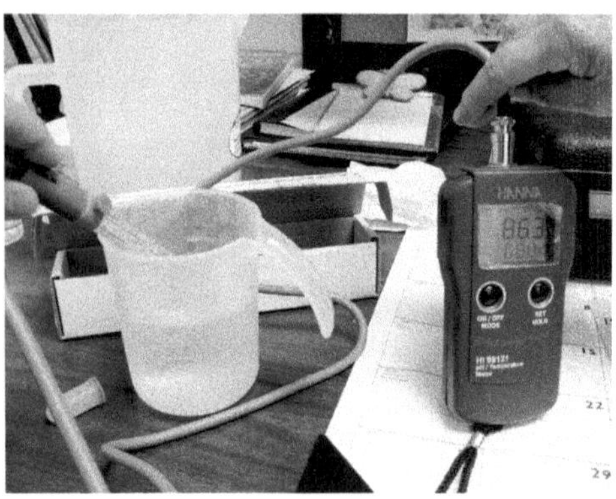

Figure 12.13. The pH of this municipality-treated water used in a greenhouse is 8.6.

Figure 12.14. Checking water quality with a pH meter.

Numerous water-testing kits are commercially available for both spontaneous and scheduled testing. The kits are readily available, reasonably priced, easy to use and interpret, and reliable. The majority of the test kits use color-changing, sensitive paper to document water hardness, pH, and iron levels. Surface water should be filtered and stored in a bulk tank before testing, whereas collecting a water sample from a faucet or a well is rather straightforward. Let the water run for a brief time; then, with the water running, rinse a clean pint or quart glass container. Fill the container with water.

Sensitive paper provides pH readings that are less precise than those of more sophisticated meters. A meter might indicate a water pH of 8.4, while the same water shows a pH of 8 when colors are matched using the sensitive paper method. Some color charts use whole numbers for pH and broad concentration categories for iron and water hardness (e.g., 50–150 ppm), but the results are adequate for analyzing pesticide spray water.

There are meters available that provide quick and accurate readings without having to carry multiple kits; these may be useful if you fill tanks from multiple sources. Whether it is a pH meter or something more elaborate that also tests for iron and hardness, one of these instruments offers a simple, efficient, and effective method of examining water sources at a reasonable cost.

SOLVING WATER QUALITY PROBLEMS

Herbicide labels recommend rates that perform across a wide spectrum of conditions: small weeds, large weeds; good water, bad water; high temperatures, low temperatures. Higher rates overcome variation in performance associated with products over wide geographic areas. When you are using lower product rates, the quality of water can play a more important role in effective weed, insect, and disease control.

Test results from each water source form the basis for your decision on whether or not to condition your spray water. The purpose of conditioning water is to maximize the effectiveness of the pesticide. Broadly defined, water conditioners are added to the spray solution or tank mix to eliminate problems associated with water hardness. A pH buffer is used to raise or lower the pH, depending on the range needed for optimum performance.

Some pesticide formulations contain water conditioners that make them compatible with a wide range of water conditions. Other products, however, perform better when adjuvants are added to overcome water quality issues (see chapter 13 on adjuvants). Correcting the pH or solving hardness problems depends on the particular pesticide label restrictions or requirements as well as the target species. The use of water conditioners is advised under the following conditions:

- It is recommended on the pesticide label.
- The pesticide label specifies the quality of the water (e.g., the water pH range) to be mixed with the pesticide.

- A pH between 4 and 7 is needed for insecticides, fungicides, and most herbicides.
- Herbicides in the sulfonylurea family, such as Steadfast®, Classic®, and Harmony®, perform best when water with a pH of 7–8 is used.
- A weak acid herbicide is used, and the water hardness exceeds 150 ppm.
- Iron levels exceed 25 ppm, and hardness plus iron exceeds 400 ppm when herbicides are used.
- A weak acid herbicide (e.g., glyphosate, glufosinate) is used, and the target weeds include velvetleaf and common lamb's-quarter (regardless of water hardness, as these and other weed species have sufficient Ca++ in and on the leaves to reduce activity of these herbicides).
- Make sure that water is perfectly clear when the Koc of the pesticide's product exceeds 800. This value is for glyphosate, for which turbidity reduces efficacy.

Numerous commercial water conditioners are available, but some have not been proven effective. The pesticide label may be very specific as to the water conditioner and application rate to use. Some manufacturers of weak acid herbicides do not recommend adding a water conditioner; instead, they may promote applying the highest rate of product indicated on the label.

The user should check the efficacy of the water conditioners in the same manner that seed variety, pesticide products, and fertilizer rates are tested. The money spent for any product—including water conditioners—should return a value to the purchaser. Ask the vendor for data showing that the water conditioner works. If none are available, consider applying the pesticide, with and without adding a water conditioner, to the targeted pest within a small test area and evaluating the differences in control as well as crop injury.

Are There Any Special Concerns About Conditioning Water? Concerns about using water conditioners may include the order of introduction into the spray tank and their action as tank cleaners (see chapters 11 and 13).

Does Introducing the Water Conditioner First or Last Make Any Difference? Few products have been evaluated to determine how—or if—the order of introduction into the tank impacts their performance. However, you can't go wrong by conditioning the water first, especially under the following conditions:

- Products have a low use rate.
- Multiple products will be tank mixed.
- The rates selected are among the lowest listed on a label.
- Problems have been observed in the past.

Do Some pH Modifiers Act as Tank Cleaners? Products that condition pH levels generally acidify the water (most herbicides are weak acids). There are a few products that raise the pH level for use with pesticides that perform best when combined with water that is alkaline, but be careful using these products because they might also "clean" the tank. Ammonia-based products increase the pH of the spray tank solution to between 10 and 12, depending on the rate used. These highly alkaline solutions tend to pull out of the tank and solubilize into the mix any residues left from previous contents. Make sure to thoroughly clean the tank, prior to raising the pH level, even when using water conditioners.

LEAVING PESTICIDES IN THE TANK: HOW LONG IS TOO LONG?

Even when water characteristics are ideal, most pesticides break down in water over time, through a chemical process called *hydrolysis*. Some break down slowly, while others become inactive within a few hours. We express how quickly a pesticide product loses its effectiveness as its *half-life*; that is, the time it takes for half of the active ingredient to break down in water. At that point the product would not provide the desired efficacy.

We need to know how long products can stay in the tank before undergoing changes that impact efficacy. The pH of the tank water often has a dramatic effect on the breakdown process. Manufacturers formulate their products to stay in slightly acidic water for 24 hours with little appreciable reduction in effectiveness. But this general statement does not hold up for all products.

If we were to assume that treated municipal water is the perfect mixer because it is safe to drink, we would be wrong. Municipalities create safe drinking water by adding products such as chlorine that disinfect by killing harmful bacteria, and these treatments often shift pH levels into the alkaline range, typically pH 7.8–8.5.

Let's assume we are using municipal water with a pH of 8.2. Consider how the pesticide products in table 12.3 react to changes in the pH. Assume that you mixed a pesticide product, but weather conditions caused a five-day job delay. Depending on the water pH, the product in the tank may have broken down and become ineffective.

Table 12.3. Product Half-Life at Different pH Values

	pH 9	pH 7	pH 5
Brand X herbicide	10 minutes	17 hours	16 days
Brand X fungicide	2 minutes	3 hours	10 hours
Brand X insecticide	24 hours	10 days	stable

Some labels bear statements that provide information on stability at specific pH levels. Here are some examples:

- "Spray preparations are stable if they are pH neutral or alkaline and stored at or below 100°F."
- "Do not let spray mixtures stand overnight."
- "Apply the spray the same day it is prepared, while maintaining continuous agitation."

The University of California, Davis, has come up with the following guidelines based on water pH:

- A pH between 3.5 and 6 is satisfactory for most spraying and short-term (12–24 hours) storage of most mixtures in a spray tank. It is not suitable for sulfonylurea herbicides.
- A pH between 6 and 7 is adequate for immediate spraying for most pesticides. Do not leave the spray mixture in the tank for more than 1–2 hours, to prevent loss of effectiveness.
- Most products mixed in alkaline water should be sprayed immediately.

Figures 12.15–12.18. The use of water for application is nearly universal for all pesticide products. The tank is filled with water (first); pesticide product is metered into the tank (second); a spreader/sticker adjuvant is added to the mixture (third); and a water conditioner is added to improve the quality of the water (fourth).

CONCLUSION

Pesticide application processes require due diligence in keeping up-to-date with today's trends, tools, and technologies. The process of managing pests takes considerable money for salaries, products, equipment, and fuel. The paybacks are better results and value from your pesticide applications.

Water quality needs to be understood and managed first. Once you manage the quality, then you can follow label directions to provide the best results possible. Conversely, starting off with water that has quality issues can mean reduced pesticide performance even if the rig is calibrated and the application is perfect. It has been proven that poor water quality can have a negative effect on certain families of pesticides. Water quality also affects the uniformity of the mix in the tank. With so much at stake, the quality of the water used in your spray solution should be evaluated and considered just as you evaluate the purchase of spray equipment and the selection of pesticide products. A small amount of effort to test your water, or have it tested, is rather cheap insurance to ensure that products perform as promised and deliver the expected results.

13

Adjuvants and Pesticide Performance

With contributions from Gregory Lindner, Bryan Young, Donald Penner, Jason Deveau, Doug Linscott, Heping Zhu, Richard Zollinger, Eric Spandl, Bill Johnson, Kiersten Wise, Aaron Patton, Chuck Champion, Nick Harre, Norm Wagoner, and Kevin Leigh Smith

Leading by one run in the bottom of the ninth, the catcher calls for a fastball on the low outside corner of the plate. The pitcher throws a strike to win the game. With seconds to spare on the clock, the quarterback tosses a perfect spiral to his receiver, who slips past the defender in the corner of the end zone for the winning touchdown. The wing-back passes the ball to the striker, who kicks the ball just out of reach of the goalkeeper's hands to help her team win an overtime thriller. The difference between winning and losing in sports is often a single well-executed play.

When the game is over, everybody will talk about the star player's perfect pitch, pass, or kick. But executing that perfect play was the result of preparation and teamwork. At every level, success in sport doesn't rely on luck but on planning, execution, and teamwork.

In many ways, the perfect pesticide application is like that single perfectly executed play. Pesticide applications and sports both require preparation, knowledge, basic skills, and commitment. Both also require teamwork. The chemicals you mix in your tank have to work together—both the pesticides and the adjuvants, which are products that enhance pesticide performance. Applicators need to be on top of their game if they are to win the battle against the insects, weeds, and diseases that are formidable opponents in successful crop production.

Of course, one can only take the analogy between sports and pesticide application so far. Protecting health, property, crops, environment, and income is definitely not a game. Pesticide applications affect people's livelihoods and can lead to costly outcomes when the pesticide fails to control the targeted pests. For these reasons, every application must be the winning play.

This chapter describes how adjuvants can enhance pesticide performance and protect against the unwanted consequences of spray drift. The goal is to define what adjuvants do and highlight their value in improving pesticide applications (see also chapters 11, 12, and 14).

UNLEASHING A PESTICIDE'S POTENTIAL

Why does a pesticide work well one time and not so well the next when everything seems to be the same? Variable weather could play a role. Or the pest could be less susceptible to the active ingredient. At other times, an inconsistent pesticide performance seems inexplicable.

In many ways, pesticide performance is similar to the way a seed corn hybrid's yield varies from one year to the next or within a given field in the same year. The genetic pedigree of that hybrid determines its yield potential, but that doesn't necessarily mean it will produce the highest-possible yields.

Performance potential is useful when discussing pesticides too. Every pesticide molecule in a jug or bag has the built-in potential to control the pests listed on its label. You can maximize performance consistency when you appropriately match the right pesticide to the targeted pests, with the right timing; mix it at the labeled rate; and apply it with calibrated equipment under favorable environmental conditions. However, even if you make every attempt to account for these factors, pesticides can still fail to produce expected or consistent outcomes. For example, a pesticide may not perform as planned because the pest has developed resistance to it.

But other factors could also have affected the pesticide's performance. An important factor is the water in the spray solution. Conditioning the water that will carry the pesticide and modifying the droplet characteristics can often improve spray quality and potency, which increases pesticide performance and consistency.

Figure 13.1. Uncontrolled weeds will reduce crop yields.

Figure 13.2. Water beading on a waxy leaf is more likely to run off.

LEAF SURFACES AFFECT PESTICIDE PERFORMANCE

Leaf surfaces are one of the more complex biological structures known to scientists. Plant surfaces harvest light, breathe, exchange water, have natural defenses, resist physical forces, and more. A leaf surface's composition (waxy, hairy, and so on) plays an important role in how a pesticide spray droplet spreads over the leaf surface and penetrates it.

The outermost part of a leaf is called the *cuticle*, which is a protective film. The composition of any given plant's cuticle affects how the spray droplet interacts with leaves:

- How well spray droplets cover the leaf
- How well spray droplets are retained on the leaf surface
- How spray droplets behave
- How pesticides move into plants from retained and drying spray droplets

The interaction between leaf and droplets is crucial to the effectiveness of spray.

1. Spray Droplets Must Contact Leaves

It is more difficult to get spray droplets to hit and remain on narrow, upright blades of grass than on wide, horizontal broadleaf plants.

2. Plant Hairs and Other Structures Keep Droplets off Surfaces

Plant hairs (trichomes) can suspend droplets above the leaf surface until they evaporate. Trichomes that point vertically up from the leaf surface are typically coated with wax. Research shows that trichomes are not a major entryway into the leaf (although some may absorb slight amounts). The bottom line: Unless a spray droplet properly contacts the leaf surface, pesticide uptake will be minimal.

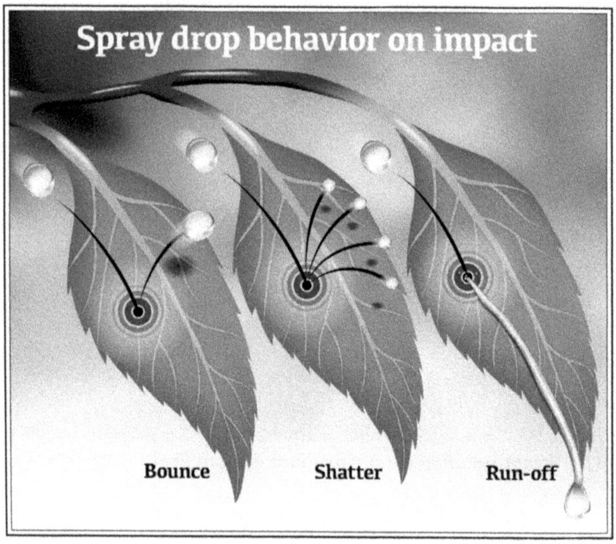

3. Leaves Must Retain Spray Droplets

Even when a pesticide spray droplet lands on a leaf, the droplet has to remain on the foliage to be effective. When a spray droplet leaves the spray nozzle, it has speed and mass, so it has momentum. At the moment of impact, the leaf surface acts like a shock absorber, bending and stretching as the spray droplet contacts it. After contacting the leaf, the droplet can remain on the surface, shatter, bounce off, or roll off the edge. How that droplet responds has as much to do with its physical properties as it does with those of the leaf surface.

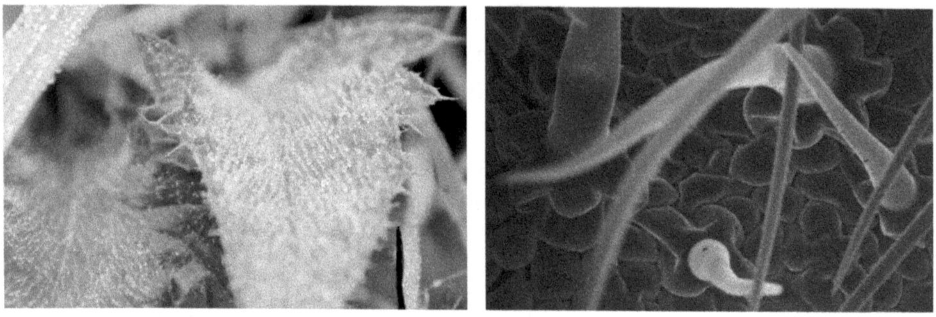

Figures 13.3–13.4. Leaves that have a dense layer of hairs (trichomes) make it difficult to get pesticide spray droplets onto the leaf surface.

SEE DROPLETS ON LEAVES

The USDA Agricultural Research Service has videos of spray droplets contacting leaf surfaces: http://www.ars.usda.gov/Services/docs.htm?docid=19299.

Figure 13.5. These spray droplets remained on the leaf surface after contact.

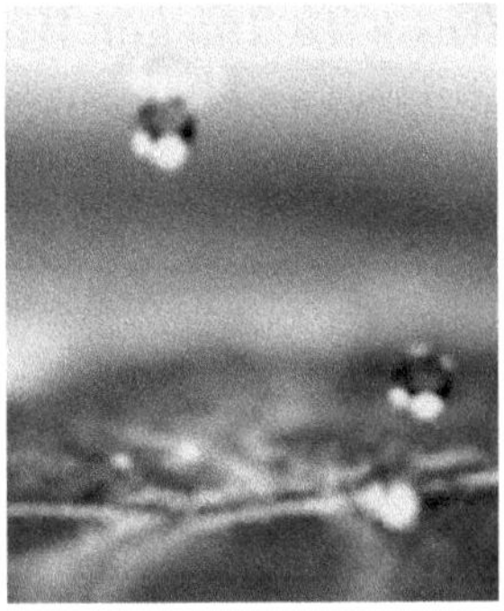

Figure 13.6. These spray droplets bounced off the leaf surface after contact.

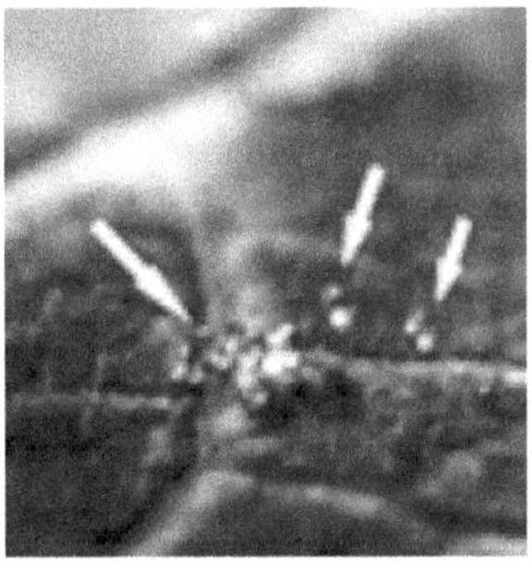

Figure 13.7. These spray droplets broke apart on the leaf surface after contact.

4. Waxes Repel Spray Droplets

The waxy cuticle on leaves can make it difficult for pesticides to enter plants (through passive diffusion) and for water to escape the plants. By itself, water can do little to carry a pesticide across the cuticle barrier and into the target.

Pesticide solubility and penetration can present one of the most significant barriers to pesticide uptake because leaf surfaces can repel water (they are hydrophobic). Polar water droplets and nonpolar waxes don't interact well; one molecule repels the other, as you can see with water on a waxed car. Less polar pesticides are more soluble in the waxy portions of the cuticle, which more readily allows dissolved pesticides to move across this barrier layer.

Just how quickly a spray solution can penetrate the cuticle depends on how much of the pesticide is in a dissolved state—solid or crystallized pesticides do not readily enter the leaf. Spray adjuvants can help improve pesticide uptake. With polar, water-soluble, organic, or inorganic pesticides, the challenge is to help the spray droplet deliver the dissolved pesticide across the impermeable outer leaf layer.

Figure 13.8. (Left) A surfactant can help pesticide spray droplets spread on waxy leaf surfaces. (Right) Surface waxes on a leaf cause water droplets to bead up.

5. Plants Protect Themselves with Thicker Wax Layers

Plants that grow during extended periods of high temperatures, low humidity, and prolonged drought can produce thicker coatings of surface wax. This process, known as *hardening off*, helps prevent plants from drying out. Hardening off makes it even more challenging for pesticides in a water carrying solution to stick to, spread over, and penetrate a plant's surface.

Plants may also alter the composition, thickness, and molecular arrangement of the wax in response to age, environment, and even leaf position on the plant. When foliar-applied herbicide applications fail to control the target during droughts, a thicker layer of wax on the plant's cuticle (which limits absorption) is commonly the culprit.

FUNGICIDES REQUIRE GOOD COVERAGE

Some fungicides have limited mobility—they don't spread across external plant surfaces well or translocate into plant tissue. Many common fungicides may penetrate leaf tissue and move only to the bottom surface of the leaf or into the leaf tip. Because of this, successful fungicide applications rely on good spray coverage. Fungicide product labels often recommend using a spray adjuvant to enhance the spreading, penetration, and mobility of the pesticide.

ADJUVANTS: OVERCOMING WATER-INDUCED BARRIERS

Once the pesticide solution is mixed and the spray is applied, you can no longer control its fate. Nothing can be added, and the only portions that can be removed are those volatile enough to evaporate as the droplets travel through the air or after they have hit the target surface.

Before spraying, though, you can incorporate one or more adjuvants into the spray mixture to influence how the spray behaves once it's left the nozzle. An *adjuvant* is a product that you add to a spray solution to improve the efficacy of a pesticide, the consistency of the solution, and/or the application safety (such as reducing drift). (See chapter 11 for the mixing order when adding adjuvants.) An adjuvant can overcome issues with:

- The properties and quality of the water in the solution (see chapter 12)
- The structure and composition of target plant surfaces
- Application equipment and conditions
- The environment in which the spray application is made

Adjuvants are not magic bullets; they can only do so much. But they can improve pesticide performance by modifying the spray application and the pesticide's interaction with the plant:

- Keeping a pesticide from binding to minerals suspended in water
- Adjusting the water pH so a pesticide is less likely to break down
- Manipulating droplet size to reduce on-target and off-site movement of pesticides
- Improving the odds that a spray droplet will stay on the leaf by reducing factors that cause droplets to bounce and roll off leaves
- Modifying or reducing surface tension to enhance a droplet's ability to be retained on or spread across the leaf surface
- Minimizing spray droplet evaporation
- Preventing spray deposit from being washed off the leaf surface
- Protecting the droplet from degrading in sunlight
- Improving a pesticide's absorption and uptake by the plant

ADJUVANT CLASSIFICATION AND NOMENCLATURE

You can't choose the proper adjuvant until you understand the full performance needs of each spray operation. Suppliers, distributors, and manufacturers have developed standardized terms to classify and describe different adjuvants, which are recognized by the American Society for Testing and Materials (ASTM International, or ASTM). Fortunately, the type names are descriptive enough that you can deduce what the adjuvant will do.

1. Water Conditioning Agents

Water conditioning agents (or water conditioners) are adjuvants that bond with dissolved hard water ions such as magnesium and calcium. They provide sacrificial molecules to keep pesticide molecules from bonding to the hard water ions. In chemistry this process is called *chelation*. Tying up these minerals allows the target plant to more readily absorb the pesticide.

The most common water conditioning agents are ammonium sulfate (AMS), dipotassium phosphate, or a low-dose alternative form of AMS. You can use dry "spray grade" AMS at a rate of 8.5 to 17 pounds per 100 gallons of water to counter the effects of hard water. Liquid AMS solutions at rates of 2.5 to 5 gallons per 100 gallons of water provide similar effects.

As the name suggests, low-dose alternatives to AMS require less product—about 2 to 6 pints per 100 gallons of water. The chemistry of these products allows them to sequester hard water cations, but product performance across brands is highly variable, and the low-dose rate is intended to be sufficient for average water sources. They may be ineffective for water that is significantly harder than average.

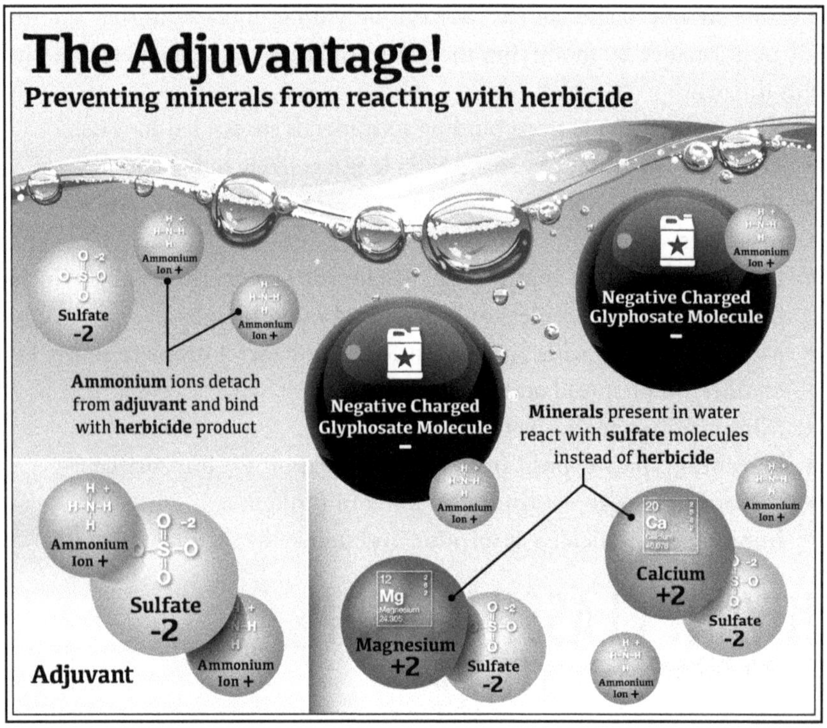

Ammonium sulfate (AMS) molecules break apart in water, allowing the sulfate to bond with the magnesium and calcium ions in hard water. For example, AMS allows the glyphosate molecule to form an ammonium salt, which target plants absorb more readily.

HOW MUCH AMS DO YOU NEED?

To determine how much ammonium sulfate (AMS) to add to overcome antagonistic ions, you must first test your water. Once you know how much potassium (K+), sodium (Na+), calcium (Ca+), magnesium (Mg+), and iron (Fe+) are in the water, you can calculate the dose this way:

(0.002 × ppm K+) + (0.005 × ppm Na+) + (0.009 × ppm Ca+) + (0.014 × ppm Mg+) + (0.042 × ppm Fe+) = pounds of AMS to add per 100 gallons of water

Let's say a water test of a farm well indicated that the water had K at 2.37 ppm, Na at 30.9 ppm, Ca at 55, Mg at 26.1, and Fe at 0.07.

Here is the calculation:

(0.002 × 2.37) + (0.005 × 30.9) + (0.009 × 55) + (0.014 × 26.1) + (0.042 × 0.07) = 1 pound of AMS per 100 gallons of water

Source: North Dakota State University

An alternative is to add a liquid or dissolved water conditioning agent that contains a chemical that will bind with calcium and magnesium. It must still be at least as effective as AMS for conditioning hard water but capable of being used at lower rates (such as 1 percent, depending on water hardness).

2. Surfactants (Wetter-Spreaders)

Surfactants (also called *wetter-spreaders*) lower the surface tension of spray droplets (measured in dynes per centimeter). This allows gravity to collapse and flatten the spherical spray droplet on the leaf surface, which allows droplets to spread over a larger area. Surfactants can also reduce the amount of bounce when a droplet hits a leaf, which allows for greater success at depositing.

Surfactants primarily work at the interfaces between the spray droplet, leaf surface, and surrounding air. However, they also can influence interfaces both inside and outside the leaf or the intended target pest. The surfactant adjuvants most commonly tank mixed with pesticides are called *nonionic surfactants* (NIS). These surfactants are often made with materials that have an affinity for waxy plant surfaces.

Organosilicones are a type of NIS that are commonly used as tank-mix adjuvants with pesticides. Organosilicones are often called *super-spreaders* or *super-wetters* because they decrease surface tension so much. They also may have penetrant properties. Organosilicones greatly increase droplet spreading and decrease droplet sizes, which may be more desirable for some active ingredients and application conditions and less desirable for others.

Surfactants are some of the most commonly recommended adjuvants on pesticide labels. This is especially true for water-soluble and systemic herbicides. An NIS primarily spreads droplets and can at times enhance the uptake of certain active ingredients. However, NIS is typically less aggressive in penetrating the leaf than adjuvants that contain oil.

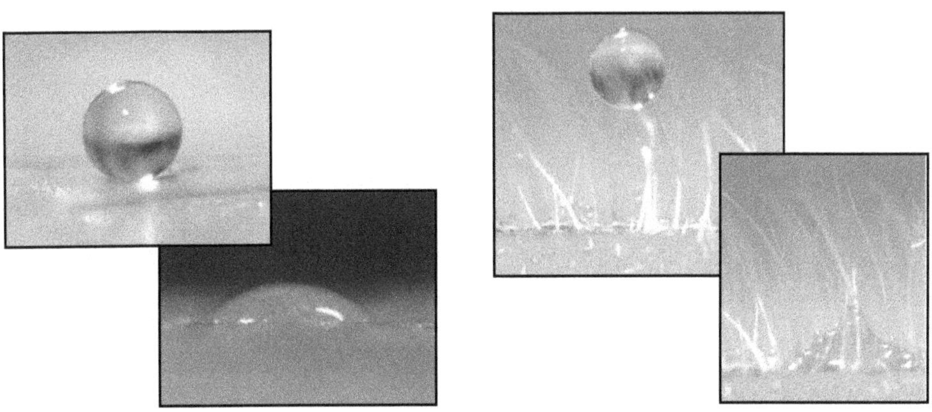

Figure 13.9. Adjuvants can reduce a spray solution's surface tension, allowing the spray droplet to spread and penetrate leaf surfaces. (Left pair) A spray droplet without an adjuvant (top) and one with an adjuvant (bottom). (Right pair) A spray droplet without an adjuvant suspended on leaf hairs (top) and one with a surfactant that allows it to reach the leaf surface (bottom).

In applications when an NIS improves penetration, the more effective approach is to match the pesticide's polarity with a surfactant of similar polarity. When you closely match a pesticide and adjuvant, like dissolves like, and you increase specific pesticides' rate of movement into the leaf. Many formulated pesticides already contain surfactants that do this very effectively.

Figure 13.10. The spray droplet on the left does not contain a surfactant. The droplet on the right contains a surfactant, which improves its coverage.

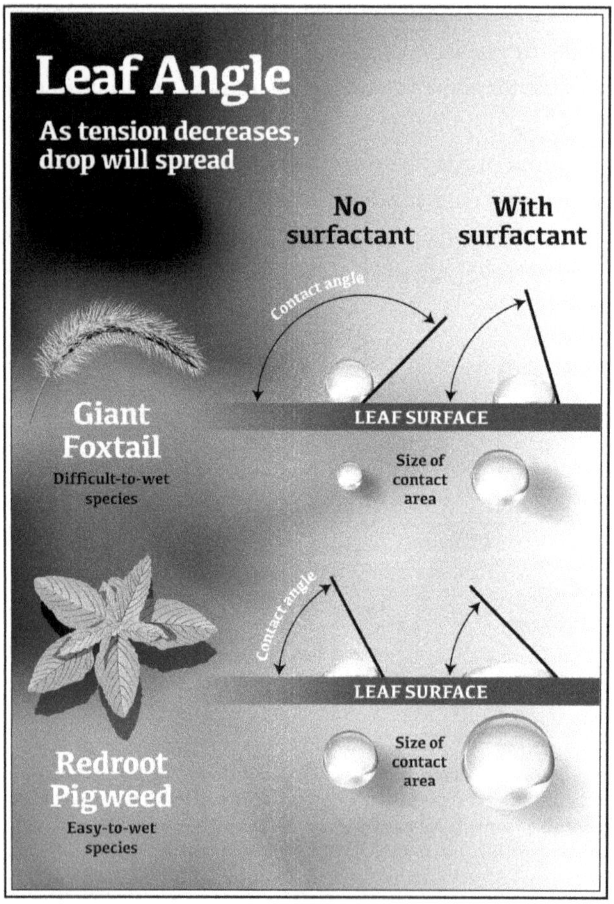

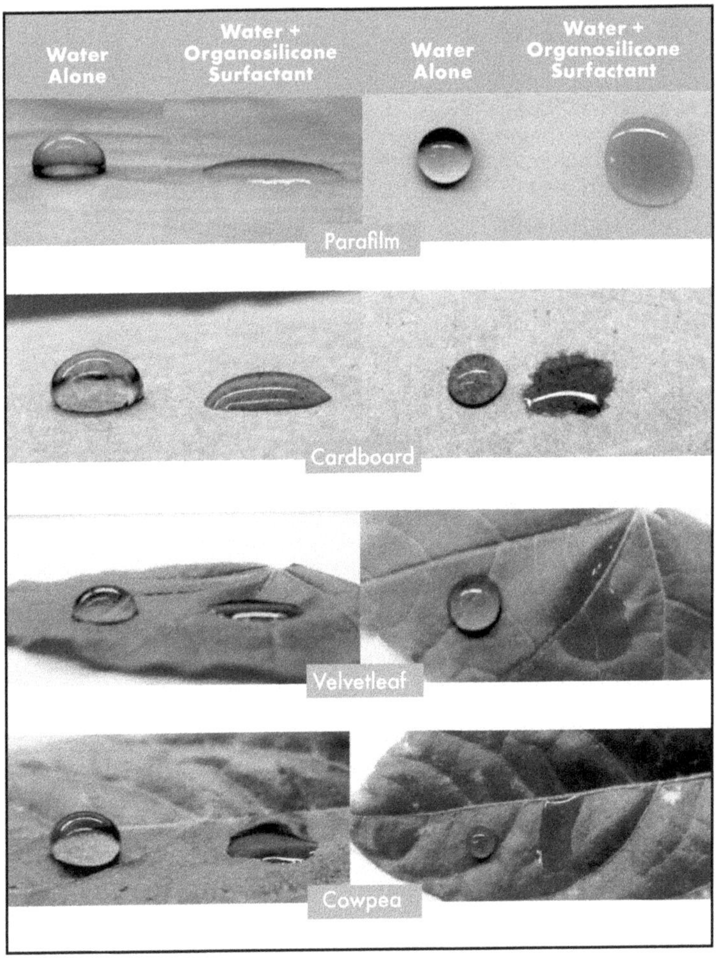

Figure 13.11. This illustration demonstrates how surfactants influence droplet spread on a variety of surfaces.

3. Oil Concentrates

Oil concentrates help pesticides overcome leaf barriers to penetrate leaves better. They soften or disrupt the waxes on leaf surfaces. There are three general groups of oil concentrate adjuvants (based on their ingredients): crop oil concentrates, modified (methylated) seed oils, and high-surfactant oil concentrates.

Crop Oil Concentrates. Crop oil concentrates (COCs) are adjuvants that contain petroleum-based (low-volatility) oils.

Modified (or Methylated) Seed Oils. Modified (or methylated) seed oils (MSOs) are adjuvants that are derived from plant seed oils. One example is biodiesel. In the United States, soybean oil is the most common MSO, although other oils can be used (such as canola, cotton, and sunflower). MSOs frequently contain 50 to 90 percent oil.

The nonoil components of these adjuvants are usually surfactants. MSOs typically do not perform as well as surfactants at promoting spreading when formulated at very high oil rates. For this reason manufacturers add more surfactants to these adjuvants to enhance wetting or spreading and to improve performance for water-soluble pesticides.

Figures 13.12–13.13. These photos demonstrate how an oil concentrate adjuvant helps a pesticide penetrate a leaf. (Left) A piece of paraffin was placed over water-sensitive paper that turns blue when it contacts water. (Right) The droplet on the left is water without an adjuvant. The droplet on the right contains an oil concentrate adjuvant. Notice the dark color that indicates where the drop containing oil penetrated the paraffin.

High-Surfactant Oil Concentrates. High-surfactant oil concentrates (HSOCs) were initially developed to improve control at lower use rates than for COCs and MSOs. HSOCs also were developed for use in tank mixes that contain glyphosate and are being applied to manage herbicide-resistant weeds.

Premium HSOC products may contain 20 to 40 percent NIS with 60 to 80 percent of the specific type of oil. These ingredients generally form more stable spray emulsions and may reduce use rates relative to comparable COC formulations that contain lower levels of surfactant. Other crop oils may only provide enough emulsifying surfactant to allow you to mix and spray the oil. The oil in these tank mixes tends to separate to the top of the spray mixture fairly rapidly once you stop mixing and agitating them.

4. Humectants (Evaporation Retardants)

Humectants, which can be synthetic or naturally derived, are designed to slow water evaporation in the air or on the leaf. Slowing the evaporation rate improves foliar uptake of some pesticides.

The longer the spray droplet stays in its liquid form, the more time certain pesticides have to penetrate the leaf surface. This is particularly true of highly polar and water-soluble pesticides in situations when solubility is critical for the pesticide to be available to move into the plant. Once a polar pesticide dries on the surface (crystallizes), it no longer diffuses as rapidly into the leaf. For pesticide products that are not rapidly absorbed, prolonging the spray droplet's drying time can be critical.

5. Compatibility Agents

Compatibility agents help incompatible products work together. They are most often used in complex tank mixtures that may include multiple crop protection products using multiple carrier systems. With so many possible product combinations, there will be occasions when products are incompatible (see chapter 11).

Always check product labels for information about the compatibility of spray materials before adding them to a spray mixture. *Compatibility* refers to the physical and chemical interactions between two or more products in a spray mixture. Incompatibility might be particularly likely when the application has a high concentration of dissolved fertilizer, water conditioners, or micronutrients.

Compatibility agents are often recommended in concentrated liquid fertilizer mixtures to allow effective mixing and to avoid rapid oil phase separation of some formulation types. Other formulated compatibility agents may allow otherwise incompatible pesticide mixtures to be prepared and sprayed to maximize user convenience.

Even with a compatibility agent, it is critical to add each product to the spray tank or inductor in the proper order (see chapter 11). For example, thicker products sometimes create mixing problems with other products. Another example is water-dispersible granules (WDGs). Whenever possible, you should always add WDGs first (that is, adding them to the water in the spray tank before other products). If you do not allow the WDG product to fully disperse in the tank before adding the next component, it can create sprayer problems because the formulation may coagulate or settle out in the mixture.

Remember, even if you perform a jar test to check compatibility (see chapter 11), that will only reveal the physical incompatibility of products in a tank mix. It will not reveal any other form of antagonism, such as products that inactivate one another or cause phytotoxicity in plants. The only way to know about potentially incompatible products for sure is to read and follow all product labels, ask the pesticide manufacturers, or contact university specialists.

6. Defoamers/Antifoamers

Defoamers/antifoamers are used to manage the trapping of air in a spray mixture while preparing, mixing, or applying it. Antifoamer agents are most useful at preventing foam from forming. Defoamer agents are generally added to a mix to reduce foam after it has formed. Some products inhibit foam from forming by reducing the surface tension of air bubbles, which allows them to revert to solution. Other products bridge and collapse the films that compose foam.

7. Scents (Masking Agents)

Scents (masking agents) conceal the noxious odors of the pesticide or other solvents in the mix. These adjuvants are concentrated fragrances that may include specific odor control agents that reduce the volatility of more objectionable compounds. Masking an odor is different from reducing a compound's volatility. Affecting volatility reduces the pesticide's potential for off-target drift.

Figure 13.14. Foam markers provide a visual clue to the applicator of where they have already treated.

8. Marking Agents

Marking agents (foam markers, spray pattern indicators) incorporate dyes or foams that allow applicators to see where they applied the spray mix. Applicators commonly use these adjuvants when spot-spraying or working with invasive species.

9. Formulated Tank Cleaners

Formulated tank cleaners, as the name suggests, clean the spray tank. There is no one-size-fits-all tank cleaner (always review product labels for recommendations). Some tank cleaners are designed to degrade or deactivate the active ingredient of a particular product. Others are detergents that help remove oily film residues from equipment surfaces, hoses, and pipes. Still other tank cleaners may release and remove mineral deposits in the spray equipment that contain pesticide residues.

Adjuvants that contain ammonia and strong detergents remove traces of unwanted pesticides in the spray tank. Ammonia, with its high alkalinity, breaks down acidic pesticide residues in tanks and hoses. The detergents have an affinity for oil that may help to pull materials off the interior surfaces of spray equipment and tank walls.

Ammonia and other materials also may degrade the remaining pesticide residues once removed from surfaces. Other more aggressive and less volatile materials may include sodium hydroxide or potassium hydroxide, which are very strong alkaline cleaners.

CAUTION

Never mix ammonia with chlorine bleach because it can produce toxic chlorine gas.

When you clean a tank or additional sprayer components (screens, strainers, lines), always make sure the cleaning solution remains in the tank long enough to effectively clean it, especially if you need to promote the degradation of pesticide residues. Always read and follow the tank cleaner label for specific directions about holding times.

If you want to remove hard water deposits and precipitated micronutrients from inside the equipment, include a chelating (or metal capture) agent. High alkalinity may reduce the performance of chelating agents since mineral deposits are less soluble under alkaline conditions. In such cases, a mild acid cleaner may also be useful as long as the equipment won't be harmed by it.

It is best to decontaminate some difficult-to-remove or highly active pesticide residues (such as those from sulfonylurea herbicides) with an aggressive alkaline cleaner or a suitable oxidizing agent. Some of these cleaning agents (including chlorine and peroxide bleaches) may be available as consumer products. However, always be sure these products are compatible with your spray equipment before you use them.

10. Stickers

Stickers help spray droplets better adhere to the target plant under windy or rainy conditions. Stickers typically incorporate latexes (the material used in house paint to form a uniform, waterproof layer) or resins that help the dry spray deposit physically stick to the leaf surface. Other products may promote spreading along with sticking (spreader-sticker). To do this, spreader-sticker products commonly contain surfactants mixed with sticky materials (such as fatty acids) to provide both adjuvant functions.

Figure 13.15. This photograph shows how a sticker adjuvant (kaolin clay) helps keep a spray droplet on a leaf surface. After the droplet dried, you can see how well it adhered to the leaf surface. *Photo provided by CPS/Timberland.*

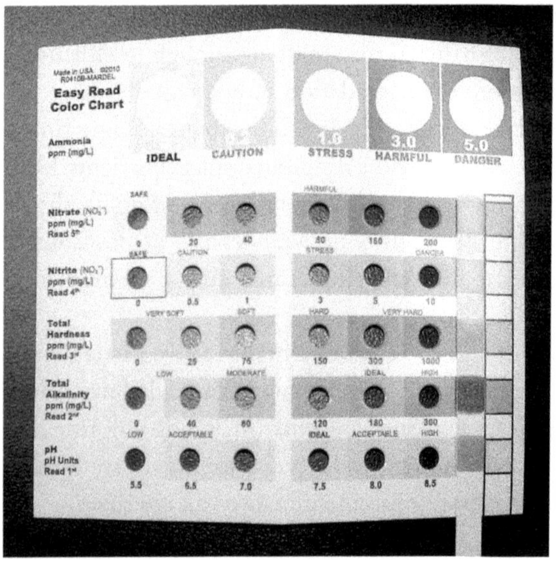

Figure 13.16. A water sampling kit uses litmus paper, which changes color to measure water pH and hardness.

11. pH Adjusters (Acidifiers, Buffers)

The pH adjusters (also called *acidifiers* and *buffers*) maintain or prolong a pesticide's stability by modifying the spray solution's final pH. Most pH adjusters are acidifiers that lower pH. Adjuvants that raise pH are less common.

Buffer and *acidifier* are often used interchangeably but are quite different. Water has little or no buffering capacity (see chapter 12), so a small amount of a pH adjuster can significantly alter the pH of the spray solution. An acidifier will lower the pH, but the pH of the resulting solution is still sensitive to any other compounds that may be added to it. To a certain point, a buffer can hold the spray solution's pH at a given level even when you add products. This is helpful when the pesticide product is particularly sensitive to pH.

12. Drift-Suppressing Agents (Deposition and Retention Aids)

Drift-suppressing agents (deposition and retention aids) help limit the drift of pesticide sprays off target. There are many different types. The performance of drift-suppressing agents depends on the spray mixture and application equipment. One type of agent will not fit all applications; each tank mixture may create different spray droplet distributions, so may require different products and use rates.

Some of these adjuvants thicken (build viscosity) in the solution to change the way spray droplets are formed. Some of the polymers that thicken the solution also help improve retention on the target by reducing the chance that droplets will bounce, shatter, or run off. However, polymers may also increase incompatibility issues. For example, they may not work well with certain drift-reducing nozzles, producing a poor spray pattern or very large droplets that may affect pesticide performance.

Figures 13.17–13.18. (Top) This photo shows a spray that does not contain a drift-suppressing adjuvant. (Bottom) This photo shows a spray that does contain a drift-suppressing adjuvant. Notice how the spray with the adjuvant remains in a "sheet" longer.

Another type of drift-suppressing agent works by reducing the percentage of smaller droplets—but it doesn't necessarily make large droplets even bigger. These adjuvants are an oil-in-water emulsion or surfactant (such as lecithin-based products), or more traditional oil adjuvants (not all oils are effective drift-suppressing agents).

The oil products work by producing many small and flexible oil-in-water emulsion droplets. When sprayed, these allow larger droplets to form without forming additional very large droplets. They may be more useful with certain nozzle designs and may not be as suitable for other nozzles, since they may also reduce the surface tension of the spray mixture or undermine air induction or Venturi nozzle designs.

Drift-suppressing agents may provide other adjuvant functions—such as improved retention and spreading of droplets. Depending on their composition, these adjuvants may also help to improve foliar uptake.

CAN YOU UNDERSTAND AN ADJUVANT LABEL?

If you read these terms on an adjuvant label, could you determine what they do?

- Nonionic low-foam wetter-spreader adjuvant
- Premium spreader-sticker
- Water conditioning agent
- Deposition and drift control aid
- Water conditioner and spreader
- High-concentrated MSO
- Nonionic surfactant, spreader, penetrator
- Alkalinity agent
- Nonionic surfactant
- Alkaline cleaner for sprayer systems
- A blend of modified vegetable oil concentrate and a silicone surfactant
- Super-concentrated foam buster
- Wetting, spreading, sticking, buffering, and drift control agent
- A premium buffered MSO concentrate
- Nonionic surfactant and buffering agent or acidifier
- Blend of nitrogen solution with premium paraffinic oil
- Buffering-sequestrant water conditioner
- Foam-marker concentrate
- 10% antifoam defoamer
- Ammonia sulfate—water conditioner—deposition aid—defoamer
- Humectant/adjuvant/surfactant
- Nitrogen solution and nonionic surfactant
- Penetrant—acidifier—surfactant
- Extender and sticker
- Vegetable oil plus emulsifier

If you figured out what these products were designed to do, then you have mastered the material to this point.

SOME ADJUVANT PRODUCTS CONTAIN MORE THAN ONE ADJUVANT TYPE

You can select an adjuvant product that serves just one purpose or a premix product that performs many functions, depending on the specific application. Although a premix may simplify selection and improve compatibility inside the container, it can be a waste of money if it is not the correct solution for the spray mixture. Premixed products may cost more because they offer added convenience. But that convenience can be valuable: Some premix adjuvants can help avoid specific compatibility problems, eliminating the need to add these materials later.

While a premixed adjuvant may be convenient, a premix isn't necessarily better than buying the individual components separately and mixing them yourself. Let your control goals dictate what you buy. Do you need an acidifier if your solution is already at the ideal pH? Do you need a water conditioner if the spray contains very little calcium or magnesium?

Another important factor is whether the ingredients in the adjuvant are at a concentration that will get the job done for your particular application. The ingredients in multifunctional adjuvants are in a fixed ratio relative to the other ingredients. If the ratio is not right, the premix product may not be the best choice for your goals. Know what spray solution problems you want to correct and match them with the right adjuvant product or products.

THE PESTICIDE LABEL: SUGGESTIONS AND SILENCE

Pesticide manufacturers design their formulations to include multiple ingredients (coformulants). These ingredients are necessary to perform the manufacturing process, deliver a stable product, and/or ensure robust performance. The specific ingredients differ based on the product form.

Water-based products typically contain a sufficient amount of a glycol antifreeze agent to stabilize the product for freezing and thawing. The antifreeze agent does not necessarily prevent freezing, but it keeps the formulation stable on thawing after a hard freeze.

Some pesticide formulations may contain surfactants to modify the wetting and spreading properties needed for foliar application, while other formulations may not. If the pesticide doesn't contain surfactants, the product label may recommend specific tank-mix adjuvants.

Other additives in pesticide formulations may include thickeners and chemical stabilizers. Water-dispersible granules (WDGs) include binders and dispersing agents to clump very fine pesticide particles to reduce dustiness and allow the resulting solid granules to redisperse into a fine particle suspension suitable for dilution in the spray application equipment.

However, including adjuvants in some pesticide formulations is impractical. For example, when a pesticide is applied at very low rates, the amount of surfactant needed to modify spray mixtures is often more than the amount of pesticide required. Sometimes the desired adjuvant may be physically incompatible with the pesticide in a highly concentrated form, which could cause formulations to become too thick to measure or pour.

In other cases, an adjuvant preferred for biological activity may not be chemically compatible with a pesticide's active ingredient, limiting shelf stability. For example, organosilicone surfactants are generally not chemically stable in soluble liquid (SL) or suspension concentrate (SC) formulations for extended storage (two or three years). However, they remain quite stable when mixed in the spray tank with many pesticides prior to application. Therefore, pesticide labels often specifically address adding adjuvants to the tank.

Adding your own adjuvants improves flexibility in terms of spray dilutions. Many pesticide products are used with multiple crops, varied climatic conditions, different application equipment types, and changing water use rates per acre. You can customize tank mixtures to assure you add enough adjuvant to modify the spray mixture's properties to meet the conditions present at the time of the application.

In general, manufacturers can provide greater value by putting some adjuvants into their formulations, but there is limited space in the package for inert ingredients. Active ingredients have a greater value than inert ingredients and adjuvants, so the primary goal is to deliver the most stable and effective concentration of active ingredient in the finished pesticide product. Getting the highest stable concentration of active ingredients in the product allows applicators to handle fewer containers or packages, but not necessarily in ways meaningful for the application.

Another very real reason why manufacturers don't add adjuvants to their products is that they want to keep their products financially competitive. In this way, pesticide products are like electronic components that are sold in packages labeled "Batteries not included." Adding a lot of adjuvants to a pesticide can drive up the cost and push it out of the market.

Even when manufacturers can add all the adjuvants to their formulated products, there is no perfect formulation that is appropriate for the entire United States and for every growing season. Applicators have to make the ultimate decision to include adjuvants. Only the applicators have access to the information, expertise, and experience about what adjuvants they need based on local conditions. The critical information applicators consider when making adjuvant decisions could include:

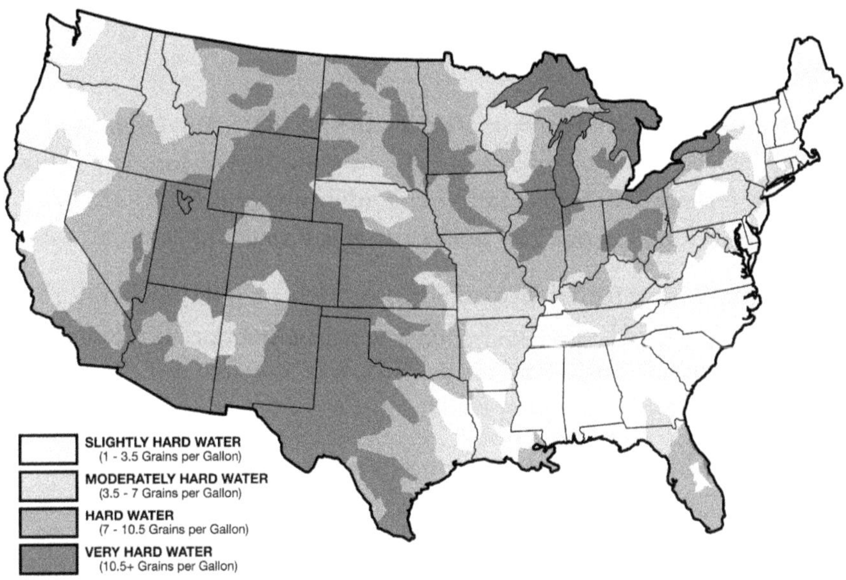

Figure 13.19. Water hardness areas in the United States.
Used with permission from H20 Distributors.

- Water quality parameters such as pH and hardness. These vary drastically around the country.
- The final volume of the spray mix. Adjuvant rates are often determined using a volume-to-volume (%v/v) calculation. Pesticide manufacturers never know where their users will be or where they farm or work. Whether applicators will be using 3 gallons or 100 gallons of spray carrier per acre depends on the industry (agriculture, lawn care, or right-of-way maintenance), farming sector (corn or tomatoes), and application equipment (helicopter, ground rig, airblast sprayer, backpack).
- The area to be sprayed. Applicators should consider adjuvant rates on an area basis (adjuvant product per acre) when the carrier is 10 gallons per acre or less to ensure enough adjuvant is present for the pesticide being applied.
- Regional, state, and local differences in targeted pests. The rates and types of adjuvant required may be different between tough-to-control, herbicide-resistant weeds and more susceptible ones. Applicators also should consider application timing and crops' tolerance for a given product.
- Weather. The current conditions can influence how much adjuvant to add. In particular, droughts can greatly influence the adjuvant category and rate. Adding an adjuvant to help reduce the droplet's evaporation rate or to allow the product to penetrate thicker plant cuticles may be necessary under dry conditions. However, these same adjuvants would be a waste of money in areas that have ample rainfall and high relative humidity.

SAMPLE ADJUVANT RATE RECOMMENDATIONS

Methylated Seed Oils and Vegetable Oils.* Methylated seed oil (MSO) or vegetable oil adjuvants may increase leaf absorption of the product. Include MSO or vegetable oil adjuvants at 0.5% to 1% v/v (2 quarts to 1 gallon per 100 gallons of spray solution).

Nonionic Surfactants. Use nonionic surfactants (NIS) at a rate of 0.25% to 1% v/v (1 quart to 1 gallon of surfactant per 100 gallons of spray solution). Surfactant products must contain at least 70 percent constituents effective as spray additives.

Crop Oil Concentrates.* Apply petroleum-based crop oil concentrates (COCs) at 1% v/v (1 gallon per 100 gallons spray solution) or 2% under arid conditions.

* Oil adjuvants must contain at least 80 percent oil quality petroleum (mineral) or modified vegetable seed oil with at least 15 percent surfactant emulsifiers.

By the time a pesticide reaches end users, a manufacturer has invested millions of dollars to understand how the product works best. A pesticide label is a national guide for using the product. Label recommendations on adjuvants fall into three categories.

1. The label may specifically recommend using an adjuvant.
2. The label may prohibit using an adjuvant.
3. The label may provide no information about an adjuvant.

When Labels Recommend Specific Adjuvants

Manufacturers do not give specific brand names for the adjuvants they recommend. Instead, they recommend a specific adjuvant category (such as NIS, COC, MSO, UAN, or AMS). For example, the label of a turf herbicide recommends using a nonionic surfactant and further states, "Use only nonionic surfactant on ornamentals." In any case, if a label recommends a particular adjuvant, you should use it.

Frequently, recommendations are too generic to assure the desired level of control under specific application conditions, or at a particular location, or for a specific tank mixture. Examples of such vague statements include:

- "To enhance foliage wetting and coverage, an approved nonionic agricultural surfactant may be added to the spray mixture as specified by the surfactant label."
- "Use only water suitable for spraying. Use an agriculturally labeled nonionic surfactant for all foliar applications. When using surfactants, follow the use directions and precautions listed on the surfactant manufacturer's label. Use the higher concentration of surfactant in the spray mixture when applying

lower spray volumes per acre. The order of addition to the spray tank is water, spray thickening agent (if used), additional herbicide (if used), and [pesticide name]. Surfactant should be added to the spray tank last or as recommended on the product label."

- "Add surfactant or another nonionic wetting agent to the mixture at label rate."
- "The addition of 1 to 2 percent dry ammonium sulfate by weight, or 8.5 to 17 pounds per 100 gallons of water (weight/volume), may increase the performance of this product on annual and perennial weeds, particularly under hard water conditions, drought conditions or when tank-mixed with certain residual herbicides. The equivalent rate of a liquid formulation of ammonium sulfate may also be used. Ensure that dry ammonium sulfate is completely dissolved in the spray tank before adding herbicides. Thoroughly rinse the spray system with clean water promptly after use to reduce corrosion."

A pesticide label may allow a variety of tank-mix adjuvants, leaving the final decision about specific adjuvant products to the applicator. In these cases, pesticide dealers or your university extension service may offer helpful recommendations based on past results in your area. This allows dealers and distributors to offer their customers more crop-specific adjuvant solutions.

Always read and follow the pesticide label's directions carefully. If you have questions, consult the local retailer for added guidance.

When Labels Prohibit Specific Adjuvants

Some pesticide products specifically state that using an adjuvant is unnecessary, that an adjuvant will make the product less effective, or that using an adjuvant may cause damage. New pesticide formulations that offer built-in spray drift control technologies may have labels that state that using other adjuvants can increase spray drift.

Product labels may specifically warn users to address specific concerns. For example, adjuvant labels may contain the following warning statements:

- "Use only nonionic surfactant on ornamentals. Do not use a crop oil concentrate with [name of product] on ornamentals."
- "Do not add surfactants, additives containing surfactants, buffering agents, or pH adjusting agents to the spray solution when [name of product] is the only pesticide used."
- "Certain spray tank additives (adjuvants, wetting agents, surfactants), liquid fertilizers, and tank mixtures containing emulsifiable concentrates may reduce the selectivity on the turfgrass. Use adjuvants and spray additives or tank-mix combinations only when your experience indicates that the tank mixture will not result in objectionable turf injury."

In some cases, adding an adjuvant might be unnecessary because the formulated product contains the needed adjuvants. In other cases, using an adjuvant may be antagonistic and lead to less-than-expected performance or even crop injury. An adjuvant may cause ancillary damage, like burning the foliage of the crop being treated with an herbicide. With newer generations of crop traits on the horizon, this may become more critical.

Adding surfactants to fast-acting herbicides increases the risk of damaging crops. For example, a turf herbicide label states, "Temporary discoloration of some turf types may result from use of surfactants or adjuvants with [herbicide name]. High temperatures and high relative humidity may increase the risk of temporary discoloration. Use of surfactants is not recommended."

When Labels Are Silent About Adjuvants

A pesticide label may neither list an adjuvant nor specifically prohibit its use. In these cases, the label is said to be silent on the matter. In the United States, under federal and state pesticide regulations, it is not illegal to add an adjuvant when a product label does not prohibit its use.

Pesticide products are often designed for a wide range of applications. Treating some crops may not require adjuvants, while other crops do. Alternately, controlling specific pest species may require using an adjuvant for some pesticides, while others don't. Due to the wide diversity in adjuvant products in the market, it is not always possible to evaluate all of them as a new pesticide product is being developed.

Amending pesticide product labels is time-consuming, since manufacturers must submit label changes to the US Environmental Protection Agency for approval. Often, labels mention adjuvants generically to allow the marketplace some flexibility for possible innovation.

So while it may be legal to use an adjuvant with a pesticide with a label that is silent, you are taking full responsibility for using it. Even the adjuvant manufacturer will specify on their labels that the surfactant "is recommended for use with those pesticides whose label recommends a nonionic wetter/spreader-type adjuvant," or "Always refer to the label on the product before using [adjuvant product name] or any other product." Working outside the scope of the pesticide and adjuvant labels means that you assume responsibility for any unintended consequences of the application. Although it carries risk, your application can produce highly desirable results if you plan it properly.

Ultimately, using adjuvants not mentioned on a pesticide label is a personal decision. You have the opportunity in each application to augment the spray mixture with an adjuvant, because you know:

- The setup of the sprayer
- The characteristics of the water source
- The prominent weeds, insects, or diseases being targeted
- The rates of pesticides and water volume used
- The environmental conditions
- Your previous experiences with the pesticide and adjuvant

Sometimes a pesticide label may give information that implies a good fit for an adjuvant. For example, a common label states, "When making applications in low relative humidity, set the equipment to produce larger droplets to compensate for evaporation. Droplet evaporation is the most severe when conditions are both hot and dry." The label's instruction to produce larger droplets could be accomplished by adding a humectant to protect small droplets from evaporation, or a deposition agent that would increase the overall droplet size. These are excellent examples of when planning plays a critical role.

The right professionals can advise whether it is safe and worth the money to add an adjuvant. These professionals include pesticide manufacturer representatives, consultants, local agronomists, and university extension scientists who may have data to support their recommendations or valuable experience with the use of adjuvants in your area.

Growers place a great deal of trust in their network of consultants for good reason: experience. Consultants generally have had more experience (year in and year out and over larger regional areas) and have seen both negative and positive outcomes from the use of adjuvants.

BUYING ADJUVANTS WITH CONFIDENCE

In the nineteenth century, pesticide adjuvants included substances like molasses, sugar, and resins that were used to "stick" sulfur, copper, and lime fungicides to grapes. Kerosene emulsions and soaps derived from whale oil improved the effectiveness of arsenic insecticides.

In the first two decades of the 1900s, insecticides and fungicides used oils. By the 1930s agricultural researchers began studying the physical properties of water droplets as they related to surface tension and contact angles. The 1940s and 1950s proved to be busy times for adjuvants after nonionic surfactants were introduced. Not until the 1980s were AMS and MSO products introduced. By the 1990s multifunctional adjuvants were released.

New adjuvant products for pesticide application are developed and introduced each year. Today there are more than 700 agricultural adjuvants manufactured by more than 40 companies in the United States alone. The US Environmental Protection Agency (EPA) does not require adjuvant products to be registered. Generally, adjuvant producers formulate their products to contain ingredients that exempt them from crop tolerance limitations set by the EPA. Not being EPA regulated doesn't necessarily mean that there are no rules governing adjuvants. Some states, including California and Washington, have extensive state regulations and registration.

American Society for Testing and Materials (ASTM) Section E35:22 also developed a standard terminology with definitions for use with adjuvant products. This enables apples-to-apples comparisons when you purchase products. Also, a trade association—the Council of Producers and Distributors of Agrotechnology (CPDA)—offers a certification service that reviews adjuvant composition and classification claims. The CPDA provides its seal of approval to an adjuvant product only after a thorough review process to assure it meets their guidelines and standards.

There are two kinds of adjuvants that commonly appear in spray mixtures. The first includes postmarket or tank-added adjuvants that are not specifically regulated by the EPA and that the grower or applicator adds at the time of application. The second kind is when that same adjuvant is added as part of a product that is offered for sale. In such cases, the product adjuvant will be called a *formulant* or *inert ingredient*, and the EPA has reviewed the adjuvant as part of the pesticide registration.

This really means that although the federal government does not have a national system for adjuvant registration like it does for pesticides, many of the same adjuvant materials are used in both the registered pesticide formulation and the tank-added adjuvant products. As a result, the EPA has in fact reviewed many of those adjuvants at one time or another.

As with any products you buy, it is always a good idea to do your homework before purchasing an adjuvant. Because so much rides on getting the pesticide applied right the first time, it is crucial that you have confidence that both the pesticide and adjuvant labels are accurate and that your final tank mix will provide the desired benefits under field use. Here are some questions to consider about registration, legal use, and efficacy.

Questions About Registration

As with any agrichemical, it is important to know whether adjuvant manufacturers will stand behind their products and whether their claims about their products' effectiveness are supported by data. The same goes for adjuvants used in medicine, health and beauty aids, and detergents. Here are two questions about registration that you should ask before selecting an adjuvant product.

1. Is the Adjuvant Registered in California and Washington?

A few states require adjuvant registration. State guidelines vary, but the trend is for more states to regulate them. California and Washington require a data review and registration process from manufacturers before they allow the adjuvants to be marketed there. Depending on each state's regulatory requirements, manufacturers can be required to submit the following information (manufacturers registered in California and Washington already address these issues):

- Documentation that the ingredients are acceptable for use on food crops
- Proof of third-party acute toxicology studies
- Evidence that the product will not harm aquatic organisms
- Evidence that the product is effective
- Proof that labels are clearly written and the directions are easy to follow
- Documentation that the labels adhere to the prescribed label formats
- Any precautionary statements to protect applicators

2. Is the Adjuvant Registered with the CPDA?

The CPDA adjuvant certification is a cooperative industry program designed to establish adjuvant product benchmarks. The program relies on ASTM standard classification guidelines.

ADJUVANTS FOR AQUATIC USES

Not all adjuvants are labeled for use on aquatic vegetation. Adjuvants approved for aquatic use must meet specific criteria. Manufacturers must demonstrate that aquatic use will not adversely affect desirable aquatic species. States like California and Washington require data about an adjuvant's possible aquatic toxicity to fish and aquatic invertebrates. An independent laboratory tests a range of concentrations of the adjuvant to accurately calculate the dose response curve, the LC_{50} (median lethal dose) for fish, and the EC_{50} (medium effective concentration) for invertebrates. **Always make sure any adjuvant you use is labeled for aquatic use.**

GENERAL INFORMATION

[Name of Product] is a spray adjuvant designed to maximize pesticide performance by improving deposition of the spray application onto the intended target. [Name of Product] improves coverage and reduces drift and evaporation of pesticides being applied by ground or air.

CA Reg. No. 1381-50004-AA
WA Reg. No. 1381-05004

These certifications provide customers confidence that they are purchasing a reputable and reliable adjuvant product. The CPDA program ensures that adjuvant products meet 17 documentable benchmarks. The manufacturers must:

- Demonstrate that the active ingredients listed on the label are present in the product and in the percentages stated in the ingredient statement.
- Verify that all ingredients were screened for acute toxicological effects on human health.
- Attest that all the ingredients have a tolerance exemption for use in food crops (under the Code of Federal Regulations 40 CFR Part 180) or be approved for food crops by the EPA. The EPA must assess all these materials and then determine that they are safe for their intended use with pesticides applied to food.

The federal code 40 CFR Part 180 organizes inert ingredients into these sections:

- 180.910—Inert ingredients used pre- and post-harvest; exemptions from the requirement of a tolerance
- 180.920—Inert ingredients used pre-harvest; exemptions from the requirement of a tolerance
- 180.950—Tolerance exemptions for minimal risk of active and inert ingredients
- 180.960—Polymers; exemptions from the requirement of a tolerance

Manufacturers that meet the CPDA standards for adjuvants may prominently display the CPDA's certification logo on their labels and in their advertising.

Some adjuvant manufacturers elect to make a stronger statement on their labels, for example, "[Name of product] is certified by the Council of Producers & Distributors of Agrotechnology (CPDA), which ensures consistency in product performance. CPDA certification assures users that the adjuvant product meets the claims indicated on the label; that ingredients used in the product meet EPA regulations for approved ingredients for use in pesticide tank mixes; and that the product is of the highest level of quality and reliability."

Using the adjuvant certification is an excellent way of knowing that the manufacturer voluntarily solicited CPDA review of their product to provide their customers confidence in the quality and utility of their product. A current list of certified products (along with their labels and material safety data sheets) is available on the CPDA website (https://cpda.com).

Question About Legal Use

Here is a question about legal use that you should ask before selecting an adjuvant product.

Is There an Exemption Statement?

Look at the label to see if any of the ingredients are exempt under 40 CFR Part 180. If one is, you should find a statement like this on the label: "Ingredients are exempt from tolerance under 40 CFR Part 180." The label may also state, "All ingredients are accepted for use under 40 CFR Part 180." If adjuvant products do not include these statements on their labels, ask the manufacturer whether the products can legally be used on food crops. Without the manufacturer's assurance, you should not use an adjuvant product on food crops.

PRINCIPAL FUNCTIONING AGENTS:	
Proprietary blend of hydroxy carboxylic acid, phosphoric acids and ammonium sulfate	52%
Components ineffective as water conditioners	48%
TOTAL	100%
All ingredients are accepted for use under CFR 40, 180.	

Figure 13.20. The "Principal Functioning Agents" section of a label gives information on whether ingredients are exempt.

Questions About Efficacy

Here are three questions related to efficacy that you should ask before selecting an adjuvant product.

1. What Do Manufacturer Representatives or Dealers Say?

It seems logical to get input from a pesticide manufacturer's sales or research representative. If these representatives don't know the answers, they have access to other professionals in their company who can answer your questions about using adjuvants with their specific products.

2. Is the Seller Reputable?

It is best to buy products from people and companies who have a vested interest in your operation. You do business with them for a number of reasons including price, expertise, trust, and prior history. It is reasonable to expect good service and well-reasoned opinions about adjuvant selection and use if you've been happy with their past recommendations.

3. Do Data Support Efficacy Claims?

Always consider your desired outcomes when choosing an adjuvant. The least expensive adjuvant may not always deliver the best value. Dealers, distributors, and adjuvant producers can often provide past field data from their customers or university trials to support their claims about the expected benefits of the adjuvants they recommend.

Adjuvant manufacturers should be able to provide data that show their products have been tested in the greenhouse and evaluated under field conditions. Contact your state's extension service to see if they have collected efficacy data in university research trials.

Without some performance data to back up your purchase decision, you are blindly betting that the cost of the adjuvants you are applying will, in fact, result in a benefit. It's always better to make an educated choice when data support using an adjuvant. Pay particular attention to data that provide:

- Side-by-side comparisons of competing products.
- Efficacy results that compare a pesticide's efficacy with and without an adjuvant.
- Data from replicated trials. Such data are better than data from demonstration plots, which are better than data drawn from testimonials. Replicated trials give scientific credence to the conclusions drawn.
- Evidence that the product provides consistent results over a number of years and environments.

It's always best to test an adjuvant by running your own field trial. It doesn't have to be large or complicated. Spray one load with the adjuvant and one load without (or with an alternative adjuvant), and then evaluate the performance within the first few weeks. You can judge yourself whether adding the adjuvant was worth the money.

Reviewing the Adjuvant Label

Adjuvants normally come with one-page labels that resemble pesticide labels. The quality of adjuvant labels has improved, largely because of the registration programs in California and Washington that require labels to give specific information on them. However, products that are not registered in those states may have much simpler labels than the one in figure 13.21.

Figure 13.21. An adjuvant label.

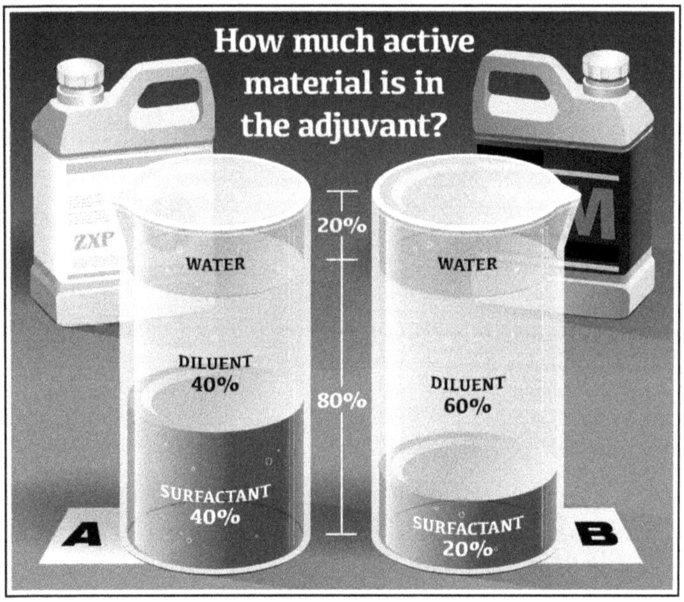

The active ingredient section can at times be confusing and misleading. However, the example in figure 13.21 indicates that 100 percent of its ingredients are active ingredients. Most of the product is an oil-based active ingredient that provides penetration properties, while the remainder is a surfactant that helps spray droplets spread over the leaf surface. This label is straightforward and easy to interpret.

In the graphic "How Much Active Material Is in the Adjuvant?," products A and B are sold as 80-20 NIS adjuvants. Both products state that active ingredients make up 80 percent of the product, and inactive ingredients make up 20 percent. You might assume that the two are similar. But look closer at the differences between these products. Product A has 40 percent surfactant load in its formulation, while product B has 20 percent surfactant. The surfactant is the ingredient that actually gets the job done in the spray tank. So while each is an 80-20 adjuvant, there are large differences in what you pay for.

Review the product label for the adjuvants you are considering to determine the reason for any difference in the cost per acre of the products, or to find out how well they may perform for your specific application.

Figure 13.22 gives another example. This adjuvant label states that the product is a 90-10 blend—the formulation is composed of 90 percent active ingredients and 10 percent carriers. As you can see, the manufacturer has added surfactants, humectants, and fatty acids to the list of active ingredients. Notice that the label indicates that 70 percent of the formulation is surfactant based. Labels such as this one make it easy to see exactly what you are paying for.

PRINCIPAL FUNCTIONING AGENTS:
Surfactants, humectants, free fatty acids 90%
Constituents ineffective as a spray adjuvant 10%
TOTAL 100%
All ingredients are exempt from tolerance under 40 CFR 180.
Surfactant content: 70%

KEEP OUT OF REACH OF CHILDREN

WARNING

Undiluted product causes eye irritation. Avoid contact with skin, eyes or clothing. Do not take internally. Wear chemical resistant eyewear, gloves, long-sleeve shirt and pants and shoes and socks. Wash thoroughly with soap and water after handling and before eating, drinking, chewing gum, or using tobacco. In addition, follow the precautionary statements on all products used with this product.

USE RATES

Follow the rates on the pesticide product label. If a rate is not specified or if they allow a wide range of use rates, follow the recommendations below:

Application	Recommended Rate	Low Rate (1)	High Rate (2)
	Pints per 100 gallons of spray mix (% of tank mix by volume)		
Ground, Aerial, CDA	2(0.25)	1(0.125)	4(0.5)
Aquatic	2(0.25)	1(0.125)	4(0.5)

Figure 13.22. A sample adjuvant label.

Rates

Adjuvant labels calculate recommended tank-mix rates using either a volume-to-volume ratio (v/v) or product rate per 100 gallons of water. A COC label tells the user to mix a 0.5 to 1.0 percent ratio per 100 gallons, which is equivalent to 4 to 8 pints of product.

Some labels provide only the percentage ratio without making the conversion to pints for you. Let's use the COC label to figure out how the 4 to 8 pints were calculated. To begin with, there are 8 pints per gallon.

100 gallons × 8 pints/gallon = 800 pints
0.5% = 0.005 and 1.0% = 0.01
0.005 × 800 pints = 4 pints
0.01 × 800 pints = 8 pints

Scientists prefer replacing volume ratios with a fixed rate per acre when the volume of water used as the spray carrier is significantly reduced, especially when using less than 10 gallons of carrier per acre. At some point, the volume of water is so low that there is simply not enough of the adjuvant present in the tank to maximize the transfer, rate, diffusion, and uptake when you use the volume-to-volume rates. In effect, you are cutting the use rate of the adjuvant as the water volume is reduced.

Some adjuvants define low-water-volume use rates. For example, the label of one COC indicates that aerial applications and low-volume sprays should use 1 to 4 pints per acre, which is a higher dose of COC than we would obtain if we calculated the rate at 0.5 to 1 percent (v/v). Similarly, an MSO concentrate indicates that when applying more than 5 gallons per acre by ground equipment, you should use 4 to 8 pints per 100 gallons. But if using less than 5 gallons per acre, you should use 6 to 8 ounces per acre. When a label does not define what constitutes a low volume of water, check with the adjuvant supplier or manufacturer to see what rates they would use given your specific needs.

The more difficult challenge is to know what exact rate to use, given that the labels provide a range. Adjuvant labels are seldom helpful. Check the pesticide label first, but this may not help. If the pesticide label calls for a specific category of adjuvant but does not provide a rate, ask your supplier for a recommendation. If you can't draw on the knowledge or experience of these individuals, then initially select the middle to high end of the rate range. After you become familiar with the product, you can adjust the rate.

You can change the rate of any adjuvant based on prior experiences. You might use the higher rate of a drift control adjuvant on a windy day but cut back on days with less wind. Spraying during drought conditions after plants have hardened may require you to use more of an oil adjuvant to deal with the thicker layer of wax on the leaf surface.

One of the last issues to address is the mixing order (see chapter 11 for more). The first rule is to follow the instructions on pesticide labels, which normally state to add adjuvants last. For example, the instructions might read: "Add pesticides as directed by labeling or in the following sequence: dry flowables or water dispersible granules, wettable powders, flowables, solutions, emulsifiable concentrates, and add [adjuvant name] last and continue agitation."

But a pesticide or adjuvant label may recommend adding the adjuvant at a different point, so read all label directions before mixing products. Some adjuvants (such as pre-emulsified, predispersed, or emulsified ones) may not be as sensitive and can be added at different points in mixing depending on convenience.

CONCLUSION

Making the winning play in a pesticide application requires pregame preparation (pest identification, growth patterns, and vulnerable life stages). The better you prepare, the more likely you are to win.

If in a football game your lineman doesn't put a solid block on the opponent's linebacker, your back might run for 5 yards. But if the lineman makes the block, that 5-yard gain could be a touchdown. A 5-yard run may seem adequate, but it is not what it might have been if the play had been properly executed. Like that 5-yard run, the results of a pesticide application may be acceptable. But adding the right adjuvant might have been like the lineman making that block. Running interference for the spray droplet, adjuvants can help the spray mixture score every time.

SECTION 3

Spraying Pesticides

14

Agricultural Spray Nozzles

With contributions from Rajeev Sinha, Debora de Oliveira Latorre, Bryan Young, Erdal Ozkan, Stephen Pearson, John Obermeyer, and Kevin Leigh Smith

Economists predict that the cost of pesticide application equipment will continue to rise as technological advances in precision application become commercially available. These breakthroughs primarily include machine learning, robotic automation, pulse-width modulation, and remote and proximal sensing capabilities. Familiar technologies such as direct inline injection, tractors and sprayers guided using computer vision and global positioning systems, and data logging continue to improve. And technologies such as remotely piloted aerial application systems that seemed futuristic only a few years ago are being rapidly adopted for pesticide applications (see chapter 15).

These rapid advances and their general adoption are expected to:

- Increase application and operational efficiency
- Permit timely and weather-appropriate applications
- Boost pesticide efficacy by improving application quality
- Save input and troubleshooting costs
- Provide environmental protections by applying product more precisely
- Improve applicator safety through equipment design
- Reduce spray loss
- Reduce application failures and required resprays

As sprayer technology evolves and advances, the need for highly trained operators has grown. Just a few years ago, if you could drive a truck, you could drive a sprayer. Back then, it was pretty much putting the equipment in gear (remember manual transmissions) and driving straight. An employee could drive a sprayer from spring through fall and then deliver fuel and propane in the winter.

Today's sprayer operators must be well versed in onboard software, electronics, digital communication, and mechanics. Operators also should have a working understanding of the legal, environmental, chemical, and agronomic aspects of their job. There is no comparison between today's sprayer operators and yesterday's applicators.

Figure 14.1. Application equipment may be computer driven, but nothing happens without nozzles.

As we explore and implement these new technologies, it is important not to overlook advances in hydraulic nozzle technology and to acknowledge their critical role. In general, a nozzle is undoubtedly one of the more important pieces of spraying equipment. It influences (among other things) the spray's quantity, uniformity, droplet size, and potential for off-target movement of pesticides.

Trade magazines are filled with stories about technologies and practices that add precision to applications, but stories about advances in nozzle materials, design, and functionality are rare. Nozzles are often regarded as old technology.

Some might think, "It's just a nozzle; how important can it be? I have used the same nozzle all my life, and it works." Such remarks clearly demonstrate how many people underestimate the humble nozzle. Many wrongly assume that nozzle designs are mere variations on one another and that they are more alike than not. As a result, operators tend to stretch a nozzle's utility, assuming a single design is appropriate for all applications. Many operators are guilty of exchanging worn sets of nozzles for new sets of the same nozzle without considering what they actually need for the job at hand.

But even the most advanced sprayers can be undermined by nozzles that are worn, damaged, or operated in ways that go against their intended use. Whether it is a pull-behind sprayer, a turf tank-hose sprayer, or an agricultural sprayer, achieving optimum performance (or getting your money's worth out of equipment and product) with these newer technologies may be compromised by having the wrong set of spray nozzles.

Nozzles are one of the most economical components in an application, which may be why many overlook their importance. Ultimately, a set of 72 nozzles on a 120-foot boom may cost hundreds of dollars but apply thousands of dollars of agrichemical products over their lifetime. The same could be said for the nozzles in an $85,000 right-of-way truck sprayer, a $20,000 ride-on turf sprayer, or a $300 backpack sprayer.

It's wrong to assume that a nozzle's importance is commensurate with its relative cost. Nozzles are the working end of application equipment that moves the diluted product out of the sprayer and onto the target. Choosing the wrong nozzle can compromise the spray job. This might mean wasting product, failing to control pests, increasing the potential for resistance, or creating off-target situations that can damage the environment, residential areas, and other agricultural operations.

Application equipment and products have evolved, and so have nozzles. Many new nozzle designs were developed during the past two decades. Today's spray nozzles are not what your grandfather used decades ago. Modern nozzles are designed and manufactured to reflect the needs and expectations of customers, industry, and regulatory agencies.

In many ways, the science behind the designs, materials, and capabilities of today's nozzles is as impressive and complicated as the chemistries being applied. This chapter discusses the current state of spray nozzle technology as it relates to today's pesticide applications. We hope this will enhance the nozzle selection process and optimal use to save applicators money and yield the most effective applications possible.

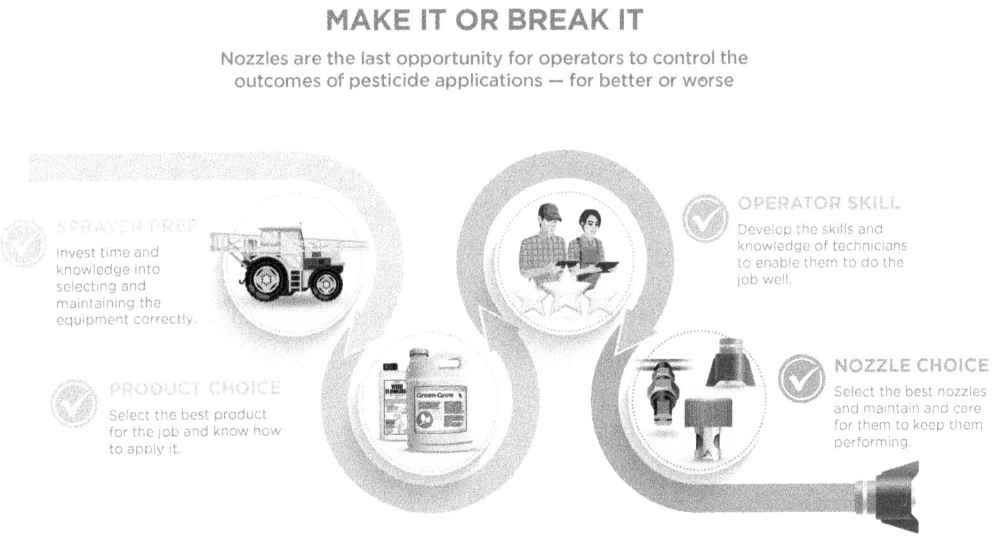

THE MANY FUNCTIONS OF NOZZLES

Nozzles are highly engineered equipment that is designed to meter, atomize, and create patterns to precisely apply the liquid in the sprayer tanks onto the targets. Applicators must carefully select the nozzle type and design, nozzle spacing/positioning, and operating pressure for the travel speed and prevalent environment conditions to deliver effective pesticide doses to targets as efficiently as possible.

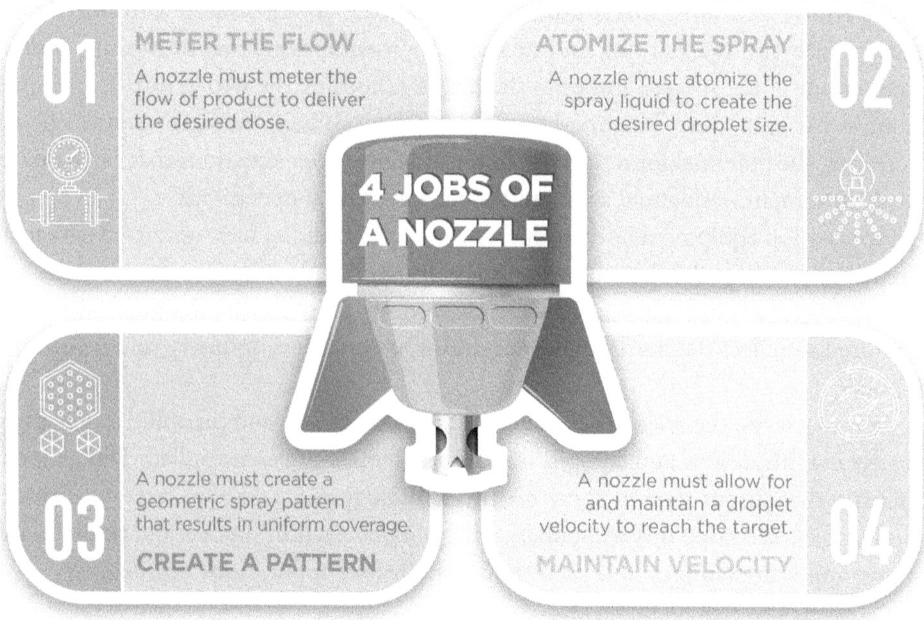

A first step toward appreciating the work spray nozzles perform is to define what nozzles need to do within milliseconds as the spray enters the nozzle body, exits into the atmosphere, and lands on the target. Nozzles must:

- Meter the flow to deliver desired doses
- Atomize spray liquid to create desired droplet sizes
- Create geometric spray patterns that uniformly cover the target
- Maintain droplet velocity to reach targets

Turn Liquid Streams into Droplet Clouds

Atomization is the process of turning a liquid stream into spray droplets. In most instances, this happens when the liquid to be atomized has a high velocity relative to the gas (surrounding air) phase. For example, in commonly used hydraulic nozzles, a high-pressure pump pushes a liquid stream through a small orifice, generating spray droplets released into ambient air.

Figure 14.2. Without a nozzle, the water comes out like a water hydrant, which would serve no purpose in the application process.

Imagine turning on an outdoor faucet connected to a garden hose. The wider you open the valve, the more water flows into the hose. Partially covering the end of the hose with your thumb decreases the orifice size, which breaks the stream into droplets.

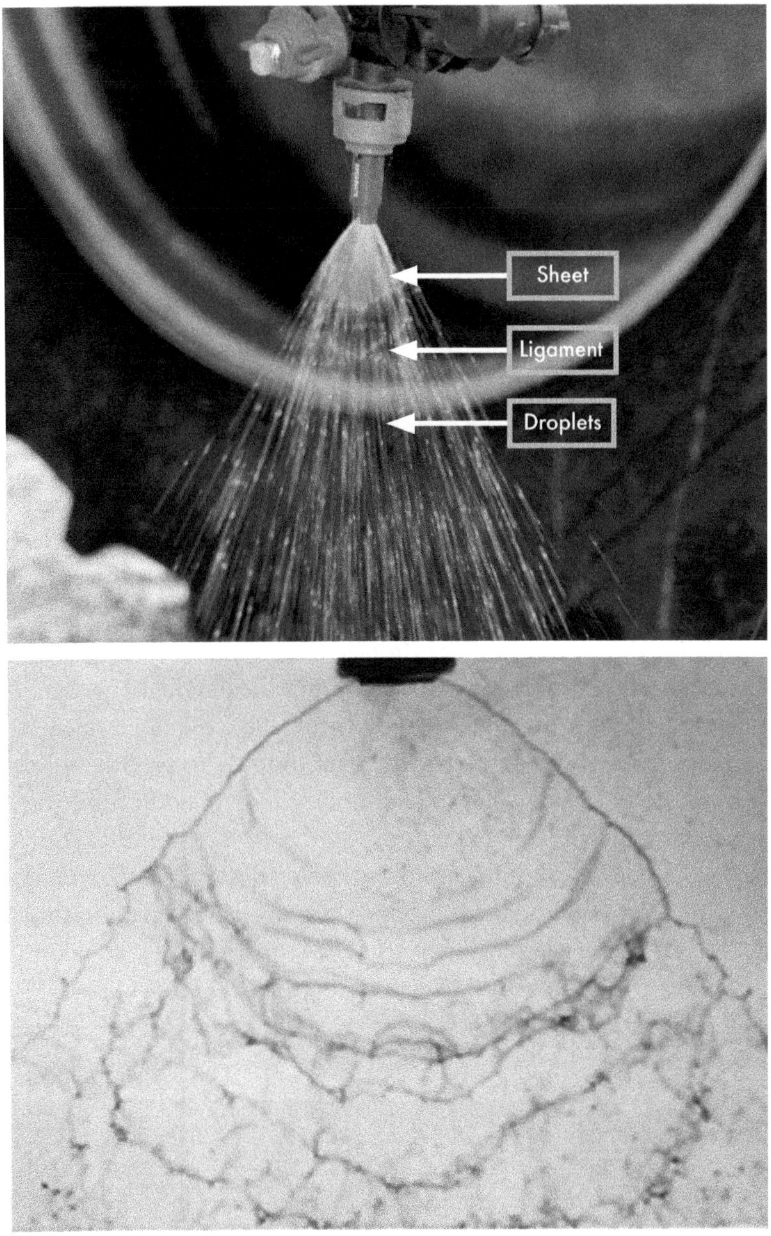

Figures 14.3–14.4. As the spray leaves the nozzle tip, the initial flow is a flat sheet of fluid. Once it leaves the exit orifice, the internal pressure no longer influences that sheet of water. As it leaves the nozzle tip and begins falling, the sheet of fluid begins to disintegrate into ligaments that will break into still smaller pieces of different sizes, called *droplets*.

WHICH IS CORRECT: *NOZZLE* OR *TIP*?

People often use the terms *nozzle* and *tip* interchangeably, but there is a difference. *Tip* refers to the part inserted into a cap. It can be made of stainless steel, plastic, ceramic, or brass. The nozzle contains the tip and gets attached to the boom through the turret. In some sprayers, the distinction between the two is often difficult, because some nozzles and tips are manufactured as a single unit (such as polymer nozzles). In such cases, the unit would be called a *nozzle*.

Create Droplets with a Specific Size and Density

Pesticide applications are most effective when nozzles produce an optimum droplet spectrum with the correct carrier rate. Therefore, you should link your nozzle selection (including type and size) to the chemistry you are using, the carrier volume you're applying, the surface the product is contacting, and the target pest.

A specific droplet size spectrum may be important to get the best activity from an herbicide, fungicide, or insecticide. Different chemistries are most effective with different droplet spectrums. Unfortunately, very few pesticide labels give information about specific droplet sizes that are best for a given application situation.

In general, fungicides require finer droplets and a higher density of droplets to provide better target surface coverage. Similarly, insecticides might need more fine or medium droplets to reach the undersides of leaves where the insects may be hiding. Systemic herbicides might work best with larger droplets to give the droplets more time on the target leaf surfaces, allowing foliar absorption to occur before the water in the spray deposit evaporates.

Think how the target can influence droplet size selection. The optimum droplet sizes for specific targets and pesticide products must be considered together. In addition, some herbicides can severely injure sensitive crops and cause crop loss to neighboring areas if there is off-target movement of the applied product.

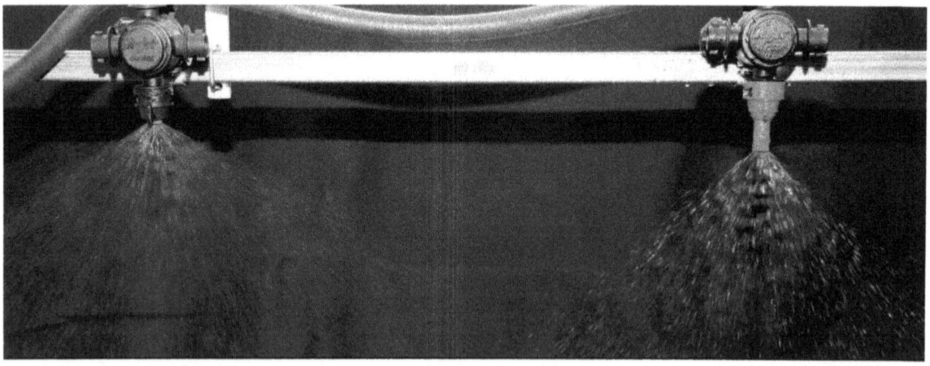

Figure 14.5. Two different nozzles working under the same pressure can produce remarkably different droplet sizes.

You can control such phenomena by paying attention to several factors, especially droplet size distribution. This helps explain why there are so many nozzle designs. Engineers are challenged to manufacture nozzles with droplet spectrums that fit largely within a narrow range of sizes (for example, very fine, fine, medium, coarse, very coarse, extremely coarse, ultra-coarse).

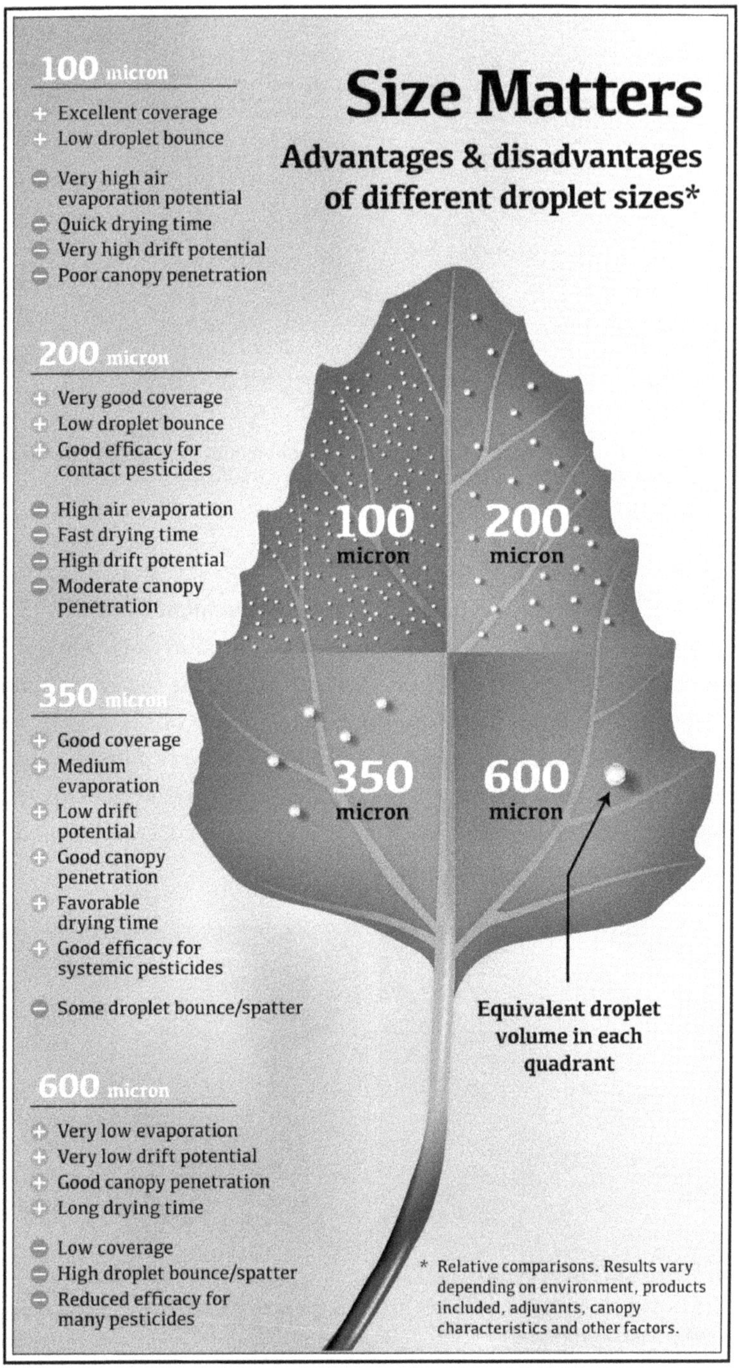

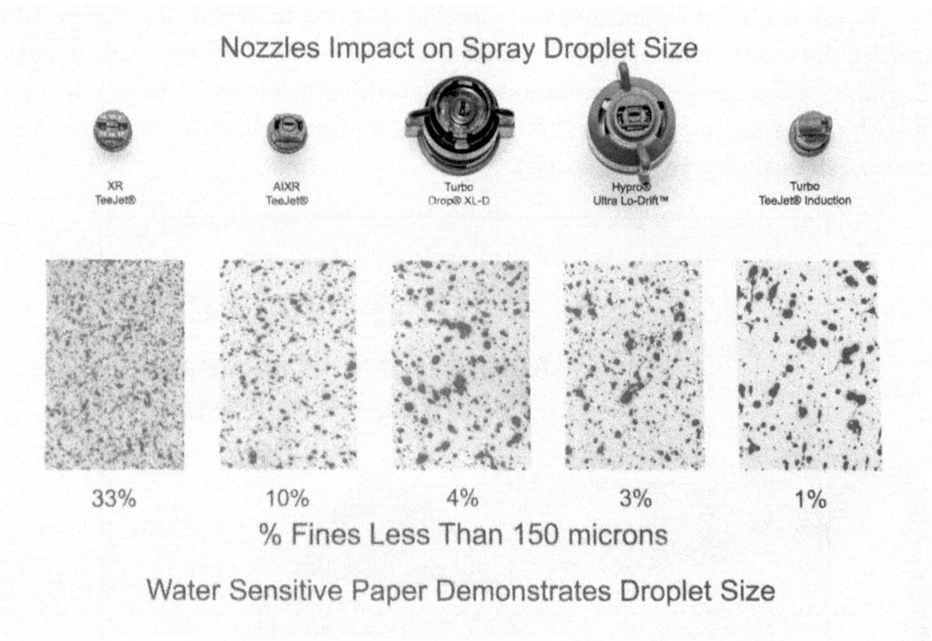

Figure 14.6. Nozzles of similar output (in this case 0.4 gallons per minute) can generate different droplet sizes that result in varying coverage on the target surface.

Form a Specific Spray Angle Width

For hydraulic nozzles, the spray angle is a static measurement of how wide the spray pattern will open up under a specific pressure. It is a function of nozzle designs, operating pressure, and the characteristics of the spray liquid. As the liquid goes through the exit orifice, the droplets conform to the physical shape of the nozzle, which then forms the angle. As the angle gets larger, the width of the spray pattern increases. That provides more surface area to form spray particles, which results in smaller droplets. Common spray angles include 65°, 80°, and 110°, but the angle selection depends on the spraying equipment and nozzle spacing.

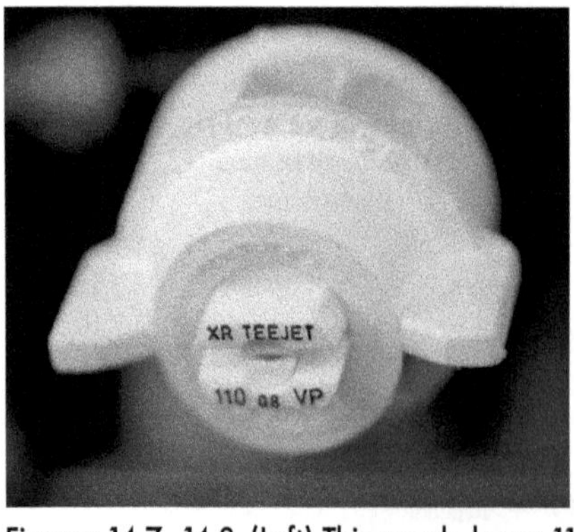

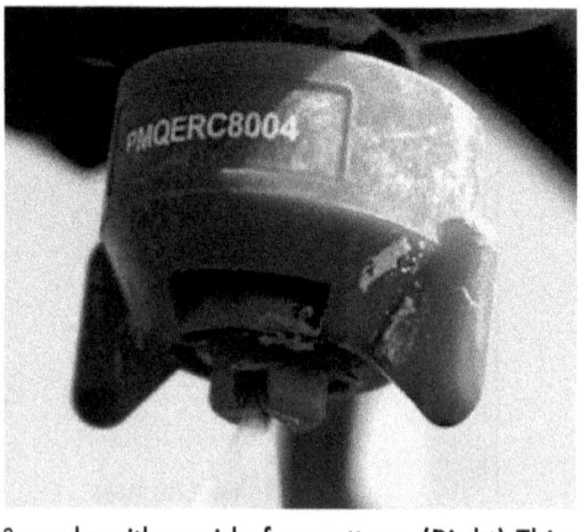

Figures 14.7–14.8. (Left) This nozzle has a 110° angle with a wide fan pattern. (Right) This nozzle has a smaller fan angle of 80°.

Figure 14.9. A turret garden hose sprayer has different spray patterns. Selecting the appropriate orifice allows you to perform various functions, including watering plants from a distance, gently misting shallow-planted seeds, and cleaning surfaces with a concentrated stream.

Create the Spray Pattern

Nozzles with different spray patterns are available on the market, and the right choice depends on the application type. The most common spray patterns are flat fan (tapered and even), cone spray, and solid stream.

- **Tapered flat fan nozzles** are designed to be used mainly in horizontal boom sprayers, because the deposition of the liquid has a tapered shape. The proper overlap of the adjacent sprays results in uniform distribution along the boom.
- **Even flat fan nozzles** do not require an overlap of adjacent nozzles for a proper distribution along the treated area. These nozzles are designed for banding pesticides (for example, over a crop row).
- **Cone spray nozzles** subdivide into hollow-cone and full-cone varieties. These nozzles are designed to be used in airblast sprayers, on specialty crops, and in directed spray applications.
- **Solid stream nozzles** are commonly used for fertilizer applications but also in some right-of-way and aquatic applications.

To picture spray patterns, consider a garden hose nozzle. If you use the hose to directly apply water without any modification, good luck with coverage and retention of the water on a target. If you attach a turret sprayer onto the hose, you can create a number of adjustable water patterns to meet your specific needs, including flat fan, cone (full or hollow), solid stream, and wide angle. Each setting forces the water through a different nozzle that allows the flow pattern and droplet spectrum to change.

Meter the Flow of the Liquid

Nozzles control the flow rate, which determines the amount of pesticide applied to a given area and the total application volume. Each nozzle is described by how many gallons per minute it will deliver at a rated pressure, based on the size of the nozzle's metering orifice. Smaller orifices have lower flow rates, and larger orifices have higher flow rates. Operators can modify nozzle flow rates by changing the pressure within the system. Higher pressures increase the flow rate, and lower pressures decrease the flow rate.

Control the Droplet Velocity

Droplet velocity measures how fast droplets are moving once they leave the exit orifice. Different nozzle designs, capacities, and operating pressures produce droplets with different velocities. Larger orifices or higher operating pressures mostly generate droplets that exit with higher velocities. Larger droplets can reach the intended target surface, but they sometimes bounce off, splatter, or run off the target site.

Smaller droplet sizes normally correspond to lower velocities. Smaller droplets lose momentum quickly as they exit the nozzle, remain as droplets in the air longer, and can either reach the target or be moved to nontarget locations. A fine droplet's initial speed is estimated to drop by 90 percent within the first 3 feet.

Weather conditions (such as wind speed) and travel speed directly influence the horizontal movement of smaller droplets. Air temperature and humidity also affect a droplet's lifetime. Higher temperatures (greater than 86°F) and lower humidities (less than 50 percent) reduce the droplet's lifetime.

Nozzles are critical to the success of any spray application. While the types and costs of application equipment vary widely across the commercial application industries and agriculture, all equipment depends on nozzles to make successful applications.

THE PUSH TO MAKE NOZZLES MORE PRECISE

Engineers studying fluid dynamics have long known that a spray entering and exiting a nozzle undergoes a complicated set of physical processes. In the early days of nozzle development, engineers were largely unable to document the processes involved in producing spray droplets with much accuracy. This resulted in a limited number of nozzles being used over a wide range of applications—more of a one-size-fits-all approach.

Only in the past few decades have advances in our understanding of fluid dynamics allowed spray nozzles to keep pace with improvements in application equipment and pesticide products. The "Factors Involved in the Evolution of Precision Nozzles" graphic indicates five factors that contributed to advances in nozzle technology.

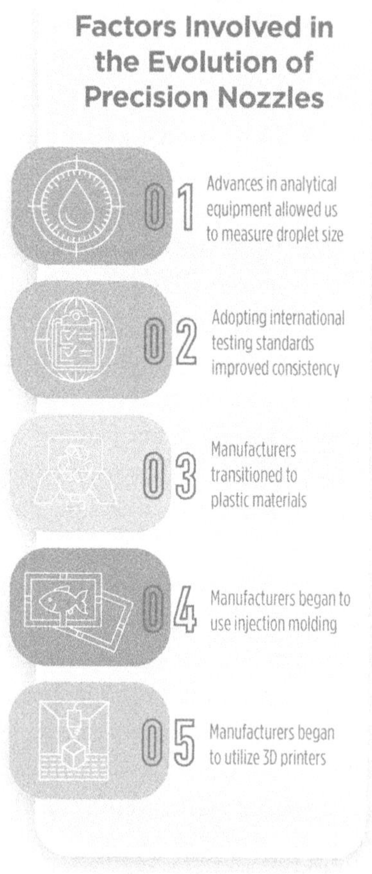

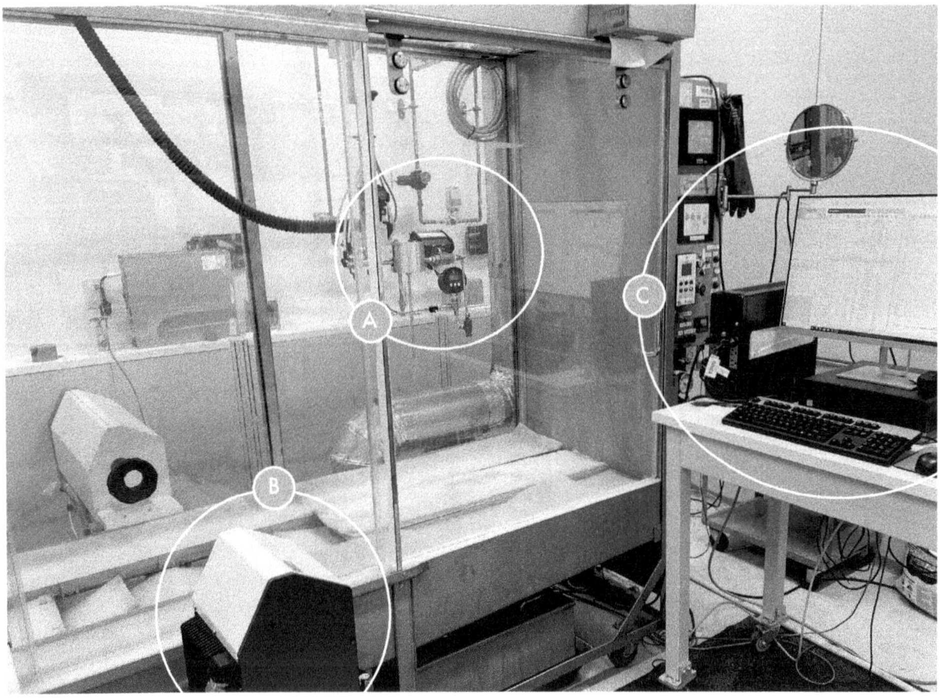

Figure 14.10. This photo shows a laser diffraction spray chamber used to analyze spray droplets and patterns. The equipment includes (A) a nozzle mount, (B) a laser diffraction-based particle size analyzer, and (C) computer software for data visualization and analysis.

Advances in Analytical Equipment Measuring Droplet Sizes

Engineers today are able to use analytical equipment that measures the droplet size spectrum, spray angle width, droplet velocity, and flow rate more precisely and in ways unavailable to past generations. Incorporating advanced tools (such as engineering software, droplet size analyzers, 3D printers, and computational fluid dynamics) helps manufacturers develop new nozzles.

Increased computer processing and data storage, advances in laser diffraction, image detection, and computational mathematical software can now produce results in an afternoon that would have taken weeks or months to complete. Rapid, accurate, and reliable measurements of nozzle parameters and spray droplet characteristics have led to nozzles being designed, manufactured, and marketed for specific applications.

Developers use three techniques to measure droplets: image analysis, laser diffraction, and phase doppler particle analysis. Image analysis and laser diffraction techniques are the mostly widely used. For image analysis, the analytical equipment images the spray pattern containing droplets of different sizes using a light source. The equipment software analyzes each image/snapshot in real time to generate droplet size data. Some droplet size analyzers can also provide other pertinent information (such as droplet velocity) by analyzing subsequent snapshots.

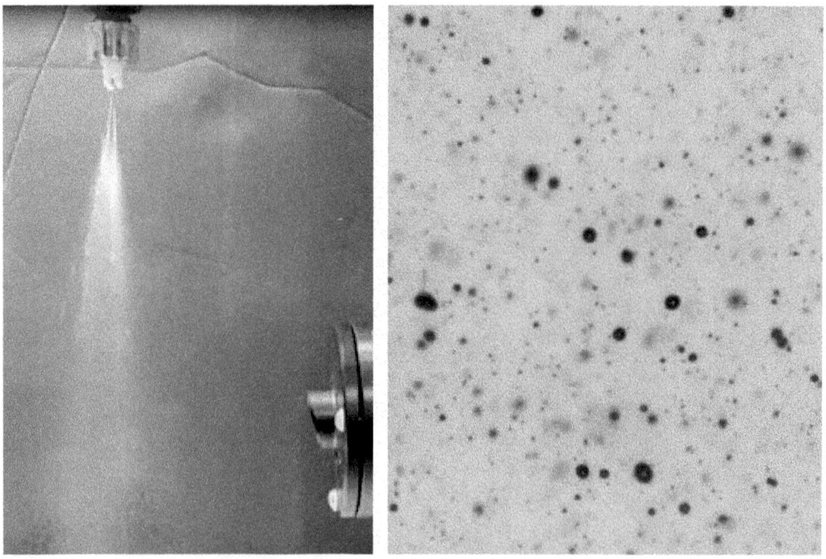

Figure 14.11. The Oxford Laser is one example of spray image analysis. It uses a camera (left) that detects droplet shadows (right). It then measures the droplet diameter in units called *microns* (one-millionth of a meter).

Acceptance of International Testing Standards

The instrumentation that developers use to define and describe nozzle output created an emerging field of science with different testing protocols and nomenclatures with different meanings. However, manufacturers were generating results using different and unrelated testing protocols, making it difficult to compare nozzles' performance across manufacturers and sometimes even within a single company. In addition, the terms used to describe those tests largely depended on how the nozzle manufacturer defined them.

To standardize the playing field, various groups formed standard-setting committees. The groups included the American Society for Testing and Materials (ASTM) International, the Institute for Liquid Atomization and Spray Systems (ILASS), the International Organization for Standardization (ISO), and the American Society of Agricultural and Biological Engineers (ASABE).

These groups built consensus, and manufacturers agreed to follow specific testing protocols when they created and marketed nozzles. Furthermore, this process standardized the definition of terms, which improved communication among universities, government, industry researchers, and end users. In the long run, pesticide application industries and applicators benefited most from the standardization process.

Transition to Thermoplastic Materials

Years ago, nozzles were largely made from stainless steel and brass and produced by a rotary cutter. Those materials are still used today in some nozzles—mostly those that handle high-temperature liquids, have special designs, or are used in scenarios where abrasion resistance is critical. Today most agricultural nozzles are made from plastics, however.

There are several advantages to using high-quality polymer-based plastics for agricultural spray nozzles. From a design and engineering perspective, plastic enables manufacturers to create new designs that have internal and external customization that cannot be duplicated in metal. From a manufacturing perspective, polymer nozzles can be produced with a high level of precision and in large quantities to meet consumer demand. From a farmer or industry perspective, polymer spray nozzles provide a cost-effective, corrosion-resistant solution offering multiple nozzle types and droplet sizes—they are also color-coded for easy identification.

Use of Injection Molding

Injection molding, using high pressure and temperature, forces melted plastic into metal molds. The metal molds provide efficient and consistent production and offer designers the freedom to create product shapes and geometries that would otherwise be difficult, cost prohibitive, or potentially impossible to produce using other methods or materials. Injection-molding resins can vary in strength, durability, heat resistance, and chemical breakdown. They can be dyed accurately and repeatably, which allows manufacturers to color-code completed parts and products.

Utilization of 3D Printers

At one time, skilled machinists were required to create prototype nozzles. After researchers tested the nozzles, the machinists would have to modify and reconstruct the prototype. Now nozzle manufacturers can design and test nozzle prototypes with 3D printers. Once prototypes meet the desired specifications, they undergo standard testing protocols. After testing, 3D printers enable manufacturers to make injection molds that can accurately mass-produce the new designs.

Nozzle manufacturers increasingly design prototypes with 3D computer simulations. Computational fluid dynamics is a field of study that allows companies to use computers to measure the impact of design changes before making prototype nozzles.

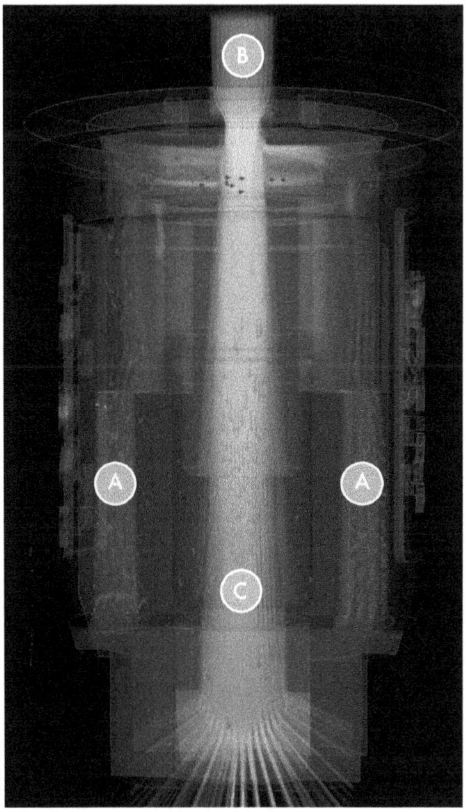

Figure 14.12. This graphic shows how computer simulations can test nozzle designs. The areas labeled (A) represent air coming from the inlet and going to the mixing chamber, where it will mix with the liquid. (B) This design has a pre-orifice with a small diameter to increase liquid velocity (the velocity is highest at the top, just below the B). The liquid and air combine in the mixing chamber (C).

CHARACTERIZING DROPLET SIZE AND SPRAY PATTERN

To better understand what nozzles do under specific work conditions, we must consider:

- The diversity of spray equipment
- The impact of droplet size on pesticide efficacy
- Application volume per area
- The shape of the plants and surfaces that will be sprayed
- Target pests
- Marketplace competition
- Applicator and customer expectations
- Potential environmental and legal penalties for off-target movement

Decades of testing and field use ultimately determine how specific droplet spectrums work for certain pesticides. A nozzle might need to produce smaller droplets for uniform coverage (such as for fungicide applications). Or a nozzle might need to compensate for spray drift concerns by producing larger droplets (such as for herbicide applications). In other cases, a set of nozzles might need to move to higher volumes because of higher application speeds. This would require a nozzle with a bigger orifice to achieve the higher flow rate. Another application might require much less volume, requiring a smaller-orifice nozzle with a lower flow rate.

COMMON MEASUREMENTS FOR NOZZLES

Here are some terms that are commonly used to describe nozzle characteristics:

- **Volume median diameter**
- **Droplet size distribution**
- **Droplet velocity**
- **Driftable fines**
- **Relative span**
- **Flow rate**
- **Spray angle**
- **Lateral distribution**
- **Pattern shape**

Droplet Size and Distribution

Today there is a wide variety of agricultural nozzle designs, some intended for specific and limited circumstances. For a pesticide label or consultant to prescribe the right nozzle for a job, we need a way to describe nozzle features and distinguish between types. This level of precision implies a level of specificity when manufacturers select nozzles to perform tasks.

Manufacturers need to be able to answer these questions about a nozzle:

- What is the volume median diameter (VMD) of the droplet spectrum?
- What is the droplet size distribution (also called the *droplet spectrum*)?
- What spray classification does it produce at different spray pressures (such as fine, medium, or coarse)?
- How uniform is the droplet size distribution (known as *relative span*) in the spectrum?
- What is the velocity of those desired droplets when they exit the nozzle?

The average size of droplets and their distribution (spanning from smallest to largest diameters) primarily depend on how a nozzle atomizes the spray liquid. As mentioned earlier, many factors influence the droplet spectrum, including operating pressure, the nozzle's internal configuration, the size and shape of the exit orifice, the spray angle, and the chemical properties of the spray liquid (such as its density and viscosity).

While different spray tips can produce the same flow rate (for example, gallons per minute), the droplet sizes making up that flow can still vary dramatically. This is why manufacturers test each nozzle type at different pressures to determine the median and average droplet sizes, and their distribution, as part of nozzle classification.

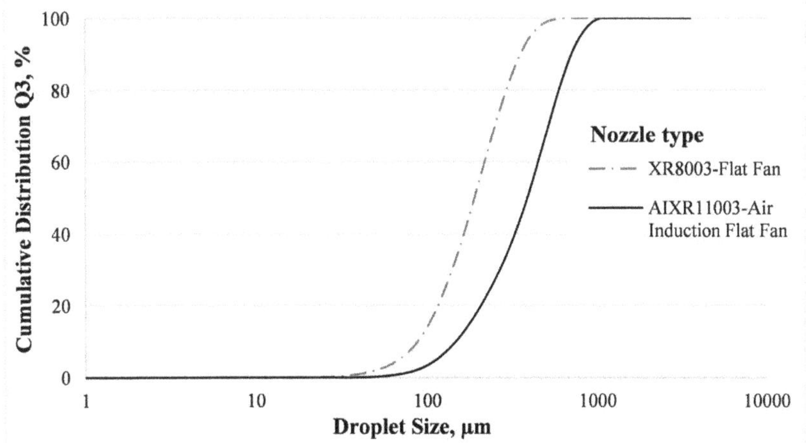

Figure 14.13. This figure illustrates changes in the cumulative distribution of particles based on two nozzle types with the same capacity (flow rate) at 40 psi. You can observe a greater cumulative distribution of small droplets for an XR nozzle compared to an AIXR nozzle.

Nozzles produce a range of droplet sizes. Knowing that range has important implications when you evaluate and select a nozzle. If you plan to use a nozzle in an area where off-target movement of spray droplets or spray drift is a concern, you should choose a nozzle that minimizes the volume of finer droplets (or driftable fines in the spectrum).

There is not one specific size that determines which droplets will drift or not. Wind velocity at application time plays the ultimate role in defining the cutoff droplet size below which droplets are considered *driftable.* However, under normal wind velocity conditions, droplets smaller than 150 microns are considered driftable.

Conversely, you might select a nozzle that produces a large proportion of small droplets (for example, fine and very fine) when applying a fungicide to a vineyard with an air-assisted sprayer or if you are spraying for a community-wide mosquito control program that must produce small droplets that remain in the air long enough to contact mosquitoes.

The software associated with different analytical techniques categorizes droplets based on three common diameter values:

1. $Dv_{0.1}$ (also called Dv_{10} or D_{10})
2. $Dv_{0.5}$ (also called VMD, Dv_{50}, or D_{50})
3. $Dv_{0.9}$ (also called Dv_{90} or D_{90})

Dv is an abbreviation for *diameter volume.* These diameters are commonly measured in microns, or micrometers (µm). There are 125,000 microns in 1 inch. For reference, the diameter of a human hair is typically 75 to 100 microns.

$Dv_{0.1}$ is a value where 10 percent of the total volume of liquid spray flux is composed of droplets with diameters smaller than this value. For example, a $Dv_{0.1}$ of 99 means that 10 percent of the droplets the nozzle produces will be smaller than 99 microns.

$Dv_{0.5}$ is a value where 50 percent of the total volume of liquid spray flux is composed of droplets with diameters smaller than this value. For example, a $Dv_{0.5}$ of 203 means that 50 percent of the droplets the nozzle produces will be smaller than 203 microns.

Some people also refer to the $Dv_{0.5}$ as the VMD (volume median diameter). Technically, the two terms are not exactly the same, but these values are very close in most cases. Therefore, they are often used interchangeably. ISO Standard 5681-2020 defines $Dv_{0.5}$ as an equivalent value to VMD.

$Dv_{0.9}$ is a value where 90 percent of the total volume of liquid spray flux is composed of droplets with diameters smaller than this value. For example, a $Dv_{0.9}$ of 417 means that 90 percent of the droplets the nozzle produces will be smaller than 417 microns.

Different nozzles that have the same flow rates may have very different $Dv_{0.1}$, VMD, and $Dv_{0.9}$ values. In table 14.1, the VMD for Nozzle 1 is 203 microns (µm), which means that 50 percent of the total volume of liquid it sprays is composed of droplets with diameters smaller than 203 microns; the other 50 percent of the droplets are larger than 203 microns. In turn, the VMD for Nozzle 5 is 782 microns. Although Nozzles 1 and 5 have the same flow rate, they yield very different droplet sizes. Nozzle 5 produces larger droplets, while Nozzle 1 produces smaller droplets.

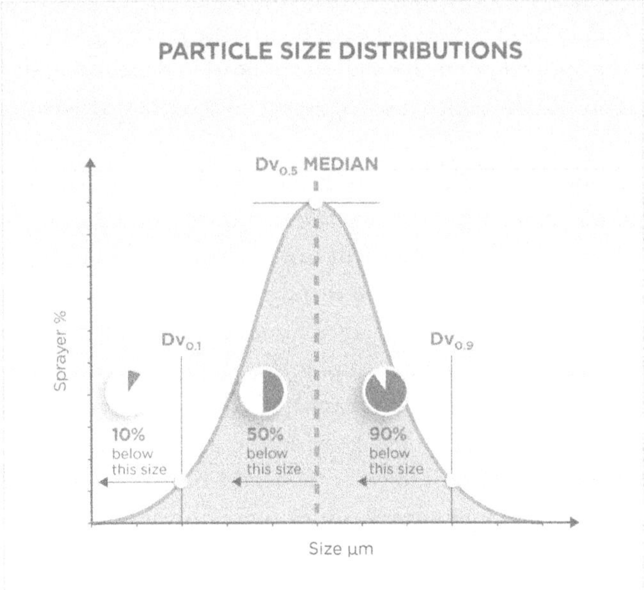

Figure 14.14. Spray clouds created from the atomization of the solution from the spray tips result in a droplet size distribution in a bell-shaped curve. Extensive research and development focuses on narrowing this curve to have relatively fewer fine droplets that may never hit the target or excessively large droplets that may bounce off target leaves or reduce spray coverage.

Table 14.1. Comparison of Nozzles with the Same Flow Rate

	$Dv_{0.1}$ (microns)	VMD (microns)	$Dv_{0.9}$ (microns)	Flow Rate at 40 PSI (gallon/ minute)
Nozzle 1	99	203	417	0.4
Nozzle 2	131	287	655	0.4
Nozzle 3	227	387	599	0.4
Nozzle 4	165	347	603	0.4
Nozzle 5	289	782	1,409	0.4

Based on the $Dv_{0.1}$ of Nozzles 1 and 2, we can observe that 10 percent of the volume of the spray consists of droplets smaller than 99 and 131 microns, respectively (the rest of the spray forms larger droplets). Smaller $Dv_{0.1}$ values indicate the driftable droplets a nozzle can produce. The $Dv_{0.1}$ values for Nozzles 3, 4, and 5 are relatively larger than for Nozzles 1 and 2. This may indicate that, under the same pressure, Nozzles 3 through 5 produce a smaller volume of driftable fines than Nozzles 1 and 2.

Among the five nozzles, there are also large differences in the $Dv_{0.9}$. Nozzle 1 presents the smallest $Dv_{0.9}$, which means that 90 percent of the volume of the spray is composed of droplets smaller than 417 microns.

The differences in droplet sizes between nozzles are neither good nor bad; they are just different. The optimal nozzle for a given application depends on the requirements of that application. VMD values provide a general overview of what to expect from the droplet spectrum.

Evaluating a nozzle's droplet spectrum includes reviewing all three droplet size distribution values. Nozzle 5, with its larger droplet spectrum across the $Dv_{0.1}$, VMD, and $Dv_{0.9}$, would be an excellent drift reduction nozzle. However, if a uniform distribution on the target surface is the goal, Nozzle 5 might not be the best choice, because the majority of the spray volume is composed of coarser droplets.

Conversely, Nozzle 1 has a fine droplet size spectrum with more of the spray volume contained in smaller droplets compared to the other nozzles. The lower Dv values for Nozzle 1 predict it would provide the greater coverage required for an effective fungicide application. However, it would be less of a candidate to apply systemic herbicides if off-target movement or spray drift is a prime concern.

Droplet Sizes Correspond to Color Codes

ISO (ISO 25358 and prior versions) and other standard-setting organizations developed droplet size classifications. The $Dv_{0.1}$ and the VMD values categorize droplets from extremely fine to ultra-coarse. Instead of providing the values for $Dv_{0.1}$ and VMD, nozzle manufacturers follow a droplet size color-coding scheme (according to the ISO standard). This color-coded scheme makes it easier for users to understand.

Table 14.2. ISO Standard Classification of Spray Droplet Size

Droplet Size Category	Initial	VMD Range (µm ISO)	Color Coding ISO
Extremely fine	XF	<60	Purple
Very fine	VF	60–145	Red
Fine	F	146–225	Orange
Medium	M	226–325	Yellow
Coarse	C	326–400	Green
Very coarse	VC	401–500	Blue
Extremely coarse	XC	501–650	White
Ultra-coarse	UC	>650	Black

It is important to highlight that droplet size varies with pressure range, nozzle capacities, spray angle, and pesticide formulations. The droplet sizes manufacturers note on their products are based on tests with water. Also, the color scheme for droplet size categories is not related to the color scheme for indicating nozzle orifice size.

Table 14.3. Classification of the Five Nozzles from Table 14.1

	$Dv_{0.1}$	VMD	Color Code (ISO 25358)	Spray Droplet Classification[1]
Nozzle 1	99	203	orange	fine
Nozzle 2	131	287	yellow	medium
Nozzle 3	227	387	green	coarse
Nozzle 4	165	347	green	coarse
Nozzle 5	289	782	black	ultra-coarse

[1] Each nozzle can deliver 0.4 gallon per minute at 40 psi. Spray droplet classification according to the ISO25358 - ASABE S572.3 standard.

In table 14.4, we can compare the droplet size of the XR TeeJet® Extended Range nozzles, which are available in two angles: 80° and 110°. The chart outlines how spray angle, pressure, and orifice size impact droplet size.

Let's use the XR8005 nozzle (taken from the TeeJet® nozzle catalog 52) as an example. This nozzle has an 80° spray angle and produces a coarse, medium, and fine droplet spectrum depending on the pressure. Notice that the XR11005, with a 110° spray angle, produces fine to medium droplets depending on the pressure.

When a spray liquid exits an orifice, a higher fan angle creates a wider pattern and area to disperse the droplets. At the same time, a greater degree of atomization leads to a greater number of smaller droplets. This can be explained by the same flow volume of spray solution being required to cover a larger (wide angle) or smaller (narrow angle) area. It takes more droplets to effectively cover a larger area with the same application volume.

Figure 14.15. These two nozzles have the same flow rate of 0.4 gallons per minute at 40 psi. The difference between them is the spray angle. The one on the left (XRC 8004) has a spray angle of 80°, while the one on the right (XRC 11004) has a spray angle of 110°. Never mix nozzles with different spray angles on the boom. Doing so results in improper pattern overlap, which leads to nonuniform spray coverage across the boom.

Table 14.4. Data for the XR TeeJet® Extended Range Nozzles

TIP PART NO. (STRAINER MESH SIZE)	PSI	DROP SIZE 80°	DROP SIZE 110°
XR8001 XR11001 (100)	15	F	F
	20	F	F
	30	F	F
	40	F	F
	50	F	F
	60	F	VF
XR80015 XR110015 (100)	15	M	M
	20	F	F
	30	F	F
	40	F	F
	50	F	F
	60	F	F
XR8002 XR11002 (50)	15	M	M
	20	M	M
	30	F	F
	40	F	F
	50	F	F
	60	F	F
XR80025 XR110025 (50)	15	M	M
	20	M	M
	30	M	M
	40	F	F
	50	F	F
	60	F	F
XR8003 XR11003 (50)	15	M	M
	20	M	M
	30	M	M
	40	F	F
	50	F	F
	60	F	F
XR80035 (50)	15	M	
	20	M	
	30	M	
	40	M	
	50	F	
	60	F	
XR8004 XR11004 (50)	15	M	M
	20	M	M
	30	M	M
	40	M	M
	50	F	F
	60	F	F
XR8005 XR11005 (50)	15	C	M
	20	M	M
	30	M	M
	40	M	M
	50	M	F
	60	F	F
XR8006 XR11006 (50)	15	C	C
	20	C	M
	30	M	M
	40	M	M
	50	M	M
	60	M	M
XR8008 XR11008 (50)	15	VC	C
	20	C	M
	30	C	M
	40	M	M
	50	M	M
	60	M	M

Driftable Fines Are Less Defined

Today's software can measure smaller droplets linked to spray drift. One of the first and most extensive attempts to correlate spray droplets to pesticide drift was performed by the Spray Drift Task Force, a group led by the pesticide industry. The task force compared exit orifice openings, pressures, nozzle designs, and nozzle heights from targets to measure how much of the total volume of a spray moved off-target and how far downwind the sprays moved.

The problem is that small droplets exiting a nozzle often do not have enough mass or energy to reach the target. Theoretically, those droplets remain in the air longer, which makes them susceptible to being moved off-target by the wind. For example, a 100-micron droplet may take 10 seconds to fall a vertical distance of 10 feet and can move 44 feet laterally at the same time when there is a 3 mph wind. By contrast, a 400-micron droplet released from the same height can fall 10 feet in 2 seconds and will only move 8.5 feet laterally when there is a 3 mph wind.

Nozzle manufacturers use the $Dv_{0.1}$ as a surrogate for spray drift potential by using a value called the *percentage of driftable fines*. The percentage of driftable fines can be expressed as an equation:

percentage of driftable fines = fraction of total spray volume contained in droplets < a given value assigned by the manufacturer for comparison

Table 14.5 shows how increasing the pressure from 20 psi to 40 psi produces higher percentages of smaller droplets (for example, 150 μm) with a selection of nozzles. For some nozzles, the pressure change causes a rather dramatic change in the percentage of smaller droplets. The opposite is also true: Reducing the operating pressure reduces the driftable fines.

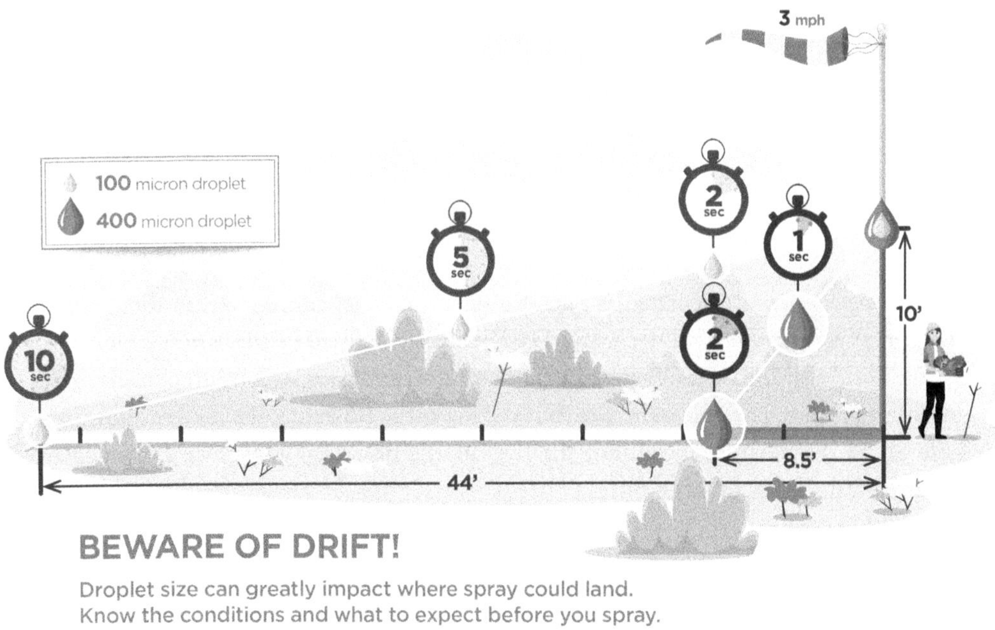

Table 14.5. Effect of Spray Pressure on Droplet Size

Nozzle Type (0.5 GPM Capacity)	Approximate Percentage of Spray Volume Less Than 150 Microns	
	At 20 psi	**At 40 psi**
XR – Extended Range TeeJet (110°)	18%	29%
TTJ60 – Turbo TwinJet® (110°)	8%	14%
TT – Turbo TeeJet® (110°)	7%	16%
TF – Turbo FloodJet®	5%	9%
AIXR – Air Induction XR® (110°)	4%	9%
AITTJ60 – Air Induction Turbo TwinJet® (110°)	2%	3%
AI – Air Induction® (110°)	5% (@ 30 psi)	7%
TTI60 – Turbo TeeJet® Induction TwinJet® (110°)	2%	4%
TTI – Turbo TeeJet® Induction® (110°)	<1%	2%
APTJ – AccuPulse® (110°)	<1%	1%

Data obtained from Oxford VisiSizer system, spraying water at 70°F under laboratory conditions.

Spray Droplet Uniformity Measured by Relative Span

A unitless value known as the *relative span* (RS) indicates the uniformity of the droplets exiting a nozzle. The formula for RS is:

$$\text{relative span} = \frac{Dv_{0.9} - Dv_{0.1}}{Dv_{0.5}}$$

A relative span is calculated as zero (0) if every droplet coming out of the nozzle is the same size (the $Dv_{0.1}$, $Dv_{0.5}$, and $Dv_{0.9}$ values would all be the same).

It is impossible to actually make an agricultural nozzle for which 100 percent of the droplets produced are the same size. But the lower the RS value, the more uniform the size of the droplets produced. For the five nozzles discussed above, we can calculate their RS values (see table 14.6).

Table 14.6. Relative Span of the Five Nozzles from Table 14.1

	$Dv_{0.1}$	$Dv_{0.5}$	$Dv_{0.9}$	$Dv_{0.9}$ - $Dv_{0.1}$	Relative Span
Nozzle 1	99	203	417	318	1.8
Nozzle 2	131	287	655	524	1.8
Nozzle 3	227	387	599	372	1.0
Nozzle 4	165	347	603	438	1.3
Nozzle 5	289	782	1409	1120	1.4

Note that Nozzle 3 has the smallest RS (1.0), meaning the diameters of all the droplets it produces are most similar. Conversely, Nozzles 1 and 2 have the greatest RS (1.8), indicating that these two nozzles (compared to the others) produce the least uniformity in droplet size, which may increase drift and/or reduce spray coverage (extremely important for contact-type pesticides). Therefore, when selecting nozzles, pay attention to both the VMD and the RS values (if the data to calculate the RS are available).

The ways that test results are displayed can depend on the type of droplet size analysis equipment and techniques used. However, most reports include the $Dv_{0.1}$, $Dv_{0.5}$, $Dv_{0.9}$, and RS. The information presented in figure 14.16 was generated using one particular droplet size analyzer (Oxford), which is also capable of determining the velocity (in meters per second, or m/s).

Figure 14.16. Data from the Oxford droplet size analyzer.

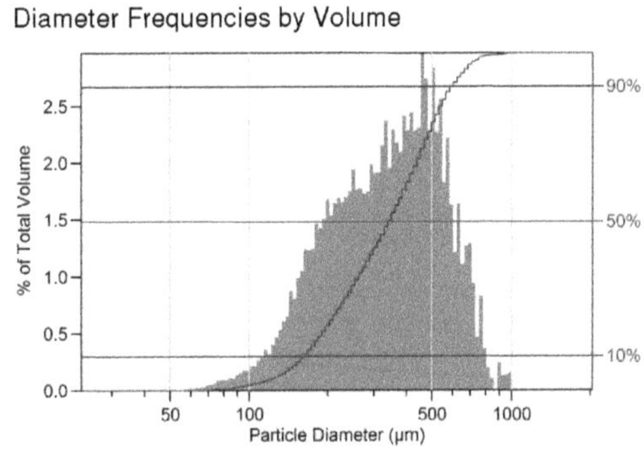

Averages:		Spread:		Volume percentiles:	
Arithmetic mean	161.8	Minimum diameter	44.2	10%	164.6
Surface mean	186.6	Maximum diameter	10005.4	50%	347.4
Volume mean	215.3	Relative span	1.26	90%	603.5
Sauter mean	286.7	Deviation	0.5	GSD	1.70
Volume-weight mean	440.5	Sphericity	0.92		
Minimum velocity	0	Maximum velocity	15	Average velocity	3.06

Flow Rate

Flow rate measures how much liquid is going through a nozzle over a given time—usually in gallons or liters per minute. At a manufacturer's site, a nominal nozzle flow rate is typically measured at 40 psi with water at 70°F. The nozzle is primed and operated continually while the output is measured for an established time interval.

Flow rates follow an international standard color-coding system (ISO 10625), in which each color represents the flow rate of a nozzle at 40 psi (2.8 bar). Manufacturers use this color code to dye plastic nozzles, making it easier for end users to identify which nozzles they are using.

In these examples, flow rate is independent of nozzle design and manufacturer. Over one minute, a red cone, a red flood nozzle, and a red air induction nozzle will all deliver approximately 0.4 gallon at 40 psi.

Figure 14.17. The left-hand nozzle is yellow, indicating a flow rate of 0.2 gallon per minute at 40 psi; the center nozzle is blue, indicating a rate of 0.3 gallon per minute; and the right-hand nozzle is red, indicating a rate of 0.4 gallon per minute.

Table 14.7. Color-Coding System for Nozzles

Flow Rate	Color Coding	Capacity (gpm @ 40psi)
01	Pure orange	0.1
0.15	Traffic green	0.15
02	Zinc yellow	0.2
025	Signal violet	0.25
03	Gentian blue	0.3
035	Purple red	0.35
04	Flamed red	0.4
05	Nut brown	0.5
06	Signal gray	0.6
08	Traffic white	0.8
10	Light blue	1
12	Raspberry red	1.2
15	Yellow green	1.5
20	Graphite black	2

This flow rate color scheme is completely separate from the colors used to classify the spray droplets produced, from extremely fine to ultra-coarse. Table 14.7 provides the current flow-rate color-coding system.

Spray Angle

Spray angle is a geometric measurement taken at the exit orifice. Higher pressures produce wider spray angles. The spray angle for a specific nozzle is determined at different operating pressures. It is calculated at its lowest pressure to define whether the spray angle is 110°, 80°, or some other angle. Using the lowest pressure to define the spray angle helps ensure that the sprays overlap correctly. Users can adjust spray overlap by changing the distance from the target.

Lateral Distribution

*Lateral distributio*n describes the uniformity of the volume that a nozzle distributes as measured by a distribution table (this is also called a *patternator*). This measurement provides a lateral footprint for the nozzle for how the volume is distributed along the spray boom. Manufacturers measure lateral distribution with a spray table that is divided into sections. Manufacturers mount a nozzle or set of nozzles on a spray boom and then collect and measure the amount of water in each of the slots.

Figures 14.18–14.19. (Above) A 110° nozzle in use. Both a 110° and 80° spray nozzle are predicated on the width of the exit orifice. (Left) Most nozzles list their spray width angles as part of their descriptions.

Figures 14.20–14.21. The front view (left) and side view (right) of two patternators show nozzles spraying into plastic compartments where their output will be measured. Lateral distribution is an important measurement to determine the proper overlap required between nozzles.

Figure 14.22. This photo shows an instrument measuring the lateral distribution of a Turbo TeeJet® nozzle with a 110° fan angle and a flow rate of 0.4 gallon per minute. The measured CV is 2.4 percent. Visually, you can tell the nozzles are spaced correctly, because the amount of water collected is almost the same from each nozzle, except for the end nozzles where there was no overlap.

When developing nozzles, manufacturers want each nozzle to maintain a specific distribution (that is, pattern) throughout the nozzle's recommended pressure range. Aside from pressure, other factors affect lateral distribution. Spray angle, spray pattern, application height, nozzle spacing on the boom, nozzle wear, clogged nozzles, nozzle quality, and a spray liquid's physical properties can all influence lateral distribution.

Manufacturers need to generate measurements pertinent to lateral distribution at different pressures, different heights, and different spacings. Critical for the design of any new nozzle is that it produces a uniform lateral distribution across different variables.

While the spray tables allow you to visually evaluate how uniformly a nozzle sprays, engineers have used a statistical parameter known as the *coefficient of variation* (CV) to measure uniformity across the treated area. The CV provides a numerical value for how much spray uniformity on the target varies across the width of the spray. A perfect CV is zero, indicating no variation across the cells collecting the spray. High CV values indicate a lot of variation; low CV values indicate more uniform distribution.

Each company sets its own internal threshold for the CV value. According to the ISO 16122-2 standard, a good-quality distribution along the treated area must present a CV value no greater than 10 percent. For example, the Julius Kühn-Institut in Germany considers that a nozzle set has an acceptable distribution when its CV is less than 7 percent.

Environmental Impacts

Manufacturers largely measure and test nozzles under controlled conditions. In actual use, changing field conditions during spray applications can positively or negatively influence nozzle performance. It is beyond the scope of this chapter to discuss these in detail; however, these factors can play important roles in modifying a nozzle's efficiency and performance:

- Sprayer travel speed
- Wind speed
- Wind direction relative to the sprayer travel direction and sensitive off-target species
- Air temperature
- Time of day for pesticide application
- Temperature inversions
- Air relative humidity
- Chemistry of the spray solution
- Boom spray height
- Pesticide label directions

CONSTRUCTION MATERIALS AND ADAPTIVE DESIGNS

After the 2,4-D herbicide was introduced in the late 1940s, it sparked the development of new nozzle designs. Since then, the rate of changes in nozzle designs has been increasing, often making earlier designs obsolete. This reflects the increasing numbers of new pesticide formulations, advances in spray equipment technology, and new methods in nozzle design and construction.

It may come as a surprise that nozzle design requires a tremendous amount of research, engineering, and technology. Even minute design changes can produce major changes in nozzle performance.

Nozzles are relatively small but highly sophisticated and important pieces of application equipment. Their compact internal spaces require tremendous accuracy with tolerances of 1/10,000 of an inch. Consider the challenge to engineers as they develop and alter nozzle designs. Do they want the nozzle to be a generalist? Or do they need a design that must contend with a very specific set of working parameters?

There are many questions that nozzle designers ask:

- Should the orifice be smaller or larger?
- What materials should the nozzle body be made of?
- What materials should the tip be made of?
- Should the spray tip be longer or shorter?
- Do multiple exit orifices affect droplet size?

- Can we place a metal or plastic platform under an exit orifice? Does the angle of that platform influence droplet size and pattern distribution?
- Where is it best to place air induction ports?
- What will a mixing chamber do to spray droplets?
- How large should the air induction port opening be?
- What screen strainer size will protect the nozzle tip from clogging?
- Can the nozzle cap fit current application equipment?

The materials that nozzles are made from affect their weight, durability, and cost. Internal plumbing affects the pressure and flow rate of the liquid as it moves through a nozzle. The size, shape, and orientation of the exit orifice affect the spray shape and spray droplets. The size and shape of the nozzle itself determine how it mechanically attaches to the sprayer and how it responds to environmental factors like wind and temperature.

All this and more goes into the nozzle development process. Engineers develop, test, and retest prototypes until the design satisfies the requirements. Let's look more closely at some of these factors.

External Dimensions

The sizes and shapes of nozzles (inside and out) influence the characteristics of the sprays they discharge. For example, some nozzles are longer to incorporate air injection ports. Some nozzles need additional girth to accommodate mixing chambers, or channels, which must be balanced against the designs of nozzle caps.

Entrance Orifices

Nozzle designs can alter initial flow rates and droplet sizes by manipulating entrance orifices. For some nozzles, liquid flows into the entrance orifice and then proceeds directly to the exit orifice. In others, designers add a rubber, ceramic, or plastic cap with one or more holes of different sizes and chambers. These modifications help set up the turbulence of the spray entering the nozzle's main body, where an internal mechanism works to determine the final size of the spray droplets when the liquid sheet exits the final orifice.

Figure 14.23. Nozzle manufacturers can make numerous modifications at the entrance orifice to affect flow rate and droplet size spectrum.

Structural Modifications Within a Hydraulic Nozzle

Hydraulic nozzles rely on energy in the form of liquid pressure to force spray liquid out through the exit orifice. The designs produce a moderate rate of flow and a wide range of spray shapes and droplet sizes.

While entrance orifices meter flow, exit orifices define spray geometry (shape) and droplet sizes (spray quality). Relatively tiny design changes can have large impacts on performance, and it takes milliseconds for the spray atomization process to complete—from the time the spray liquid moves into the entrance orifice to when it is expelled out of the exit orifice.

In recent times, most nozzle designs have focused on producing consistently larger spray droplets to reduce the likelihood of spray drift. Such nozzles often feature internal pre-orifices that meter the amount of liquid flowing into the nozzle. After these areas, there is a reduction in the internal operating pressure, producing larger spray droplets than conventional flat fans at the same boom pressure.

Structural modifications can include different cavities, chambers, and tunnels that the spray liquid is forced through. Altering the flow and pressure inside nozzles helps determine the droplet spectrum as liquid sheets pass through the exit orifices. In drift reduction nozzles, exit orifices are larger than the pre-orifices, which reduces spray pressure and produces larger droplets.

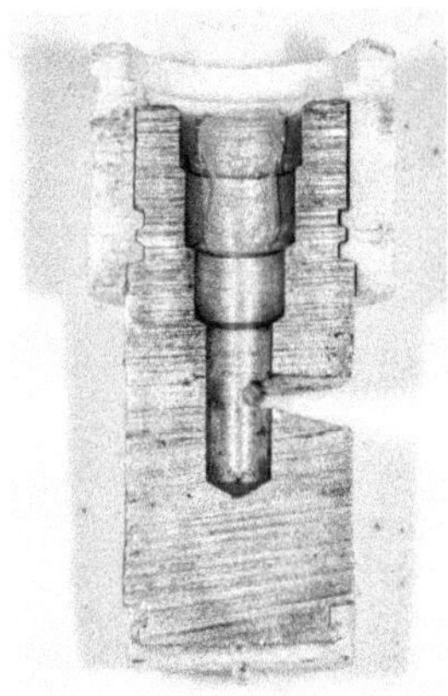

Figure 14.24. This nozzle cross section reveals it has increasingly smaller chambers before a liquid exits the orifice. This intentional compression of liquid forms specific droplet sizes

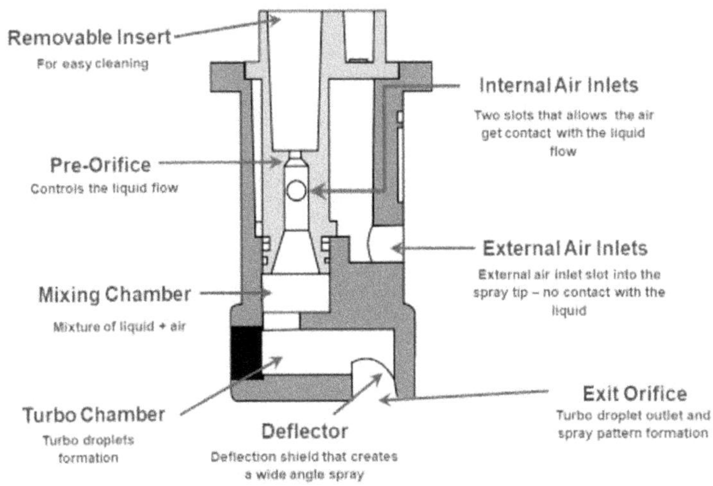

Figure 14.25. A cross section of a nozzle and all the parts that allow it to deliver unique characteristics.

One way to modify a nozzle internally is by using an *air induction* or *Venturi* design. Such nozzles pull in air, which reduces pressure on the liquid and produces larger droplets that exit the nozzle. Some manufacturers suggest that air bubbles trapped in the resulting droplets promote shattering and secondary coverage when droplets contact targets. Depending on the design, these nozzles must be operated at higher pressures than conventional tips.

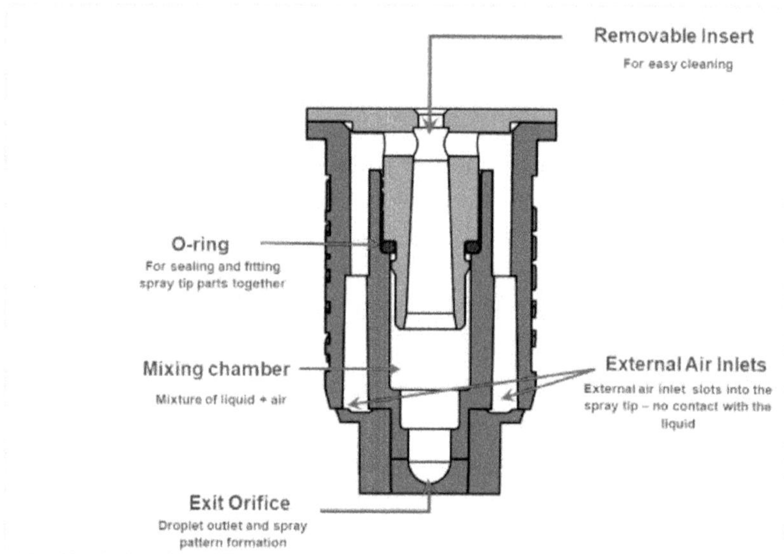

Figure 14.26. This illustration shows air inlets that bring air into the nozzle to increase droplet size.

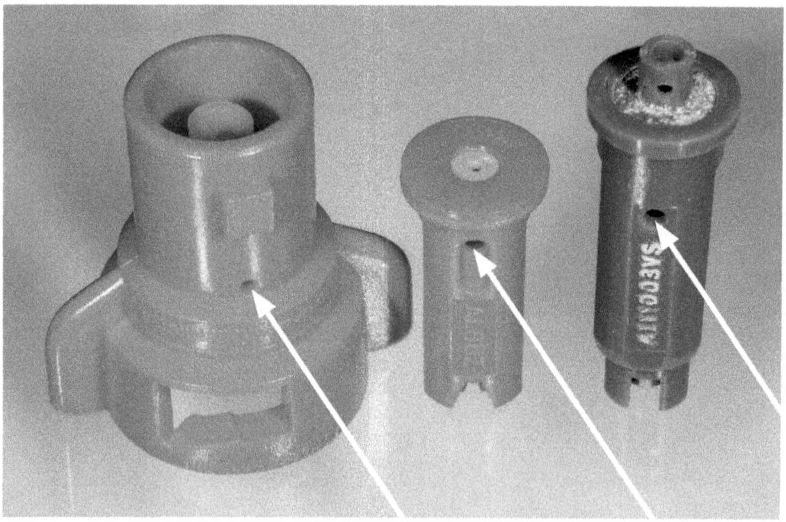

Figure 14.27. The arrows point to the air induction ports on these nozzles.

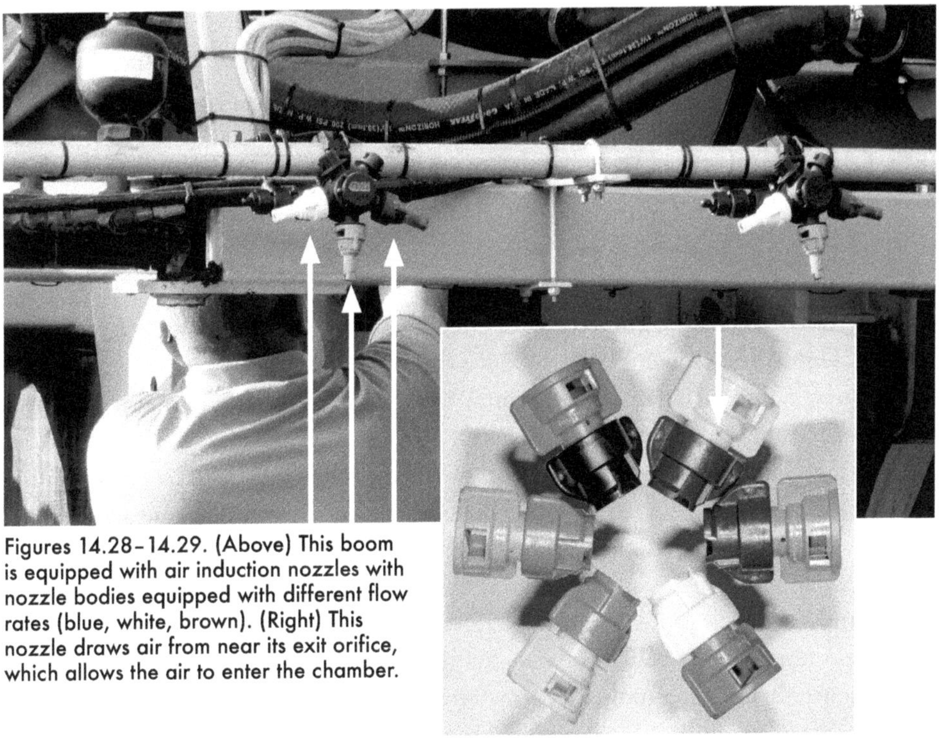

Figures 14.28–14.29. (Above) This boom is equipped with air induction nozzles with nozzle bodies equipped with different flow rates (blue, white, brown). (Right) This nozzle draws air from near its exit orifice, which allows the air to enter the chamber.

A second type of internal modification is called the *turbulence chamber*. The chamber is a cavity between the pre-orifice and exit orifice. When spray liquid passes through the pre-orifice, it enters the chamber, where turbulence slows the liquid, reduces its energy, and facilitates the creation of larger droplets upon leaving the exit orifice.

Structural Modifications to the External Exit Orifice

As previously discussed, the exit orifice appears to be a simple modification with one or more slots cut into the nozzle. Given that the exit orifice can define the spray pattern, width, flow rate (for some nozzles), and droplet cloud, it should not be surprising that this part of the nozzle has undergone numerous design changes over time.

Figure 14.30. Many manufacturers have modified the exit orifice to influence spray angle width and droplet size.

Materials Used in Nozzle Tips and Bodies

Materials used to make nozzle tips include brass, plastic (polymer), stainless steel, and ceramic. There are many questions engineers consider when deciding which materials to use:

- Will the tip spray abrasive or corrosive materials?
- What pressures will the nozzle be operated under?
- How small will the nozzle be?
- How many units will be sold?
- What will be the unit cost with one material over another?
- How difficult will it be to produce the nozzle?

The materials used to manufacture spray tips and nozzle bodies can be divided into two categories: metals and nonmetals. Metal nozzles, such as those made of brass and stainless steel, are manufactured using computers that control a laser that cuts the nozzle out of a block of the material. Once the nozzle is made, it then goes to a machine that cuts the exit orifice slot and drills the hole. Spray tips manufactured by lasers are said to be more precise because of the computer-driven technology.

Ceramic and plastic tips are manufactured through an injection molding process. Manufacturers heat the solid material to a liquid phase and then force the material into specifically designed molds. The material in the mold is allowed to cool and solidify. As the material cools, it shrinks slightly.

EVALUATING NOZZLES FOR RESISTANCE TO ABRASION AND CORROSION

Manufacturers test for nozzle wear using an accelerated test. They do this by continuously passing an abrasive material (like quartz and clay) through the nozzle for defined period. At the end of the test, manufacturers evaluate the nozzle to see if the flow rate and distribution pattern have changed. In this way, they can obtain results in a few days or weeks depending on the material being tested.

Corrosion is more difficult to test in the laboratory. The amount of corrosion is largely determined by the concentration of the material in the spray. If corrosion occurs, it happens over a longer period than the effects observed from the abrasion test. Characteristics associated with corrosion include rust and the nozzle material becoming brittle.

Nozzle manufacturers work with pesticide manufacturers to ensure the compatibility of pesticide products with the common materials used in nozzle construction.

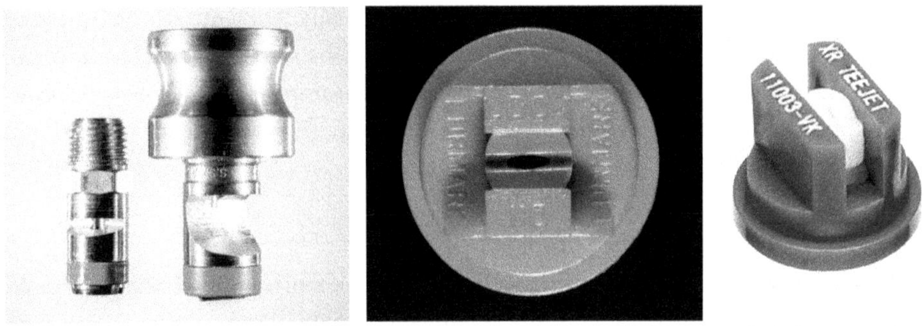

Figures 14.31–14.33. (Left) An all-stainless-steel nozzle with a plastic color-code belt. (Center) A polymer tip built into a plastic nozzle. (Right) A ceramic insert built into a polymer nozzle.

Compared to metal tips, plastic and ceramic tips are less precise from one nozzle to the next, so they have slight variations in flow rate and droplet size. However, the precision gap between metals and nonmetals has become less noticeable over time because of refinements in the manufacturing processes. The advantage of polymer and ceramic tips is that they are mass produced, which can lower the cost.

It is a common misconception that all nozzles are made from plastic. When one looks at a nozzle, what stands out is the colored polymer component (that is, the plastic nozzle body), but the tip inserts can also be made of stainless steel or ceramic.

These hybrid nozzles provide the best of both worlds by using a tip that has an insert with a longer life. At the same time, they allow new innovations or design modifications to be incorporated into the plastic body, because it is relatively easier to mold plastics than other common materials. Using plastic also reduces manufacturing costs and allows manufacturers to color-code their nozzles, which can help growers save time in the field when switching from one nozzle to the next during spraying.

Here are some brief descriptions of the materials manufacturers commonly use to fabricate spray nozzles and tips.

Brass (BR, B, or VB)

Brass was the first metal used to manufacture nozzles. It is a soft metal that can easily be machined into nozzles. Brass nozzles work better under low pressures.

The softness that makes brass easy to machine makes brass nozzles susceptible to abrasive materials during spraying. As pressures increase, abrasive materials will enlarge the exit orifice and

Figures 14.34–14.35. A brass nozzle like this one is commonly used by the pest control industry with manual pump-up sprayers. The nozzle shown here can be set to four positions depending on the desired spray pattern.

change the flow rate, fan angle, and droplet size. Generally, brass nozzles are not used in agriculture because of the higher pressures used and the abrasive pesticides applied. In addition, brass nozzles can be easily damaged when improperly cleaned. Furthermore, it becomes harder to read the nozzle size and spray angle information imprinted on the nozzle as they age.

Polymer (P, VP)

Plastics have advanced to the point where many of the historical drawbacks associated with them have been virtually eliminated. Advanced polymers are more resistant to abrasive and corrosive materials, which helps improve the serviceability and longevity of polymer nozzles.

Polymer nozzle tips have excellent wear resistance—they demonstrate less wear than stainless steel tips. Another advantage to using polymer spray tips and bodies is the flexibility to make slight design changes. This allows the manufacturer to build specific characteristics into the nozzle on a very small scale. Manufacturers can tweak molds to add different designs, including internal chambers or air induction ports.

Using polymer materials and an injection-molding process can turn out many nozzles in a shorter time and save money.

Stainless Steel (SS, VS)

Stainless steel can withstand abrasive and corrosive materials. It also provides excellent wear life. The main problem with stainless steel nozzles is their cost. Manufacturers often insert stainless steel tips into polymer bodies to help reduce costs.

Ceramic (VK)

Ceramic is a very hard material made from a quartz powder that contains a high percentage of aluminum oxide (alumina). The material is mixed with water and a binder, then is placed in a mold, heated in an oven, and cured.

Figure 14.36. The VP printed on this nozzle indicates that the nozzle tip is made from a polymer (aka plastic). The V means the nozzle follows the ISO flow rate color code.

Figure 14.37. VS or SS printed on the nozzle body indicates the insert is made of stainless steel.

Figure 14.38. VK printed on a nozzle indicates that it has a ceramic tip. The ceramic tip shown here is molded into a plastic nozzle body.

Table 14.8. Comparison of Different Nozzle Materials

Material	Ease of Manufacturing	Degree of Resistance		Damage Potential from Cleaning[1]	Relative Cost
		Abrasion	Corrosion		
Brass	excellent	poor	good	high	medium
Polymer	good-excellent	excellent	excellent	high	low
Stainless Steel	good-excellent	excellent	excellent	medium	high
Ceramic	good	excellent	excellent	medium[2]	high

[1] Rating based on how easy it is to damage the nozzle when cleaned.
[2] Ceramic nozzles edges are brittle and chip easily during mechanical cleaning.

Ceramic tips, like most stainless steel tips, are inserted into polymer nozzle bodies. Ceramic tips are the hardest and most durable tips on the market. They hold up well to corrosive and abrasive materials. However, some growers express concerns that ceramic is too fragile and could break if nozzles are dropped on a hard surface like a concrete floor.

Ceramic tips are widely used in the hollow-cone and disc and core nozzles used in tree fruit orchards and vineyards. Many of the fungicides applied to fruit trees are formulated as wettable powders and applied by air-assisted sprayers under high pressures. Some markets (like Europe and South America) have a greater interest in using ceramic nozzles on horizontal boom sprayers because of their durability.

NOZZLE DESIGNS IN THE MARKETPLACE

Historically, flat fan spray patterns were intended for horizontal booms. Applicators could adjust the nozzle spacing, boom height, and tapered volume of the spray to create an overlap resulting in a uniform swath across the field. Specialty crops, such as vegetables, may use overlapping flat fans for broadcast applications, or banded applications where flat fans are arranged only over the row.

In some cases, hollow cones, which have patterns that are intended to meet but not overlap, are still used on horizontal booms in specialty crops. Without some form of drift mitigation in these cases, the applications are very prone to off-target movement. Directed applications, such as hand wands or drop hoses, tend to employ flat fans or hollow cones, depending on the application.

Airblast (that is, orchard) sprayers, which employ air to direct spray to a three-dimensional canopy, often use hollow cones. With the increased use of wraparound towers and vertical booms, airblast sprayers may also employ flat fan nozzles. The point is that there are always exceptions to how nozzles can be used to best apply the pesticide.

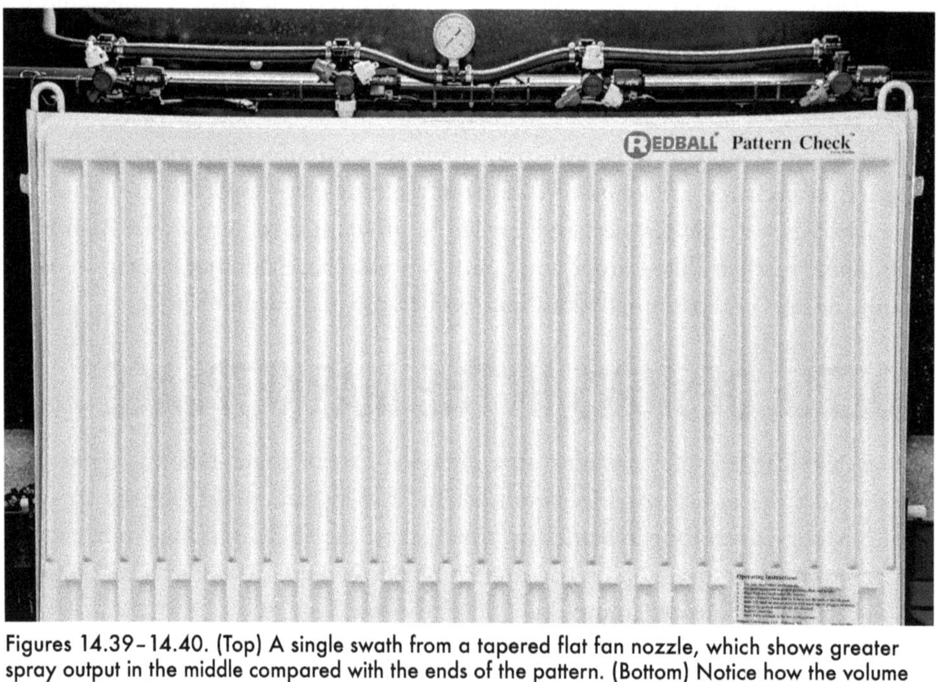

Figures 14.39–14.40. (Top) A single swath from a tapered flat fan nozzle, which shows greater spray output in the middle compared with the ends of the pattern. (Bottom) Notice how the volume tapers (reduces) along the outer edges. This is why overlap between nozzle spray patterns is required.

Nozzles That Primarily Fit on Spray Booms

Tapered Flat Fan Spray Pattern

Flat fan spray nozzles are the most commonly used nozzles in agriculture. As the name implies, these nozzles initially produce a two-dimensional triangular spray pattern. In cross section, the fan distributes more liquid in the middle and tapers toward each edge. In addition, spray droplet size distribution is not homogeneous at all locations within the fan. For these reasons, these nozzles are never intended to be operated singularly or to deliver the liquid as spot-spray applications.

To produce even distribution when using multiple nozzles on a boom, the neighboring nozzles must overlap. Furthermore, there are fewer, coarser droplets toward the edges of the boom compared to in the middle. An overlapping pattern not only makes the flow uniform but also yields a more homogeneous droplet spectrum.

Broadcast tapered flat fan nozzles are mounted to a linear boom. They are popular because equipment fitted with booms can spray large areas at fast speeds. Correctly overlapping the spray patterns of these nozzles results in a uniform spray across the swath of the boom.

The proper overlap for 110° flat fan nozzles should be at least 50 percent, so that the sprays from adjacent nozzles intersect each other at the midpoint of their sprays. Each nozzle on the boom thus delivers 50 percent of the target product to the spray area—the other half is delivered by the adjacent nozzles on its right and on its left. The resulting spray in the swath of the boom provides 100 percent coverage.

Figure 14.41. For tapered nozzles, the jets need to be offset 15° to avoid collision. Adjacent jets need to overlap but not collide.

Typically, this can be achieved at a 20-inch boom height for a boom with 20-inch nozzle spacing. Such spacing provides a very robust spray swath with maximum coverage and a low coefficient of variation. Keeping variation low is critical to avoid skips or streaks in the application and to maximize efficacy.

Tapered flat fan nozzles must be a certain height above the target to give the spray pattern enough time to develop after the spray leaves the exit orifice. Tapered flat fans are most commonly available in 80° and 110° spray angles and normally can operate within a pressure range of 15 to 100 psi, depending on the nozzle design. One advantage of these nozzles is that they are forgiving (within limits) as the boom moves up and down, because the overlap prevents skips or streaks.

There are five types of flat fan nozzles. One example is called an *extended-range flat fan* (XR). Many nozzles need a minimum nozzle pressure to create the desired pattern. XR nozzles maintain a pattern at lower pressures compared to older flat fan nozzles. For example, a standard tapered flat fan nozzle commonly has an operating pressure range of 30 to 60 psi, while the XR can produce a pattern when the pressure is as low as 15 psi.

XR flat fan nozzles work well with spray equipment outfitted with automated rate controllers. Spray patterns remain mostly the same over a considerable pressure range; however, changing pressures may alter droplet sizes (depending on the intensity of the pressure change). Spray solution characteristics (including higher viscosity, lower surface tension, and so on) may also affect the fan angle. Therefore, applicators must find the right combination of these parameters (along with boom height and nozzle spacing) to achieve a uniform spray application.

A second type of tapered flat fan nozzle is called an *off-center flat fan* (OC). With these spray tips, the exit orifices are often wider to one side; they are mounted at the ends of booms. OC nozzles essentially spray half a pattern toward one side (toward the center of the spray boom) instead of down the center like other tapered flat fan nozzles (see figures 14.42–14.44). It's as if a tapered fan nozzle's spray pattern is cut in half. Applicators should not use OC nozzles for broadcast applications across a field. They should only be used along field perimeters.

A third type of tapered flat fan nozzle is called a *dual (or multi) flat fan*. As the name suggests, these nozzles produce one fan pattern that sprays in the sprayer's travel direction (forward), while a second fan sprays in the other direction (backward). Manufacturers incorporated these designs into specific nozzles to improve both spray coverage and foliar canopy penetration. For example, spraying a fungicide on wheat heads or an herbicide on grassy weeds can be challenging, because they are vertical targets. Attacking the target at different angles increases the chance that the pesticide will contact the foliage and that the foliage will retain the pesticide.

Some dual fan nozzles produce symmetrical fans that use the same angle (for example, 60°). In such cases, the spray patterns are mirrored, and the spray angle is the same backward and forward. For example, the Turbo Twin Jet® (TTJ60) nozzle is symmetrical and has a 60° angle between the two sprays. Each spray is 30° backward or forward from the vertical, and each individual spray forms an 80° or 110° wide spray angle.

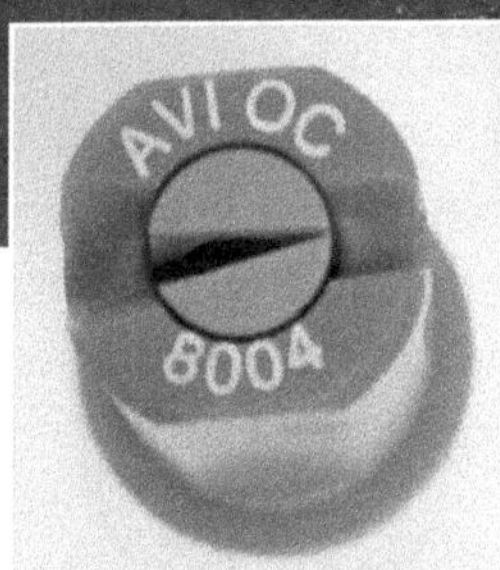

Figures 14.42–14.44.
The left-hand nozzle here is an off-center flat fan nozzle, showing its distinctive spray pattern.

Figures 14.45–14.46.
The two exit orifices on this nozzle create a symmetrical flat fan pattern with one spray directed forward and the other backward. The exit orifices are identical in terms of droplet size and flow rate (top and bottom).

Figures 14.47–14.48. This asymmetrical TeeJet® AI3070 nozzle shows two different spray patterns.

Other dual spray nozzles are designed to have asymmetrical angles (for example, 30° and 70°). For example, the TeeJet® AI3070 nozzle has a forward pattern of 30° to the perpendicular and backward pattern of 70°. This type of nozzle is ideal for fungicide applications because it provides superior coverage of all sides of cereal crop heads (such as wheat) at the top of the canopy.

With these nozzles operators may choose to alternate tips forward and backward along the boom to produce four different angles. There are also nozzle bodies that can accomplish the same thing by holding two flat fans of different angles. These nozzles may also come with a pre-orifice for air induction, generating relatively coarser droplets in both fans.

A fourth type of tapered flat fan is called a *flooding nozzle* or *deflector flat fan*. This type of nozzle is designed to deliver high volumes, especially when operators apply products directly to the soil (as with preemergence liquid fertilizer applications). The spray pattern is not defined by the exit orifice but by a stream of fluid being pushed against a flat metal or polymer deflection plate.

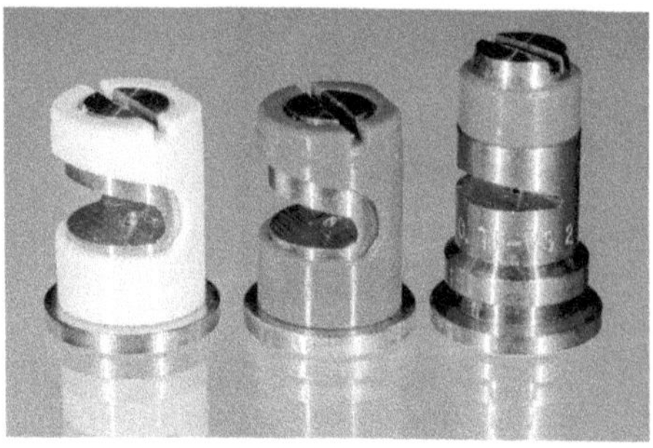

Figure 14.49. Examples of flood jet nozzles.

Figure 14.50. Flood jet nozzles produce much larger drops than other nozzles.

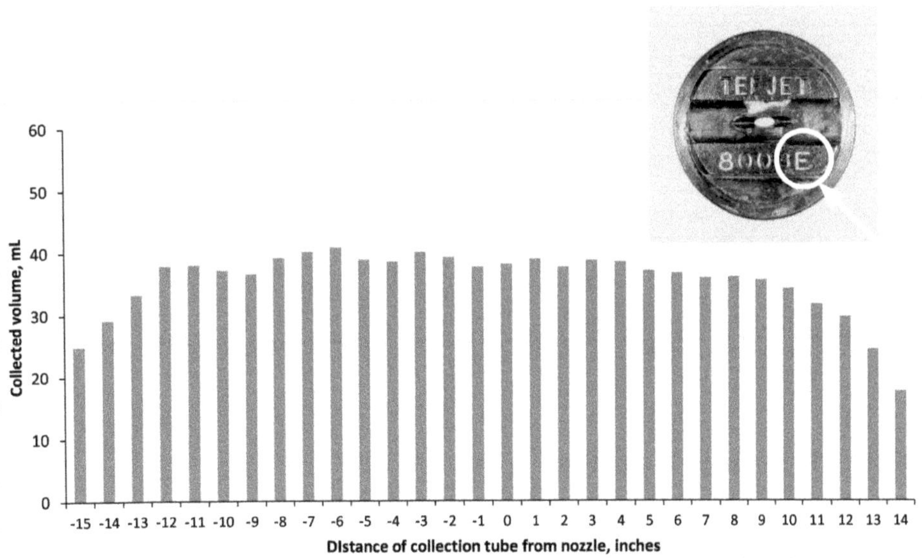

Figures 14.51–14.52. An even flat fan nozzle is identified by the letter E in part of its description. The illustration shows an even pattern across the length of the spray width.

These high-volume nozzles have larger exit orifices that don't clog easily. Usually, these nozzles do not provide as even a distribution along the boom as other flat fan nozzles. However, new deflector-type flat fan spray tip designs produce larger droplets with a more uniform size and a lower coefficient of variation.

A fifth type of tapered flat fan is called an *even flat fan nozzle*. The flow from even flat fan nozzles does not taper toward the edge of the long axis of the fan pattern. These nozzles are generally used for banding applications over a specific area rather than broadcast. For instance, even nozzles may be suspended above planted rows when the pesticide may be too expensive to use as a broadcast application. Alternatively, applicators may align even nozzles with the alleys and not the plant rows to apply herbicides to the weeds between rows of crops, not to the crop itself.

Even flat fan nozzles provide equal distribution over the entire spray swath without the need to overlap adjacent nozzles. Even flat fan nozzles are intended to be used individually. Any overlap between adjacent even nozzles would result in overapplication of the spray solution.

Using an even flat fan nozzle requires the operator to keep the height above the target rather precise. If the boom is too low above the target, then the pattern will be narrower than desired and result in overapplication of the product (a smaller surface area will be covered with the same amount of product). If the boom is too high above the target, then the pattern will be wider than desired and result in underapplication of the spray solution (a greater surface area will be covered with the same amount of product). Fortunately, nozzle manufacturer catalogs and websites provide the appropriate boom height for a given spray band width when using even flat fan nozzles.

Figures 14.53–14.54. These photos look into the nozzles toward the exit orifices of a conventional tapered flat fan (top, stainless steel nozzle) and an even flat fan (bottom, brass nozzle). Notice the differences in the shape of the exit orifices and the width of the cuts of the tapered flat fan (top nozzle) and the even flat fan (bottom).

Other Hydraulic Nozzles

Nozzles in the "other" category include cone, solid stream, and boomless nozzles. Some boomless equipment may come with rotary atomizers, which are present on some new drone application units. However, this chapter does not discuss rotary atomizers (see chapter 15).

Cone Nozzles

Cone nozzles can be divided into two types: nozzles that produce a hollow-cone pattern and nozzles that produce a full-cone pattern, which can be produced mostly by swirling the liquid.

Cone nozzles may be offered in a single-piece design or in a two-piece disc and core design. Cone nozzles are used mostly for insecticide and fungicide applications to field and fruit crops when foliage penetration and complete coverage of the lower leaf surface is critical for effective application. The major difference between the two is that with the hollow cone, most of the spray droplets are concentrated at the edges

Figure 14.55. This illustration shows a hollow cone (left) and full cone (right). The illustration of the hollow cone shows spray droplets concentrated at the edges of the cone; however, this is somewhat misleading. As the droplets are dispersed along the circular pattern, many small droplets are propelled into the center of the nozzle.

Figures 14.56–14.57. These photos show examples of hollow-cone nozzles (left), which show the larger exit orifices to allow high-volume applications. The nozzle with the larger opening can apply 2.5 gallons of water per minute, while the smaller exit orifice can apply 1.25 gallons (right).

of the cone. With the full cone, the spray droplets are spread somewhat uniformly across the cone pattern. The cone angles for nozzles may vary between 30° and 170°.

Operators can control the flow rates for both nozzles by selecting orifice/disc inserts of different sizes or volume capabilities. For a disc insert, the full-cone disc presents a slot in the middle (as shown in figure 14.55). Manufacturers have developed a newer type of hollow-cone nozzle with an air induction system for use in airblast sprayers when applicators need drift control. Some applicators combine two nozzle types, using regular hollow-cone nozzles at the bottom part of an airblast sprayer and air induction hollow-cone nozzles at the bottom.

These nozzles were designed to work separately from one another. However, some applicators use them on horizontal linear booms. When they do, the distribution across the boom shows a lot of variability, with a coefficient of variation of 20 percent or more.

Solid Stream Spray Pattern

As the name implies, solid stream nozzles produce a liquid stream. Some solid stream nozzles have multiple exit orifices to help distribute the product. There are a variety of sizes, including one to seven individual streams.

This nozzle forms a solid stream for applications that do not require atomization of spray liquid (or when atomization is not desirable) for an effective application, such as liquid fertilizer applications on wheat after emergence. A spray cloud of droplets would effectively cover the wheat foliage and result in leaf burn. The solid stream design avoids thorough leaf coverage and results in minimal leaf injury.

Figure 14.58. This photo shows the differences in the number of openings in solid stream nozzles.

Figure 14.59. A straight-stream jet nozzle positioned on a boom.

Figure 14.60. A straight-stream jet nozzle delivering product in the field.

For instance, applicators use one-stream jet nozzles to apply fertilizers, insecticides, and biologicals directly into the furrow close to the planted seed. Three-stream (or similar) tips produce a narrower coverage area and project liquid more directly onto the soil surface. Users commonly use such nozzles for directed spray applications. Seven-stream tips produce a wider coverage area, which makes them ideal for broadcast applications on both standing crops and bare soil. Due to the greater number of streams and the wider spray pattern, these tips are ideal for higher boom heights and higher ground speeds.

Boomless Nozzles

Boomless nozzles deliver uniform spray patterns across broad areas where application with a boom sprayer would be difficult—if the boom could hit obstructions (such as brush) or the terrain is so uneven that the boom height would be too variable. An applicator might mount a single boomless nozzle on the back of an ATV. This would be easier than using a 10-foot-wide fixed boom for herbicide applications in pastures and golf courses.

Figures 14.61 – 14.62. Boomless nozzles used (left) to control invasive weeds in rights-of-way and (right) to control brush in a ditch bank.

NOZZLE TIP SCREENS, CAPS, AND TURRETS

Screens and caps are accessories that play important roles in the pesticide application process.

Screens Help Prevent Nozzle Plugging

Applicators often use water from local sources that contain solid particles and debris, especially for right-of-way applications. When the applicators place screens (that is, filters and strainers) before the nozzles' exit orifices, it can help remove those suspended materials in the liquid stream that could plug the exit orifices. Line strainers and tip strainers are very important to the systems and must be properly sized and placed and be in good condition. Such preparations can help prevent spray tips from clogging and assure the tank mixture is properly filtered.

Applicators can select nozzle screens by the nozzle sizes and the types of products (herbicides, fungicides, etc.) they are applying. Properly sized screens can remove solid objects (like grains of sand) found in mixing water. They can also capture undissolved solid pesticides that were inadequately agitated in the spray tank. Screens can extend nozzle life by capturing these abrasive materials before they pass through the orifices.

There are several common mesh sizes for nozzle screens used in pesticide applications: 50-, 80-, 100-, and 200-mesh. A 50-mesh screen has 50 openings per linear inch, an 80-mesh screen has 80 openings per linear inch, and so on.

The screen openings get much smaller the more openings it has. The openings on a 100-mesh screen are 150 microns, while the openings on a 50-mesh screen are 300 microns. This means that 100-mesh screens capture and retain smaller solid particles before the spray liquid enters the nozzle.

Figures 14.63–14.64. Inline strainers filtering particulate matter in well water. The photo with the nozzle body shows what the nozzle body strainer (screen) filtered out so that the Greenleaf tip would not plug or have the spray pattern compromised by the particulate.

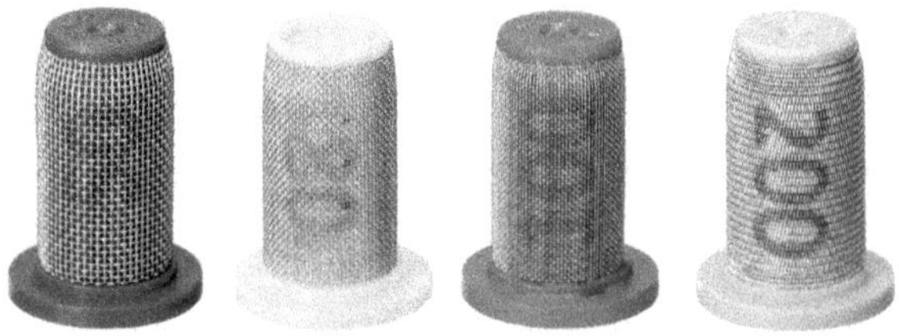

Figure 14.65. Screens are color-coded to help identify mesh sizes. More recently manufactured screens are also marked with their mesh size for quicker identification. From left to right, this photo shows a 50-mesh screen (blue), a 80-mesh screen (yellow), a 100-mesh screen (green), and a 200-mesh screen (pink).

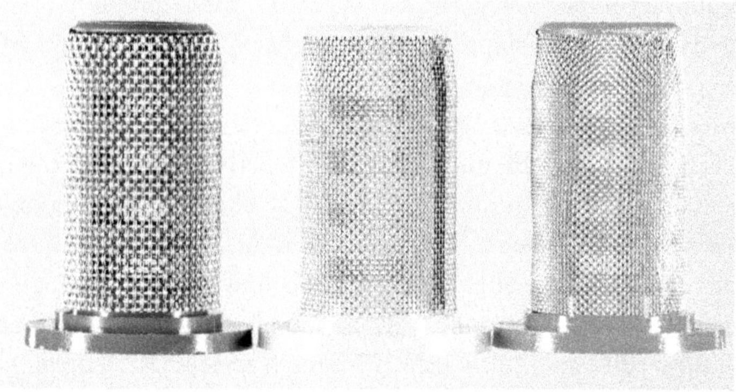

Figure 14.66. A screen's width determines what size materials will pass through it. A 50-mesh screen (left; blue) has larger openings than an 80-mesh screen (center; yellow), which has larger openings than the 100-mesh screen (right; green).

Manufacturers determine screen size recommendations by considering the pre-orifice nozzle flow rate and exit orifice size. Nozzles with lower flow rates have smaller screen sizes (higher mesh numbers), while higher flow rates have larger screen sizes (lower mesh numbers). For example, a nozzle with a low flow rate of less than 0.2 gallons per minute (gpm) will generally be paired with a 100-mesh screen. The lower flow rate means the nozzles have smaller pre-orifice and exit orifice sizes, which means they are easier to plug. Nozzles with flow rates of 0.2 gpm or greater have 50- or 80-mesh screens. Such screens are recommended because smaller particles can pass through the exit orifices.

COLOR CODES ARE NOT LINKED

It is important to note that the colors assigned to screen mesh size, nozzle flow rates, and spray droplet size categories are independent of one another. Each one follows a different ISO standard that determines the color coding.

Dual nozzles can use different screen sizes from single nozzles because each exit orifice is half the size of that of a single fan nozzle with the same flow rate.

Nozzle tip screens do not alter the flow rate unless the screen becomes clogged with debris. Problems with clogging exit orifices often arise when operators use finer mesh screens than the nozzle manufacturers recommend (such as using 100-mesh rather than 50- or 80-mesh). For example, a general recommendation for applying complex pesticide formulations (such as a suspension concentrate, capsule suspension, etc.) is a 50-mesh or coarser screen. If an applicator uses a finer screen (such as 100-mesh), the screen can get clogged and slow the flow rate. If applicators use these finer, larger-numbered screens, they may have to clean them more often, meaning downtime in the field.

Figure 14.67. Each nozzle type should have a recommendation for the size of screen needed, given here in parentheses under the nozzle number.

To avoid excessive clogging, it is also important to correctly select the screens for the line strainers. Self-cleaning line strainers minimize clogging but do require minimum flow rates—6 gallons per minute for ¾- and 1-inch sizes, 8 gallons per minute for 1¼- and 1½-inch sizes through the bypass line. Operators also should make sure the tank mixture is prepared according to the manufacturer's recommendations to avoid incompatibility.

Opinions differ on whether screens at the spray tips are necessary. Some commercial application companies no longer use tip screens ahead of the nozzles. Instead, they have reduced the mesh sizes of the screens on the inline filters before and after the pump. They save time by not having to clean so many screens—there could be 96 screens on a 120-foot boom with nozzles placed on 15-inch centers.

Figures 14.68–14.69. (Left) This stainless steel screen uses a spring ball valve to reduce dripping out of the nozzle. This is important if a nozzle body doesn't have a rubber diaphragm valve above the tip as shown in the right-hand photo. As a general rule, do not combine a spring ball valve screen with a nozzle body that has a diaphragm valve.

WARNING! NOZZLE SCREEN COLORS CHANGED

Since screens do not wear out as quickly as spray tips, applicators may have screens they purchased more than 10 years ago. Using screens purchased before 2016 is problematic because the ISO changed the colors assigned to screen mesh sizes that year. Be wary of screens that do not have the number marked on them. You could unintentionally use the wrong screen with a nozzle. While today's green screen is the same mesh size as the pre-2016 screen, the old red and blue screens changed. Color-coded screens that do not have numbers may be older screens.

Spray Boom, Turrets, and Caps

There are generally two categories of booms: wet and dry. The categories are based on how the liquid is supplied to the nozzle body. A wet boom has a standard stainless steel or plastic pipe, and the nozzle bodies are attached directly into the boom. Wet booms come with predrilled holes (for example, 15 inches apart) where applicators will mount the nozzle bodies. If an applicator wants 30-inch centers, they can shut off every other nozzle.

A dry boom has a steel, aluminum, or carbon fiber boom structure with rubber hoses (for example, EPDM) that connect individual nozzle bodies, which are mounted on the boom structure (see figures 14.71–14.73). They are called *dry booms* because no liquid flows through the boom itself.

Figure 14.70. This 90-foot wet boom has 15-inch centers, which means there are 72 evenly spaced holes in which to place the nozzle bodies.

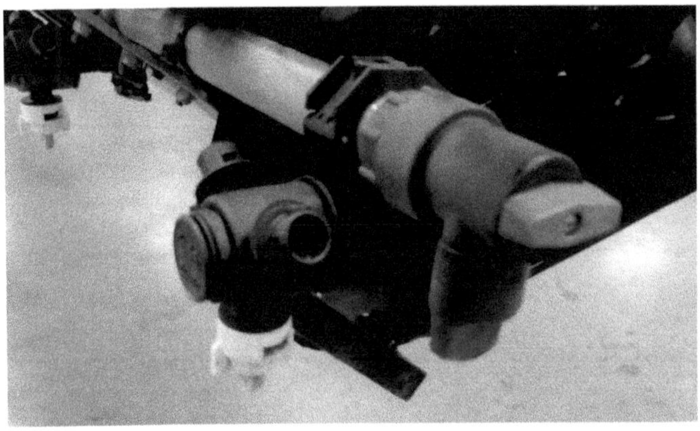

Figures 14.71–14.73. Different cap designs hold and lock specific nozzles in place.

One popular adaptation for wet and dry booms is a nozzle turret. Turrets can be made to hold three to six nozzle caps. Applicators can rotate the nozzle turret to select the proper spray tip to deal with different scenarios: to manage drift, provide a greater application volume, or generate a finer droplet spectrum for greater spray coverage. Because turrets make it easy to switch from one nozzle to another, applicators can change spray tips quickly to optimize each application.

Figure 14.74. By twisting the turret, applicators can quickly switch nozzles when crops or environmental conditions change. When installing nozzle tips on turrets, make sure to place each tip in the same order on each turret. This will speed the transition to a new nozzle and reduce any chance for error in selecting the wrong nozzles.

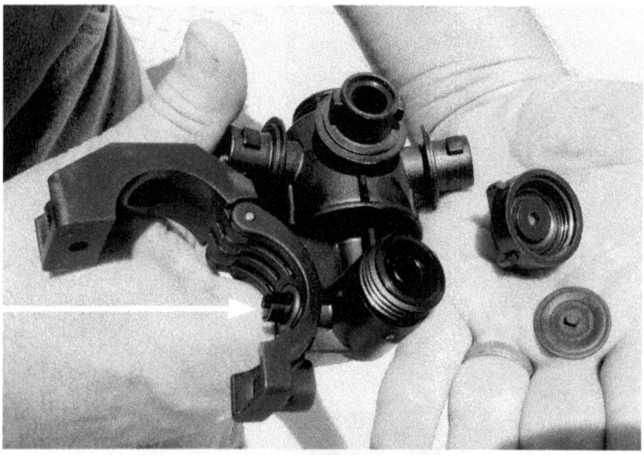

Figure 14.75. The stem on this new turret (arrow) fits into the hole in a wet boom, which is then clamped in place.

Figure 14.76. This photo shows a cap with a built-in plastic tip before being placed on the turret.

Figure 14.77. Different cap designs hold and lock specific nozzles in place.

Spray caps hold the nozzles and (if used) filter screens. Caps are specifically shaped parts that hold nozzles in place, attach the nozzles to the nozzle bodies, and ensure proper seals to prevent any leaks. Manufactures have created spray caps of numerous designs as nozzle types, equipment booms, and other application equipment have changed over time. More caps are specifically designed for different spray nozzles to fit in.

This means nozzles are made to fit into certain caps, which are then attached to certain boom components. There are a few standards that designate the dimensions of spray bodies. There are differences, but spray bodies from different manufacturers are more similar than dissimilar with regard to external dimensions. Nozzles must be designed to adapt to the bodies that applicators commonly use.

It is important to realize that a cap's color is independent of the color of the nozzle inserted into it. Manufacturers color their caps for applicators who want to match the cap color to the nozzle flow rate color. There are a couple of reasons to use colored caps matching the nozzle colors. First, many stainless steel and polymer spray tips are difficult to read once they are on a sprayer. By using caps with a color similar to the nozzles, applicators can quickly ascertain the flow rates of each nozzle. Second, it is easier to see the colors when the spray equipment gets dirty.

Many nozzles are available where the nozzle tips and caps are molded into one unit. Two examples are the TeeJet® AIC and XRC. The *C* in each name indicates *combination/combo.* When combo nozzle tips and caps are available and meet your needs, this is usually the best way to purchase the nozzle tips and caps for easy assembly.

DEMYSTIFYING THE NOZZLE SELECTION PROCESS

Spray nozzles have evolved from the more simplistic designs of yesteryear to more complicated and diverse designs. With improved manufacturing and engineering processes, modern nozzles can operate reliably over greater pressure ranges and produce consistent droplet sizes and spray patterns for a wide range of flow rates.

With literally hundreds of designs available, selecting the nozzle best suited to your task can seem daunting. It's easy to feel overwhelmed when confronted with the technical language and seemingly complicated multicolored tables filled with numbers.

The problem gets worse if you go straight to the catalogs and haven't considered what you want those nozzles to accomplish. The good news is that with a little up-front effort, you will soon be an expert at reading and interpreting the information from nozzle manufacturers' tables and charts.

Still, there will always be a segment of growers who prefer to rely on the nozzle recommendations of retailers, salespeople, consultants, or fellow farmers in the neighborhood. This faithful approach assumes that your advisor isn't selling something and that they have done the legwork required to understand the best nozzle choice for your unique situation. Wouldn't it be prudent to have a working understanding of how to choose a nozzle, if only so you can ask them a few important questions to keep them honest?

Perhaps you feel that replacing a worn set of nozzles with exact replicas is the fastest and safest approach. After all, if those worn nozzles performed well in the past, it is logical to assume they will continue to perform well in the future. Buying the exact set of nozzles that worked for you in the past may be fine, but make sure that these application parameters have not changed:

- You are spraying the *same* pests.
- You are operating the *same* sprayer.
- You are applying the *same* products or similar chemistries.
- You are applying the *same* volume (gallons per acre).
- You are using the *same* speeds and pressures.
- You are spraying under the *same* weather conditions.
- You have noted that everything bordering the field remains the *same.*
- You are operating under the *same* regulatory requirements.

If all these factors remain the same, there is no harm in saying, "I'll take another set of these same nozzles, please." But even if your situation has remained the same, there may be new nozzle designs that offer better performance. You may not be

WHAT YOU HATE TO SEE

One crop consultant tells the story about the problems improper nozzle choices can create:

> **I recently went to a field complaint where the applicator claimed the lack of control was due to weed resistance. The grower had used an air induction nozzle to apply a postemergent contact herbicide. The manufacturer's catalog clearly indicated that this nozzle design was not appropriate for contact herbicides. So the true culprit was not resistance; it was poor coverage that resulted from droplet sizes that were too large. There was nothing else I could say to make the situation any better.**

taking full advantage of new designs and technology that can reduce the range of droplet sizes. Having fewer droplets in the largest part of the spectrum reduces runoff and waste. Having fewer droplets in the smallest part of the spectrum reduces the potential for off-target drift. This leaves you in the Goldilocks zone of efficient and optimal control.

Altering One Variable Changes All Others

Replacing worn-out nozzles is an opportunity to evaluate whether the current nozzles are still the best option. When purchasing new nozzles, remember to consider the nozzle spacing and spray fan angle, the amount of spray carrier, and the travel speed.

Nozzle Spacing and Spray Fan Angle

Nozzle spacing and the spray fan angle (as influenced by nozzle design and pressure) determine the optimal boom height. Nozzle spacing and boom height collectively determine the degree of spray overlap. Excessive overlap isn't a problem, but additional nozzles do mean added expense, cleaning time, and potential for plugging.

The closer the nozzle spacing, the more nozzles there are on the boom, and the smaller each nozzle must be to achieve the correct gallon-per-minute output per nozzle. This could mean more potential driftable fines as the nozzle spacing decreases. An operator could negate this slight change in driftable fines by using a lower boom height.

Conversely, gaps in the pattern from relatively widely spaced nozzles and low boom height could lead to sublethal applications or flat-out misses. Some nozzles have an attack angle of 15°. This permits a lower boom height and improves the spray pattern compared to nozzles pointed straight down.

Be aware: Some nozzles don't produce the fan angle they advertise. It's always prudent to visually inspect sprayers during calibration to ensure sufficient overlap. Make sure to use water in the calibration and not the pesticide solution. Use gloves even though it is only water in the tank. Although it will be hard to calculate the exact overlap in the field, you can have a great idea. To check the spray pattern, use an LED strobe light in dark or low-light conditions to more easily determine the spray pattern and droplet size.

Amount of Spray Carrier

Use the amount of spray carrier necessary to achieve spray coverage. This improves the odds of achieving threshold coverage. A nozzle that emits a higher flow rate tends to produce fewer fines than another nozzle with the same design but a lower flow rate.

Travel Speed

Travel speed can influence boom height. As a sprayer's travel speed increases, the boom height can fluctuate more based on field topography changes, or the boom can bounce over uneven field surfaces. Manual or automatic boom height corrections will not be as responsive at faster speeds.

As the speed increases, boom height usually varies more. If you can maintain the boom height as the sprayer speed increases, then there should not necessarily be more drift. To avoid problems, drive as slow as is economically feasible to maintain a stable boom height and allow for coverage across the field or treated area. Typically, sprayer travel speeds for perimeter application passes around fields are slow to maintain minimal boom height and low spray pressure to allow operators to reduce physical spray drift and watch for spray obstacles or sensitive areas along field edges.

Many people still believe nozzle choice is solely a matter of droplet size, carrier volume, travel speed, and pressure. Those are all critical selection criteria, but what is more important is how these factors interact. Altering one spray variable affects the other variables.

Assume you are having trouble controlling white mold in soybean with a nozzle that generates a volume median diameter (VMD) in the very coarse category. The retailer asks, "What carrier volume do you use?" You answer, "Ten gallons per acre." The retailer follows up, asking if can you bump your rate to 15 or 20 gallons per acre. You reply, "I can make 15 gallons work, but 20 gallons would slow me down too much."

"Can you slow down a little?" the retailer asks. "Pick a speed that allows you to get the job done in a reasonable time. At these faster speeds you are traveling, the booms are higher."

The retailer asks one last question, "Is there anything around you that means you are worried about off-target spray drift?" If the answer is yes, then a medium VMD nozzle usually works well for reducing spray drift while still providing satisfactory spray coverage for contact-type pesticides. You are lucky that the nozzle being used was versatile enough to make those changes without having to purchase a new set.

In a perfect world, the best application for the above example would be achieved by using more water, driving at slower speeds, having nozzles that produce smaller droplets, and lowering the booms to improve spray movement into the canopy. But is there ever a time that's perfect for spraying? Many of the questions that determine how the sprayer is calibrated and pesticides are applied properly rely on the applicator's knowledge and management skills.

Without additional information and modifications, the charts nozzle manufacturers provide are only one piece of the whole application puzzle. The charts provide the technical parameters. In other words, if you do this, then this will happen. But you still must decide the nozzle type, droplet size, application speed, and carrier volume by considering different factors. For example, you know what is around the property being treated—such as ponds, lakes, streams, sensitive crops, and beehives. You must know the context of your application before you can pick a nozzle specific to that pesticide, to that crop, and to the targeted site.

	Droplet	PSI	GPM	GALLONS PER ACRE BASED ON 15" NOZZLE SPACING 4 MPH	5 MPH	6 MPH	8 MPH	10 MPH	12 MPH	14 MPH	16 MPH	18 MPH	20 MPH
AM11001	C	15	0.06	6.1	4.8	4.0	3.0	2.4	2.0	1.7	1.5	1.3	1.2
	C	20	0.07	7.0	5.6	4.7	3.5	2.8	2.3	2.0	1.8	1.6	1.4
	M	30	0.09	8.6	6.9	5.7	4.3	3.4	2.9	2.4	2.1	1.9	1.7
	M	40	0.10	9.9	7.9	6.6	5.0	4.0	3.3	2.8	2.5	2.2	2.0
	F	50	0.11	11.1	8.9	7.4	5.5	4.4	3.7	3.2	2.8	2.5	2.2
	F	60	0.12	12.1	9.7	8.1	6.1	4.8	4.0	3.5	3.0	2.7	2.4
	F	70	0.13	13.1	10.5	8.7	6.5	5.2	4.4	3.7	3.3	2.9	2.6
	F	80	0.14	14.0	11.2	9.3	7.0	5.6	4.7	4.0	3.5	3.1	2.8
	F	90	0.15	14.9	11.9	9.9	7.4	5.9	5.0	4.2	3.7	3.3	3.0
AM110015	C	15	0.09	9.1	7.3	6.1	4.5	3.6	3.0	2.6	2.3	2.0	1.8
	C	20	0.11	10.5	8.4	7.0	5.3	4.2	3.5	3.0	2.6	2.3	2.1
	M	30	0.13	12.9	10.3	8.6	6.4	5.1	4.3	3.7	3.2	2.9	2.6
	M	40	0.15	14.9	11.9	9.9	7.4	5.9	5.0	4.2	3.7	3.3	3.0
	F	50	0.17	16.6	13.3	11.1	8.3	6.6	5.5	4.7	4.2	3.7	3.3
	F	60	0.18	18.2	14.5	12.1	9.1	7.3	6.1	5.2	4.5	4.0	3.6
	F	70	0.20	19.6	15.7	13.1	9.8	7.9	6.5	5.6	4.9	4.4	3.9
	F	80	0.21	21.0	16.8	14.0	10.5	8.4	7.0	6.0	5.3	4.7	4.2
	F	90	0.23	22.3	17.8	14.9	11.1	8.9	7.4	6.4	5.6	5.0	4.5

Figure 14.78. Nozzle manufacturers publish tables that list the critical calibration factors: nozzle spacing, gallons per acre, travel speed, droplet size, nozzle flow rate, pressure, and in some cases the initial spray angle. Using Greenleaf Technology's AirMix® nozzle as an example, you can see that these factors are interrelated–changing one affects the others. You need to know how your sprayer operates before looking at such tables.

Criteria for Selecting New Nozzles

As we have established, no single nozzle can do it all. You'll need a few nozzles to handle common situations, weather changes, and specialty applications. There are many ways to select a nozzle for each situation, but we propose these 10 criteria.

Criterion 1: Droplet Size

Some pesticide labels prescribe droplet sizes, while others do not. For example, one herbicide label says to use "nozzles and pressures that generate Medium (about 250 to 300 microns) spray droplets as reported by the nozzle manufacturer and in accordance with ISO 25358. Do not use nozzles and pressures that result in Coarse sprays. Fine sprays should also be avoided to minimize spray drift risk."

When you search a nozzle table for a spray droplet, it's best to focus on the classification symbols (VF, F, M, C, VC, XC, UC). There are colors associated with droplet size, but they are easily confused with the colors associated with flow rate. In table 14.9, a medium (M) nozzle has a yellow color code. However, yellow can also denote a 0.2 (XR8002 and XR 11002), which is a nozzle that emits 0.2 gpm at 40 psi. In other words, if a nozzle is constructed of yellow plastic, that does not mean it creates a medium droplet size.

Looking at the XR flat fan nozzle in the TeeJet® catalog, we find that a medium droplet size is influenced by pressure and the orifice size that dictates flow rate (gallons per minute per nozzle). Table 14.9 clearly shows that changing the orifice size and pressure can change the droplet size category from fine (F, orange), to medium (M, yellow), to coarse (C, green), to very coarse (VC, blue), to extremely coarse (XC, white). Based on the herbicide label requirements, an operator could produce a medium spray droplet by utilizing several different nozzles with different orifice sizes (or flow rates) at different pressures.

Table 14.9. XR TeeJet® Flat Fan Nozzle Droplet Sizes

XR TEEJET® (XR)

TIP PART NO.	PSI 15	20	30	40	50	60
XR8001	F	F	F	F	F	F
XR11001	F	F	F	F	F	VF
XR80015	M	F	F	F	F	F
XR110015	M	F	F	F	F	F
XR8002	M	M	F	F	F	F
XR11002	M	M	F	F	F	F
XR80025	M	M	M	F	F	F
XR110025	M	M	M	F	F	F
XR8003	M	M	M	F	F	F
XR11003	M	M	M	F	F	F
XR80035	M	M	M	M	F	F
XR8004	M	M	M	M	F	F
XR11004	M	M	M	M	F	F
XR8005	C	M	M	M	F	F
XR11005	M	M	M	M	M	F
XR8006	C	C	M	M	M	M
XR11006	C	M	M	M	M	M
XR8008	VC	C	C	M	M	M
XR11008	C	M	M	M	M	M
XR8010	VC	C	C	M	M	M
XR11010	C	C	C	M	M	M
XR8015	XC	VC	VC	C	C	M
XR11015	VC	VC	C	C	C	M
XR11020	XC	VC	VC	C	C	C

Table 14.10. Turbo TeeJet® Nozzle Droplet Sizes at Different Pressures

TURBO TEEJET® (TT)

TIP PART NO.	PSI 15	20	25	30	35	40	50	60	70	80	90
TT11001	VC	C	C	M	M	M	M	M	F	F	F
TT110015	VC	VC	C	C	C	M	M	M	M	F	F
TT11002	VC	VC	C	C	C	M	M	M	M	F	F
TT110025	VC	VC	C	C	C	M	M	M	M	F	F
TT11003	XC	VC	VC	C	C	M	M	M	M	F	F
TT11004	XC	VC	VC	C	C	M	M	M	M	F	F
TT11005	XC	VC	VC	C	C	M	M	M	M	F	F
TT11006	XC	VC	VC	C	C	M	M	M	M	F	F
TT11008	XC	VC	VC	VC	C	M	M	M	M	M	F
TT11010	UC	XC	XC	VC	VC	VC	C	C	C	M	M
TT11012	UC	XC	XC	XC	VC	VC	VC	VC	C	C	C

But if you compare the driftable fines of an XR with those of a Turbo TeeJet® (TT) in table 14.10, you will see that both produce a similar droplet size range; however, the TT nozzle produces far fewer driftable fines than the XR, which means the TT provides a lower risk of spray drift for herbicide applications.

When pesticide labels are silent on preferred droplet sizes, you need to use your judgment when selecting nozzles. This means you need to know if the product is a contact spray or a systemic product, as this will influence the droplet size you desire and the nozzle you select. Research shows that larger droplets better penetrate the canopy because they maintain velocity, while smaller droplets lose momentum and fail to penetrate the canopy as well.

Contact products tend to require smaller droplets, because they increase the odds of hitting (and being retained on) smaller and hard-to-wet targets and cover more out-of-sight surfaces. Finally, the finer the spray, the more droplets there are for a given volume, which has the potential to increase deposit density and improve the odds of contacting targets that may have diseases and some insects.

But as we have established, it is hard to control the fate of small droplets. They are more prone to drift, and this can compromise coverage uniformity. Large droplets have a greater chance of penetrating the canopy. The question becomes how much coverage we need in the canopy because large droplets don't provide adequate coverage.

Systemic (translocated) pesticides do not require thorough coverage on target surfaces, which may permit the use of coarser droplets. Coarser droplets are less prone to evaporate or to be moved off target by wind or turbulence. Once coarser droplets deposit on a surface, they may remain in liquid form longer than finer deposits, and this can improve the chances for a pesticide's active ingredients to penetrate the target tissue.

Unfortunately, current nozzles do not provide uniform droplet sizes. For this reason, nozzles will always discharge some unwanted droplets. For example, we know that soil-active herbicides are relatively insensitive to droplet size, so it would make sense to use coarser droplets to ensure they stay on target. On the other hand, crop residues left in the field may prevent coarser sprays from reaching the target, so finer droplets would improve coverage. Fortunately, medium-coarse spray classifications have enough variability and range in droplet sizes to give us the best of both worlds.

Herbicides are generally more forgiving than fungicides and insecticides when it comes to coverage, because many herbicides have systemic activity in the target weeds. That means a droplet need only contact some part of the target weed for the active ingredient to physically penetrate the tissue. For such uses, you can use a coarse spray or carrier (with sufficient volume) without the risk of missing the target. Additionally, coarser droplets form deposits that remain in liquid form longer than finer droplets, which provides more time for the active ingredient to penetrate the plant.

Fungicides are often formulated to work as protectants on leaf surfaces. Many diseases infect leaves as airborne spores. That means fungicides must already be in place and spread evenly over the leaf to ensure the spore contacts the fungicide residue. To improve the odds of fungicide-to-spore contact, the fungicide may be best applied as fine and medium spray droplets.

Insecticides also tend to be formulated as contact products, but spray coverage may not be as crucial as with contact fungicides. Many insects are mobile, and there are better odds that they will walk on or eat the product. A medium-coarse spray droplet may be sufficient for threshold coverage.

½ = 8

Consider the volume of liquid in a 500 µm diameter droplet. Now consider the volume in a droplet half that diameter: 250 µm. You might think that by cutting the diameter in half, you get twice as many droplets, but you'd be wrong. Volume is cubic, so there is enough volume in a single 500 µm droplet to make eight 250 µm droplets! So there are really two ways to improve the odds of achieving good coverage:

1. **Raise the carrier volume applied per acre.**
2. **Reduce the average droplet size.**

Criterion 2: Carrier Volume

The general convention is that more water translates to better coverage. That is true up to a point, but there's a diminishing return as sprays saturate surfaces and the product simply runs off.

A conventional hydraulic nozzle meters the flow of bulk spray liquid, converts it into a range of droplet sizes, and disperses them in a specific geometric shape. This means operators performing a broadcast spray can control carrier application volume in four ways:

1. Alter travel speeds.
2. Change pressures.
3. Choose a nozzle with a different flow rate.
4. Shut off every other nozzle with the nozzle spacing (for example, on 10-inch or 15-inch spacing).

Labels often list the minimum carrier volume. An instruction from one insecticide label reads: "Use a minimum of 5 to 10 gallons per acre, increasing volume with crop size and/or pest pressure." The maximum carrier application rate you can use depends on your schedule, of course (refills take time). But as the label indicates, the size and density of the crop and the degree of pest densities you are dealing with should be the primary considerations when choosing carrier volume.

Criterion 3: Travel Speed

Travel speed often involves making a compromise between *timely* applications and *precise* applications. If you find you're always rushed and you abandon best practices just to get it done, then you should step back and look critically at your spraying strategy. What is the bottleneck? How can you address it?

In any application it is nearly impossible to maintain a constant speed from start to finish—at a minimum, you must slow down to navigate end row turns. The faster

Table 14.11. Application Rates for the AIXR11003 Tip

TIP PART NO. (STRAINER MESH SIZE)	PSI	DROP SIZE	CAPACITY ONE TIP IN GPM	CAPACITY ONE TIP IN OZ/MIN	APPLICATION RATE FOR 20" SPRAY TIP SPACING — GALLONS PER ACRE (GPA)								TURF APPLICATION GALLONS PER 1000 SQ. FT.			
					4 MPH	5 MPH	6 MPH	8 MPH	10 MPH	12 MPH	15 MPH	20 MPH	2 MPH	3 MPH	4 MPH	5 MPH
AIXR11003 (50)	15	XC	0.18	23	13.4	10.7	8.9	6.7	5.3	4.5	3.6	2.7	0.61	0.41	0.31	0.24
	20	VC	0.21	27	15.6	12.5	10.4	7.8	6.2	5.2	4.2	3.1	0.71	0.48	0.36	0.29
	30	VC	0.26	33	19.3	15.4	12.9	9.7	7.7	6.4	5.1	3.9	0.88	0.59	0.44	0.35
	40	C	0.30	38	22	17.8	14.9	11.1	8.9	7.4	5.9	4.5	1.0	0.68	0.51	0.41
	50	C	0.34	44	25	20	16.8	12.6	10.1	8.4	6.7	5.0	1.2	0.77	0.58	0.46
	60	M	0.37	47	27	22	18.3	13.7	11.0	9.2	7.3	5.5	1.3	0.84	0.63	0.50
	75	M	0.41	52	30	24	20	15.2	12.2	10.1	8.1	6.1	1.4	0.93	0.70	0.56
	90	M	0.45	58	33	27	22	16.7	13.4	11.1	8.9	6.7	1.5	1.0	0.77	0.61

you drive your sprayer, the more the rate controller will increase the spray pressure to achieve the desired carrier volume. As pressure increases, the droplet size distribution will change. For example, an XR11004 nozzle identified as producing an average medium droplet spray (M, color-coded yellow in table 14.9) at 15 psi could generate fine spray (F, color-coded orange) in the pressure range starting at 50 psi.

In table 14.11, we can see that when applying 10.1 gallons per acre at a speed of 10 mph at 40 psi, the AIXR11003 tip will produce a coarse spray. But if the speed is increased to 12 mph, the pressure increases to 60 psi, and the nozzle will now produce a medium spray.

As we've discussed, faster travel speeds carry the potential challenges of higher or bouncing booms. This is a function of the boom linkage design. Even with autoleveling systems, this can still be a concern. High or changeable boom heights increase drift potential and produce less uniform deposition under the boom.

THE POOR HUMBLE NOZZLE

Everyone wants the perfect droplet size—large enough to stay on course and defy evaporation but small enough to increase droplet counts and improve coverage and retention. Fortunately, we can hedge our bets a little when we consider that each nozzle produces a range of droplet sizes. Ultimately, the humble nozzle can only do so much. While it represents the last point of control for a sprayer operator, it is still only a part of the overall spray strategy. So it is not only what tool you use but how you use it.

Criterion 4: Nozzle Angle

The key to effective application is to get as close to 100 percent uniformity in volume distribution across the boom as possible. This simply means that whatever is below the nozzles is going to get the same rate and the same combination of droplets. This consistency always improves performance.

Twenty years ago, 80° nozzles were the standard flat fan angle choice. The spray pattern for an 80° flat fan nozzle, like all flat fan nozzles, tends to have large droplets on the outsides of the pattern and small droplets in the center. To get consistency across the pattern, a boom may have an 80° nozzle and a 110° nozzle set up to obtain a minimum of 30 percent overlap with neighboring nozzles (ideally 50 percent). For a 110° nozzle with a 1:1 ratio of nozzle spacing and boom height (e.g., both at 20 inches), the output of neighboring nozzles begins to overlap halfway between the nozzles and the target.

As manufacturers began to design nozzles that reduce drift by producing coarser droplets, operators found that a second option for reducing drift was to lower the height of the boom. Getting the boom closer to the target promoted more droplets reaching the canopy. Lower boom heights also have the advantage of keeping the finer drops from being blown by the wind out of the treatment area. But without sufficient overlap, low booms (or booms that sway low) may create gaps in coverage by reducing the overlap.

A 110° angle provides a wider swath by using different orifice designs. A wider swath can also be achieved by using higher pressures to push the spray outward. For an effective application, a 110° spray angle nozzle would also need a 100 percent overlap to ensure an even distribution of the spray solution, as discussed earlier.

A 110° angle makes the application more consistent when spraying rolling ground when the boom goes up and down. Moreover, a 110° angle would allow the operator to set the boom closer to the target for a 100 percent overlap. This creates a scenario for mitigating off-target movement of spray droplets. For these reasons, the 110° angle is preferrable to the 80° angle in many situations.

But as in most things, we may be circling back. With the rise of selective spraying, a more targeted application has a value. Nozzle manufacturers are exploring the utility of 80° flat fans (or even narrower fans) on closer boom spacing to more accurately target small weeds with little product waste.

Criterion 5: Optimum Spray Pressure

Working in the middle of the pressure range for a given nozzle can help compensate for changes in speed (and resulting pressure) during the application. If a nozzle is working at pressures near its higher or lower range limits, it may negatively affect the spray geometry and create gaps. It also may push the spray droplet out of the desired size range. In some cases, it may do both.

Certain extended-range or variable orifice nozzles are designed to handle a larger range of pressures without requiring the operator to switch to a different nozzle size. Pulse-width modulation is available, which removes pressure from the flow equation and instead employs a duty cycle of intermittent shutoffs to compensate for speed changes.

We also recommend that you operate spray nozzles in the middle of their pressure range. This ensures optimal nozzle performance. Operating at the low end of the pressure range can result in a spray pattern that inadequately forms (mainly for the old nozzle models). Operating at the higher end of the pressure range can increase wear and reduce nozzle life.

THE PHYSICS RULE OF 4X SPRAY PRESSURE

Spray pressure controls the flow rate through a nozzle. Pressure and flow share an inverse-square or square root relationship. What this amounts to is that to double the flow, you need to quadruple the pressure. Not only is there diminishing return for lots of pressure, but it also affects droplet size. This is why only minor flow changes are achievable using pressure. For large changes in flow rate, it is better to switch to a different nozzle.

Criterion 6: Distance to Target

Nozzle design, pressure, and boom height all determine the overlap of a spray pattern. Table 14.12 from the Pentair Hypro nozzle catalog lists minimum and optimum boom heights for 80° and 110° spray angles for different nozzle spacings.

The optimum boom height for a nozzle is linked directly to the nozzle angle and spacing. For nozzles with a 80° fan angle, setting the boom height too low can create underapplication zones between the nozzles. If the spray pattern for the 80° fan does not have enough distance to fully spread, it will instead place much of the spray in a narrow band.

The 110° angle allows more flexibility in application heights and drift-mitigation measures. A 110° flat fan nozzle follows roughly the 1:1 ratio for spacing and height. For every inch of space between nozzles, the nozzle needs to be 1 inch above the target. For example, for 110° flat fan nozzles spaced 20 inches apart on the boom, the boom would be placed 20 inches above the target. (In comparison, 80° flat fan nozzles 20 inches apart would need a boom height of 30 inches.)

While operators set the boom heights in the cab, we encourage everyone to inspect the overlap at the target height. That inspection is useful partially because many operators are unable to operate their booms at less than 24 inches because of sway, partially because of where the spray intercepts the canopy, and partially because patterns change with heights due to angles and wind speeds.

Table 14.12. Optimal Boom Height Above the Target

Spray Angle	15-inch Nozzle Spacing Height (in inches)		20-inch Nozzle Spacing Height (in inches)		30-inch Nozzle Spacing Height (in inches)	
	Minimum	Best	Minimum	Best	Minimum	Best
80°	13-14	22	17-19	30	26-28	43
110°	10-11	15	15-18	20	20-22	30

Criterion 7: Nozzle Material

Many of today's pesticides and fertilizers are abrasive products formulated with inert ingredients such as clay, dry materials suspended in water, and nitrogen (for example, 28 percent and 32 percent). Given that brass is too soft for many of the abrasive products used today, most industries rely on three spray tip materials: polymer, stainless steel, and ceramic.

Criterion 8: Single or Dual Spray Pattern

A single spray pattern has been the norm for years. Newer nozzle designs have forward and backward spray patterns. Some operators have observed that double fans are better suited to applications when the spray target is vertically oriented and close to the upper part of the canopy.

Criterion 9: One-Piece Cap/Nozzle or Separate Nozzle and Cap

Purchasing an all-in-one cap is all about convenience. All-in-one caps also ensure the nozzle is lined up in the body with the proper orientation. If you don't have a nozzle body turret, then purchasing the cap and nozzle separately might be a better choice given that the cap may be able to hold several different spray tips.

Newer caps (whether separate or all-in-one) only allow a certain position for the nozzle. Each flat fan nozzle is slightly off horizontally from the boom direction. However, if the fan patterns are pointed straight down and parallel to the boom, then the droplets from adjacent fans would collide instead of bypassing one another. Rotating the nozzles just a few degrees relative to each other (say, 15°) means the droplets are no longer colliding with each other. The lugs built into newer caps (separate and all-in-one) automatically align the nozzles so the opposing spray particles will not collide but still overlap to form the overall boom pattern.

Make sure to use the right caps for the nozzles. Some caps (such as those used for hollow-cone nozzles) can fit most nozzle types; however, they do not automatically offset the spray pattern at 15°. If you use the wrong cap, you would need to manually adjust nozzles one by one along the boom.

Figure 14.79. TeeJet® offers a cap installation and removal tool that makes removing nozzles much easier than by hand.

Figure 14.80. Notice the 15° offset from the boom frame direction between the spray pattern for adjacent nozzles.

Criterion 10: Nozzle Type

All of the previous discussion considered equipment speed, carrier volume, spray angle, droplet distribution, and other factors separately to help explain the inner workings of nozzles. Nozzle manufacturers also provide tables that list their nozzles and the applications each was designed to perform, so applicators have a starting point to optimize their applications.

Working Through a Nozzle Chart

Once you select a couple nozzle designs based on your target needs, it will be time to add your specific application information. The following box shows an example of the conditions placed on a nozzle by an applicator. The applicator chooses an extended-range flat fan nozzle (XR TeeJet®) spray tip to provide good-to-excellent coverage on weeds. The intended travel speed is 8 mph, although it may not be possible to maintain this speed at all times. The product label recommends using a nozzle that will provide medium spray droplets.

According to the catalog, this nozzle comes with a 110° fan angle and a stainless tip, which were two of the desired application criteria. Using table 14.13, how do we go about determining the specific size of the XR flat fan nozzle that will satisfy the other application criteria given in the box? We outline the steps to follow here.

Step 1. Find the 8 mph travel speed at the top of the table, go down the column, and then identify the nozzle sizes that can deliver 15 gallons per acre (gpa). The 110° nozzles that can deliver this rate are 11004 (red), 11005 (brown), and 11006 (gray).

Step 2. Now let's add the next requirement, which is for the droplets to be categorized as medium. Which of the three nozzles from Step 1 are capable of producing medium droplets at the 15 gpa application rate?

- Nozzle 11004: Produces medium droplets at 40 psi
- Nozzle 11005: Produces medium droplets at close to 30 psi
- Nozzle 11006: Produces medium droplets at close to 20 psi

Table 14.13. XR TeeJet® Flat Fan Nozzle

TIP PART NO. (STRAINER MESH SIZE)	PSI	DROP SIZE 80°	DROP SIZE 110°	CAPACITY ONE TIP IN GPM	CAPACITY ONE TIP IN OZ/MIN	APPLICATION RATE FOR 20" SPRAY TIP SPACING: GALLONS PER ACRE (GPA)								TURF APPLICATION GALLONS PER 1000 SQ. FT.			
						4 MPH	5 MPH	6 MPH	8 MPH	10 MPH	12 MPH	15 MPH	20 MPH	2 MPH	3 MPH	4 MPH	5 MPH
XR8001 XR11001 (100)	15	F	F	0.061	7.8	4.5	3.6	3.0	2.3	1.8	1.5	1.2	0.91	0.21	0.14	0.10	0.08
	20	F	F	0.071	9.1	5.3	4.2	3.5	2.6	2.1	1.8	1.4	1.1	0.24	0.16	0.12	0.10
	30	F	F	0.087	11	6.5	5.2	4.3	3.2	2.6	2.2	1.7	1.3	0.30	0.20	0.15	0.12
	40	F	F	0.10	13	7.4	5.9	5.0	3.7	3.0	2.5	2.0	1.5	0.34	0.23	0.17	0.14
	50	F	F	0.11	14	8.2	6.5	5.4	4.1	3.3	2.7	2.2	1.6	0.37	0.25	0.19	0.15
	60	F	VF	0.12	15	8.9	7.1	5.9	4.5	3.6	3.0	2.4	1.8	0.41	0.27	0.20	0.16
XR80015 XR110015 (100)	15	M	M	0.092	12	6.8	5.5	4.6	3.4	2.7	2.3	1.8	1.4	0.31	0.21	0.16	0.13
	20	F	F	0.11	14	8.2	6.5	5.4	4.1	3.3	2.7	2.2	1.6	0.37	0.25	0.19	0.15
	30	F	F	0.13	17	9.7	7.7	6.4	4.8	3.9	3.2	2.6	1.9	0.44	0.29	0.22	0.18
	40	F	F	0.15	19	11.1	8.9	7.4	5.6	4.5	3.7	3.0	2.2	0.51	0.34	0.26	0.20
	50	F	F	0.17	22	12.6	10.1	8.4	6.3	5.0	4.2	3.4	2.5	0.58	0.39	0.29	0.23
	60	F	F	0.18	23	13.4	10.7	8.9	6.7	5.3	4.5	3.6	2.7	0.61	0.41	0.31	0.24
XR8002 XR11002 (50)	15	M	M	0.12	15	8.9	7.1	5.9	4.5	3.6	3.0	2.4	1.8	0.41	0.27	0.20	0.16
	20	M	M	0.14	18	10.4	8.3	6.9	5.2	4.2	3.5	2.8	2.1	0.48	0.32	0.24	0.19
	30	F	F	0.17	22	12.6	10.1	8.4	6.3	5.0	4.2	3.4	2.5	0.58	0.39	0.29	0.23
	40	F	F	0.20	26	14.9	11.9	9.9	7.4	5.9	5.0	4.0	3.0	0.68	0.45	0.34	0.27
	50	F	F	0.22	28	16.3	13.1	10.9	8.2	6.5	5.4	4.4	3.3	0.75	0.50	0.37	0.30
	60	F	F	0.24	31	17.8	14.3	11.9	8.9	7.1	5.9	4.8	3.6	0.82	0.54	0.41	0.33
XR80025 XR110025 (50)	15	M	M	0.15	19	11.1	8.9	7.4	5.6	4.5	3.7	3.0	2.2	0.51	0.34	0.26	0.20
	20	M	M	0.18	23	13.4	10.7	8.9	6.7	5.3	4.5	3.6	2.7	0.61	0.41	0.31	0.24
	30	M	M	0.22	28	16.3	13.1	10.9	8.2	6.5	5.4	4.4	3.3	0.75	0.50	0.37	0.30
	40	F	F	0.25	32	18.6	14.9	12.4	9.3	7.4	6.2	5.0	3.7	0.85	0.57	0.43	0.34
	50	F	F	0.28	36	21	16.6	13.9	10.4	8.3	6.9	5.5	4.2	0.95	0.63	0.48	0.38
	60	F	F	0.31	40	23	18.4	15.3	11.5	9.2	7.7	6.1	4.6	1.1	0.70	0.53	0.42
XR8003 XR11003 (50)	15	M	M	0.18	23	13.4	10.7	8.9	6.7	5.3	4.5	3.6	2.7	0.61	0.41	0.31	0.24
	20	M	M	0.21	27	15.6	12.5	10.4	7.8	6.2	5.2	4.2	3.1	0.71	0.48	0.36	0.29
	30	M	M	0.26	33	19.3	15.4	12.9	9.7	7.7	6.4	5.1	3.9	0.88	0.59	0.44	0.35
	40	F	F	0.30	38	22	17.8	14.9	11.1	8.9	7.4	5.9	4.5	1.0	0.68	0.51	0.41
	50	F	F	0.34	44	25	20	16.8	12.6	10.1	8.4	6.7	5.0	1.2	0.77	0.58	0.46
	60	F	F	0.37	47	27	22	18.3	13.7	11.0	9.2	7.3	5.5	1.3	0.84	0.63	0.50
XR80035 (50)	15	M		0.21	27	15.6	12.5	10.4	7.8	6.2	5.2	4.2	3.1	0.71	0.48	0.36	0.29
	20	M		0.25	32	18.6	14.9	12.4	9.3	7.4	6.2	5.0	3.7	0.85	0.57	0.43	0.34
	30	M		0.30	38	22	17.8	14.9	11.1	8.9	7.4	5.9	4.5	1.0	0.68	0.51	0.41
	40	M		0.35	45	26	21	17.3	13.0	10.4	8.7	6.9	5.2	1.2	0.79	0.60	0.48
	50	F		0.39	50	29	23	19.3	14.5	11.6	9.7	7.7	5.8	1.3	0.88	0.66	0.53
	60	F		0.43	55	32	26	21	16.0	12.8	10.6	8.5	6.4	1.5	0.97	0.73	0.58
XR8004 XR11004 (50)	15	M	M	0.24	31	17.8	14.3	11.9	8.9	7.1	5.9	4.8	3.6	0.82	0.54	0.41	0.33
	20	M	M	0.28	36	21	16.6	13.9	10.4	8.3	6.9	5.5	4.2	1.0	0.63	0.48	0.38
	30	M	M	0.35	45	26	21	17.3	13.0	10.4	8.7	6.9	5.2	1.2	0.79	0.60	0.48
	40	M	M	0.40	51	30	24	19.8	14.9	11.9	9.9	7.9	5.9	1.4	0.91	0.68	0.54
	50	F	F	0.45	58	33	27	22	16.7	13.4	11.1	8.9	6.7	1.5	1.0	0.77	0.61
	60	F	F	0.49	63	36	29	24	18.2	14.6	12.1	9.7	7.3	1.7	1.1	0.83	0.67
XR8005 XR11005 (50)	15	C	M	0.31	40	23	18.4	15.3	11.5	9.2	7.7	6.1	4.6	1.1	0.70	0.53	0.42
	20	M	M	0.35	45	26	21	17.3	13.0	10.4	8.7	6.9	5.2	1.2	0.79	0.60	0.48
	30	M	M	0.43	55	32	26	21	16.0	12.8	10.6	8.5	6.4	1.5	0.97	0.73	0.58
	40	M	M	0.50	64	37	30	25	18.6	14.9	12.4	9.9	7.4	1.7	1.1	0.85	0.68
	50	M	F	0.56	72	42	33	28	21	16.6	13.9	11.1	8.3	1.9	1.3	0.95	0.76
	60	F	F	0.61	78	45	36	30	23	18.1	15.1	12.1	9.1	2.1	1.4	1.0	0.83
XR8006 XR11006 (50)	15	C	C	0.37	47	27	22	18.3	13.7	11.0	9.2	7.3	5.5	1.3	0.84	0.63	0.50
	20	C	M	0.42	54	31	25	21	15.6	12.5	10.4	8.3	6.2	1.4	0.95	0.71	0.57
	30	M	M	0.52	67	39	31	26	19.3	15.4	12.9	10.3	7.7	1.8	1.2	0.88	0.71
	40	M	M	0.60	77	45	36	30	22	17.8	14.9	11.9	8.9	2.0	1.4	1.0	0.82
	50	M	M	0.67	86	50	40	33	25	19.9	16.6	13.3	9.9	2.3	1.5	1.1	0.91
	60	M	M	0.73	93	54	43	36	27	22	18.1	14.5	10.8	2.5	1.7	1.2	0.99
XR8008 XR11008 (50)	15	VC	C	0.49	63	36	29	24	18.2	14.6	12.1	9.7	7.3	1.7	1.1	0.83	0.67
	20	C	M	0.57	73	42	34	28	21	16.9	14.1	11.3	8.5	1.9	1.3	0.97	0.78
	30	C	M	0.69	88	51	41	34	26	20	17.1	13.7	10.2	2.3	1.6	1.2	0.94
	40	M	M	0.80	102	59	48	40	30	24	19.8	15.8	11.9	2.7	1.8	1.4	1.1
	50	M	M	0.89	114	66	53	44	33	26	22	17.6	13.2	3.0	2.0	1.5	1.2
	60	M	M	0.98	125	73	58	49	36	29	24	19.4	14.6	3.3	2.2	1.7	1.3
XR8010† XR11010†	15	VC	C	0.61	78	45	36	30	23	18.1	15.1	12.1	9.1	2.1	1.4	1.0	0.83
	20	C	C	0.71	91	53	42	35	26	21	17.6	14.1	10.5	2.4	1.6	1.2	0.97
	30	C	C	0.87	111	65	52	43	32	26	22	17.2	12.9	3.0	2.0	1.5	1.2
	40	M	M	1.00	128	74	59	50	37	30	25	19.8	14.9	3.4	2.3	1.7	1.4
	50	M	M	1.12	143	83	67	55	42	33	28	22	16.6	3.8	2.5	1.9	1.5
	60	M	M	1.22	156	91	72	60	45	36	30	24	18.1	4.1	2.8	2.1	1.7
XR8015† XR11015†	15	XC	VC	0.92	118	68	55	46	34	27	23	18.2	13.7	3.1	2.1	1.6	1.3
	20	VC	VC	1.06	136	79	63	52	39	31	26	21	15.7	3.6	2.4	1.8	1.4
	30	VC	C	1.30	166	97	77	64	48	39	32	26	19.3	4.4	2.9	2.2	1.8
	40	C	C	1.50	192	111	89	74	56	45	37	30	22	5.1	3.4	2.6	2.0
	50	C	C	1.68	215	125	100	83	62	50	42	33	25	5.7	3.8	2.9	2.3
	60	C	M	1.84	236	137	109	91	68	55	46	36	27	6.3	4.2	3.1	2.5

All three nozzles can produce medium droplets while the sprayer is traveling at 8 mph and delivering an application rate of 15 gpa. However, since we want to buy only one set of this nozzle type, which one would be the best choice? To answer this question, we need to consider the travel speed range (6–10 mph) at which the nozzle can be used while delivering medium droplets and meet the 15 gpa requirement.

- Nozzle 11004 (Size 4): Can be operated when the travel speed is 6, 7, or 8 mph at pressures from 15 to 40 psi (the label does not show data for 7 mph, but if the nozzle can satisfy the conditions at 6 mph and 8 mph, it should also be acceptable at 7 mph).
- Nozzle 11005 (Size 5): This nozzle also can be operated when the travel speed is 6, 7, or 8 mph at pressures from 15 to 40 psi.
- Nozzle 11006 (Size 6): This nozzle can be operated when the travel speed is 8, 9, or 10 mph at pressures from 20 to 40 psi (the label does not show data for 9 mph, but if the nozzle satisfies conditions for 8 mph and 10 mph, it should also be acceptable at 9 mph).

In this case, all three nozzles (sizes 4, 5, and 6) satisfy all the requirements when they are operated between 20 and 40 psi. The choice you make will be based on the upper and lower limits of the travel speed you expect the sprayer to operate at.

If the travel speed range will be 6 to 8 mph, size 4 or 5 will be better. However, size 4 will have an edge, because at 6 mph, the spray pressure with size 5 must be 15 psi (the lowest operating pressure for this type of nozzle) to get the desired medium droplet size. It is always a good practice to not operate nozzles at the minimum and maximum pressure settings. If, however, the travel speed range will be 8 to 10 mph, the size 6 nozzle (11006) will be the best choice.

A final, but important, suggestion when selecting a nozzle: Before you select a nozzle, check the manufacturer's catalog or website for any additional information, such as whether the nozzle is approved for a PWM system.

KNOW WHAT YOU ARE LOOKING FOR BEFORE SEARCHING THE NOZZLE CATALOG

Before you look at a nozzle catalog, think of the criteria you need to select the right nozzle. Given the desired characteristics in the table below, what nozzle size can best satisfy these spraying conditions?

Criterion	
Nozzle type	XR Flat Fan
Tip material	Stainless steel preferred
Spray angle	110°
Overlap	100%
Nozzle spacing	20 inches
Speed	8 mph
Application rate	15 gallons per acre
Product	Herbicide
Droplet size	Medium

What Happens If Nozzle Spacing Differs from the Catalog?

It is important to understand that nozzle selection charts usually show only one particular nozzle spacing. In the previous example from the TeeJet® catalog, the information in the table for the air induction dual spray pattern was calculated using 20-inch spacing between each nozzle. Some manufacturers provide tables that also list 15-, 20-, and 30-inch spacings.

Sometimes they instead provide a conversion factor to help you convert the values in the table for your own boom spacing. If they do not provide a conversion factor for your spacing, you can determine your own conversion factor. To convert, divide the spacing in the table by the spacing on your boom. For example, let's say the spacing in the table is 20 inches and the spacing on your boom is 15 inches:

$$\text{conversion factor} = 20 \div 15 = 1.33$$

You then take the application rate from the table and divide it by the conversion factor to calculate the adjusted rate.

$$\text{adjusted rate} = 15 \div 1.33 = 11.28$$

This 11.28 gpa rate is the value you will look up in the table for nozzle ratings.

There is a second way to determine the rate, which some applicators prefer to use to calculate which nozzle meets their needs.

$$\frac{\text{gpa} \times \text{mph} \times \text{spacing}}{5{,}940} = \text{gpm}$$

gpm: gallons per minute per nozzle
gpa: gallons per acre of total spray carrier
mph: sprayer travel speed in miles per hour
spacing: nozzle spacing on the boom in inches

So, if your desired nozzle output is 15 gpm, the sprayer travel speed is 5 mph, and the nozzle spacing is 15 inches, your equation is:

$$\frac{15 \times 5 \times 15}{5{,}940} = 0.19 \text{ gpm}$$

In table 14.14, look down the column headed "Capacity One Tip in GPM." The second nozzle (AI370-02VP) nozzle puts out 0.2 gpm, and at 40 psi this nozzle will give a very coarse (VC) droplet spectrum.

Most major nozzle manufacturers have also developed apps that can provide exact nozzle flow rates required for any given set of application parameters. They can also identify a specific set of nozzle recommendations for the given application parameters.

Table 14.14. An Alternative Way of Calculating Nozzle Output

TIP PART NO. (STRAINER MESH SIZE)	PSI	DROP SIZE	CAPACITY ONE TIP IN GPM	CAPACITY ONE TIP IN OZ/MIN
AI3070-015VP (100)	20	XC	0.11	14
	30	VC	0.13	17
	40	VC	0.15	19
	50	C	0.17	22
	60	C	0.18	23
	70	M	0.20	26
	80	M	0.21	27
	90	M	0.23	29
AI3070-02VP (100)	20	XC	0.14	18
	30	VC	0.17	22
	40	VC	0.20	26
	50	C	0.22	28
	60	C	0.24	31
	70	M	0.26	33
	80	M	0.28	36
	90	M	0.30	38

To find these apps, simply visit your preferred application store and search for "spray nozzle calculator" or some other keywords related to nozzle size selection. You may also want search for the name of the nozzle company from which you are interested in buying the nozzles.

However, not all apps are user friendly. They sometimes do not account for droplet size requirements when recommending nozzles. Although the apps and catalog tables may expedite the nozzle selection process, it is best to understand the procedure and the math that nozzle manufacturers use to generate the values they list in tables and use in their apps.

Criteria for Selecting Fertilizer Nozzles

You may recall that manufacturers use clean water to test nozzle output under different pressures. Water weighs 8.3 pounds per gallon, but products like 28 percent and 32 percent UAN weigh approximately 11 pounds per gallon. To spray viscous materials like fertilizers, operators must increase the pressure to compensate for the thicker material and lower flow rate under the same pressure.

To account for the difference in viscosity between water and fertilizer, you want to use a conversion factor (derived by determining the square root of the spray solution's specific gravity) to make the fertilizer solution equivalent to water. Table 14.15 shows conversion factors for a range of solutions. Some catalogs will contain this information.

Assume you want to apply 28 percent UAN at a rate of 15 gpa. Use this formula to calculate a rate equivalent to water.

$$\text{target gpa of fertilizer} \times \text{conversion factor} = \text{gpa of water}$$

$$15 \text{ (target rate of fertilizer)} \times 1.13 \text{ (conversion factor from table)} = 16.95 \text{ gpa of water}$$

In other words, to apply 15 gpa of 28 percent UAN, the operator must use the same pressure required to apply 16.95 gpa of water.

Let's assume that the operator decided to use a StreamJet 3® stream pattern nozzle, the travel speed is 10 mph, and the operating pressure is 50 psi. What nozzle would you choose to apply 15 gpa of 28 percent UAN fertilizer, which is equivalent to 16.95 gpa of water? The choices would be the SJ3-05-VP (brown) nozzle at a pressure between 50 to 60 psi and the SJ3-06-VP (gray) nozzle at a pressure between 30 and 40 psi.

Table 14.15. Conversion Factors for Different Fertilizer Solutions When Selecting Nozzles

Weight of Solution	Specific Gravity	Conversion Factor
7.0 lbs./gal.	0.84	0.92
8.0 lbs./gal.	0.98	0.96
8.34 lbs./gal.	1.00 (water)	1.00
9.0 lbs./gal.	1.08	1.04
10.0 lbs./gal.	1.20	1.10
10.65 lbs./gal.	1.28 (28% UAN)	1.13
11.0 lbs./gal.	1.32	1.15
12.0 lbs./gal.	1.44	1.20
14.0 lbs./gal.	1.68	1.30

Table 14.16. Application Rates for StreamJet 3® Nozzles

TIP PART NO. (STRAINER MESH SIZE)	PSI	CAPACITY ONE TIP IN GPM	APPLICATION RATE FOR 20" SPRAY TIP SPACING										
			3 MPH	4 MPH	5 MPH	6 MPH	8 MPH	10 MPH	12 MPH	14 MPH	16 MPH	18 MPH	20 MPH
SJ3-015-VP (100)	20	0.11	10.9	8.2	6.5	5.4	4.1	3.3	2.7	2.3	2.0	1.8	1.6
	30	0.13	12.9	9.7	7.7	6.4	4.8	3.9	3.2	2.8	2.4	2.1	1.9
	40	0.15	14.9	11.1	8.9	7.4	5.6	4.5	3.7	3.2	2.8	2.5	2.2
	50	0.16	15.8	11.9	9.5	7.9	5.9	4.8	4.0	3.4	3.0	2.6	2.4
	60	0.17	16.8	12.6	10.1	8.4	6.3	5.0	4.2	3.6	3.2	2.8	2.5
SJ3-02-VP (50)	20	0.14	13.9	10.4	8.3	6.9	5.2	4.2	3.5	3.0	2.6	2.3	2.1
	30	0.17	16.8	12.6	10.1	8.4	6.3	5.0	4.2	3.6	3.2	2.8	2.5
	40	0.20	19.8	14.9	11.9	9.9	7.4	5.9	5.0	4.2	3.7	3.3	3.0
	50	0.21	21	15.6	12.5	10.4	7.8	6.2	5.2	4.5	3.9	3.5	3.1
	60	0.22	22	16.3	13.1	10.9	8.2	6.5	5.4	4.7	4.1	3.6	3.3
SJ3-03-VP (50)	20	0.24	24	17.8	14.3	11.9	8.9	7.1	5.9	5.1	4.5	4.0	3.6
	30	0.27	27	20	16.0	13.4	10.0	8.0	6.7	5.7	5.0	4.5	4.0
	40	0.30	30	22	17.8	14.9	11.1	8.9	7.4	6.4	5.6	5.0	4.5
	50	0.33	33	25	19.6	16.3	12.3	9.8	8.2	7.0	6.1	5.4	4.9
	60	0.35	35	26	21	17.3	13.0	10.4	8.7	7.4	6.5	5.8	5.2
SJ3-04-VP (50)	20	0.30	30	22	17.8	14.9	11.1	8.9	7.4	6.4	5.6	5.0	4.5
	30	0.36	36	27	21	17.8	13.4	10.7	8.9	7.6	6.7	5.9	5.3
	40	0.40	40	30	24	19.8	14.9	11.9	9.9	8.5	7.4	6.6	5.9
	50	0.43	43	32	26	21	16.0	12.8	10.6	9.1	8.0	7.1	6.4
	60	0.47	47	35	28	23	17.4	14.0	11.6	10.0	8.7	7.8	7.0
SJ3-05-VP (50)	20	0.36	36	27	21	17.8	13.4	10.7	8.9	7.6	6.7	5.9	5.3
	30	0.45	45	33	27	22	16.7	13.4	11.1	9.5	8.4	7.4	6.7
	40	0.50	50	37	30	25	18.6	14.9	12.4	10.6	9.3	8.3	7.4
	50	0.55	54	41	33	27	20	16.3	13.6	11.7	10.2	9.1	8.2
	60	0.59	58	44	35	29	22	17.5	14.6	12.5	11.0	9.7	8.8
SJ3-06-VP (50)	20	0.42	42	31	25	21	15.6	12.5	10.4	8.9	7.8	6.9	6.2
	30	0.54	53	40	32	27	20	16.0	13.4	11.5	10.0	8.9	8.0
	40	0.60	59	45	36	30	22	17.8	14.9	12.7	11.1	9.9	8.9
	50	0.66	65	49	39	33	25	19.6	16.3	14.0	12.3	10.9	9.8
	60	0.70	69	52	42	35	26	21	17.3	14.9	13.0	11.6	10.4

ADJUVANTS MAY INFLUENCE FLOW RATE AND DROPLET SIZE

Nozzle manufacturers' tables use water as the spray carrier. Consider that we are already adding formulated chemistry to water that carries its own set of inert and active ingredients. Any pesticide product and adjuvant you add to the spray mixture may change the physical properties of the liquid you are spraying (see chapter 13). The sprayer is no longer pumping just water out to the nozzles. How does that affect droplet size and flow rate?

Herbicides produce different droplet sizes compared to water. Some herbicides reduce the percentage of fine droplets, but many increase the percentage of fines. Glyphosate and glufosinate, with high surfactant load, have shown to move the spray quality as much as two categories lower than shown in the nozzle charts.

Adding adjuvant at rates of 0.5 to 1 percent represents a small fraction of the total amount mixed. However, most drift-reduction adjuvants will also increase the viscosity of the spray solution. You might notice that when using a drift-reduction adjuvant, the sprayer's operating pressure is a little higher to achieve the same output compared to when no drift-reduction adjuvant is used. Other drift-reduction/deposition aids do not influence viscosity or spray angle.

The bigger effect adjuvants have may be on spray angle and droplet size. Some adjuvants change the physical properties of the liquid and subsequent behavior of the liquid when it atomizes and forms the spray cloud—this will affect droplet size.

Nonionic surfactants often create smaller droplets, because the lower surface tension of the droplets allows the spray to break more easily into smaller droplets. Oils and drift-reduction agents (guar and polymers) can create larger droplets and affect spray viscosity. In some cases, adding an adjuvant will change the droplet spectrum to a relatively smaller or larger spectrum. Adjuvants like ammonium sulfate and other water conditioners do not affect either flow rate or droplet size.

In many cases, pesticide manufacturers have already tested how adding adjuvants affects control of the targeted pest. Just be aware that when you are adding products like adjuvants and some herbicide formulations, you may change droplet sizes. Some new herbicides have a specific website with nozzles and adjuvants approved for tank mixtures that will not drastically alter droplet size and product efficacy.

MAINTENANCE AND REPLACEMENT

There is not a lot of maintenance involved with screens and nozzles. Most of the required maintenance is detecting when a set of nozzles is worn and needs to be replaced.

Cleaning Screens and Nozzles

The size of the mesh is normally slightly smaller than the nozzle's exit orifice, so the mesh catches debris before the nozzles do. Products that easily dissolve in water (like glyphosate) leave no solid particles to clog screens. Conversely, when you apply products that are formulated with clay particles (like atrazine), it is a good practice to check the screens at the end of the day, especially when using dry product formulations. We always recommend that operators follow nozzle manufacturer catalogs to find the correct mesh for nozzle strainers.

If your nozzle tip screens are clogging up, that might be a sign that you need to examine the mixture or the agitation process. It might also be that the partners you used in the mixture are physically incompatible or that you're using the wrong tip screens for the mix. In other cases, clogged tip screens may indicate that you need to set up the equipment differently. Most spray equipment has a series of inline screens. Reducing the mesh size of these inline screens may help reduce the solid residues plugging up the tip screens and exit orifices.

There is also the possibility that the mixing process is not allowing enough time for the products to disperse completely before they are pumped through the sprayer components. Using a screen (strainer) that is too small may compromise the flow of certain pesticides that contain particles as part of their formulation and delivery of their active ingredients.

The maintenance required for nozzles is more difficult than screens. Frankly, it is hard to tell when a nozzle is partially plugged when they are behind the sprayer. Rate controllers really do not help when a nozzle or two is completely or partially plugged. If one or two nozzles are not working, the controller will increase the pressure to force more volume out of the remaining nozzles, resulting in an uneven spray distribution along the treated area and probably changes in droplet size.

For these reasons, it is important to know the normal operating pressure when the boom and all nozzles are functioning correctly. If you observe a change in spray pressure when operating the sprayer under normal conditions, you should try to diagnose the cause of the disparity, because it could be a sign of plugged nozzles (increased psi) or a damaged or missing spray tip (decreased psi).

Always check the nozzles when flushing the boom. Do a walkaround when spraying water and tank cleaner to check for spray abnormalities. If a nozzle is plugged, shut the sprayer down, use appropriate personal protective equipment, and clean the outside of the nozzle with a nylon-bristle brush to remove caked-on soil and mud.

Most of the time, the blockage is a small piece of debris such as sand, rust, or dried chemical inside the nozzle. This means you will have to remove the nozzle body to

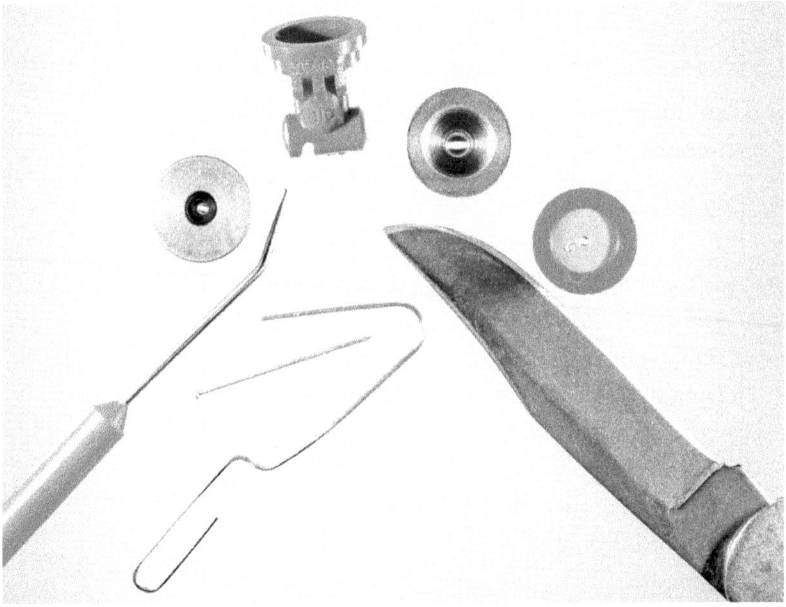

Figure 14.81. Do not use these objects to clean nozzles. All nozzle materials can be damaged when cleaning them. Small grooves etched into the exit orifice or damage to the walls can change the spray angle, flow rate, and droplet size.

examine the exit orifice. Remember that the obstruction may not be at the exit orifice but may be in the strainers leading up to the nozzles. Do not insert paper clips, knives, and other objects into exit orifices, because they *will* damage the nozzle and alter the pattern. Clean plugged nozzles only with a soft brush, water, or air pressure.

Checking Nozzles for Wear

Solid inerts such as clay particles, dry formulations suspended in the liquid stream, and sand particles in the mixture can abrade the sides of exit orifices, which allows more fluid to flow through. Worn or damaged spray tips can negatively affect how uniform the spray is across a boom. Instead of an even distribution, worn or damaged nozzles can show peaks and valleys from the fluid collected from nozzle to nozzle. Earlier, we discussed coefficient of variation (CV) as a measure of uniformity; this can be tested using equipment known as a *spray table/patternator*, as shown in figure 14.82.

NOZZLES WORN FROM USE

As a nozzle naturally wears, the exit orifice opening gets larger. Larger openings increase the flow rate and affect the spray patterns, droplet sizes, and coefficients of variation.

NOZZLES DAMAGED FROM CLEANING

If you use abrasive materials or objects to clean nozzles, it can alter the spray pattern.

Operators frequently ask, "How long does a set of nozzles last?" While manufacturing processes and materials have greatly improved, all nozzles wear out and need replacing at some point. Various factors affect the useful life of any nozzle:

- The quality of the water used to fill the spray tank
- Use of abrasive products
- Use of filters
- Spray pressure
- Nozzle tip material
- Acres applied
- Cleaning practices
- Regularity of cleaning and maintenance, particularly end-of-season maintenance

These and other factors influence how long a set of nozzles will last, but it also means the specific answer to how long nozzles last is not simple. It "all depends."

There are proven and unproven strategies that are used to evaluate nozzle wear. But evaluating wear is almost impossible, because the human eye cannot detect differences in flow rates between nozzles. Rather than try to predict nozzle lifespans, there are a number of practices that operators have developed with regard to replacing nozzles.

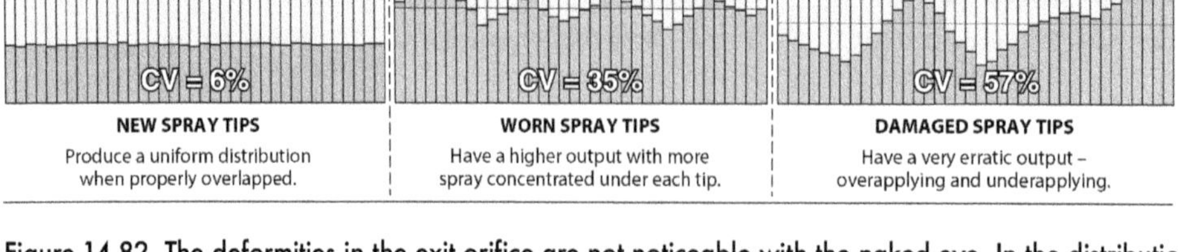

Figure 14.82. The deformities in the exit orifice are not noticeable with the naked eye. In the distribution table, it is possible to see the uneven distribution instead of an inverted V pattern found on tapered flat fan nozzles. The dark (blue) shapes at the bottom show the amount of water collected underneath each nozzle. New nozzles (left) show a nearly straight line for the water collected for each nozzle, illustrating a very low and acceptable CV of 6 percent. Notice the discrepancies in the amount of water collected from worn (middle) and damaged (right) spray tips. Instead of an even distribution across the spray boom, some nozzles put out more volume than others.

Replace Nozzles Annually

Changing nozzles each year is an option that some operators may consider excessive. The principle guiding this strategy is that it is cheaper to replace nozzles than to spend time checking each one. Compared to the costs of the pesticide and fertilizer products, equipment, and labor, a set of nozzles is a relatively cheap investment. This is an economic decision.

Wear Rating: Excellent Strategy!

Replace Nozzles Based on Acres Sprayed

Some operators replace their nozzles after every 25,000 acres sprayed. Larger farms and commercial companies may have to change nozzles yearly. For smaller operations, it will take many years to reach 25,000 acres. This strategy might work for some who know their operations well. All the variables that cause nozzles to wear are specific to each operation.

Wear Rating: Moderate to Poor Strategy

Look at Spray Patterns

Some suggest that you turn on the booms and look at spray patterns. The problem is that you simply cannot detect differences in flow rates. However, you can sometimes see different patterns, but it may be difficult to see the subtle changes resulting from damaged or worn nozzles. Looking at your spray patterns does have other advantages (for example, noticing clogged tips), so don't think this activity is not worth your time.

Wear Rating: Moderate to Poor Strategy

Compare Old Nozzles to New Ones

It seems as if placing two or three new nozzles on a section of the boom and comparing them to the old nozzles should work. However, the human eye cannot see small differences in flow rate, which makes this method suspect.

Wear Rating: Poor Strategy!

THE HUMAN EYE IS A POOR WEAR DETECTOR

Your eye is not able to see a 50 percent difference in the flow rate between two nozzles.

Measure Nozzle Output

The only sure way to tell if a nozzle is worn is to use a collection device and measure the nozzle's output. We often write about measuring the flow rate of a set of nozzles, but in fact, a turret nozzle assembly may have up to five spray tips. On a 120-foot boom with nozzles spaced 15 inches apart, there are 96 turrets with up to 480 nozzles. This is why you should check nozzles in the winter when there is more time for maintenance.

Wear Rating: Excellent Strategy!

A 2022 study measured different methods for determining nozzle flow rates and methods that produced consistent results (https://sprayers101.com/measure-nozzle-flow). A calibration tool that stood out in the test was the Innoquest SpotOn Digital Calibrator. This calibration device was consistently accurate, was easy to use, and provided results quickly.

To use the device, you set up the sprayer for a pressure of 40 psi. Place the device at a slight angle under the nozzle (roughly 15°) to collect the liquid. The meter has a lower and upper sensor. When the liquid reaches the first sensor, it activates the stopwatch inside the system. When the water level gets to the second sensor, it turns off the stopwatch, and the screen displays the flow rate in gallons per minute.

Compare the digital readout value to the output of the nozzle expected at 40 psi. With worn nozzles, it is typical to replace a nozzle if the flow rate is 10 percent or more above or below the predicted flow rate listed in the manufacturer's catalog. For example, a 04 flat fan nozzle at 40 psi has a theoretical flow rate of 0.4 gallons per

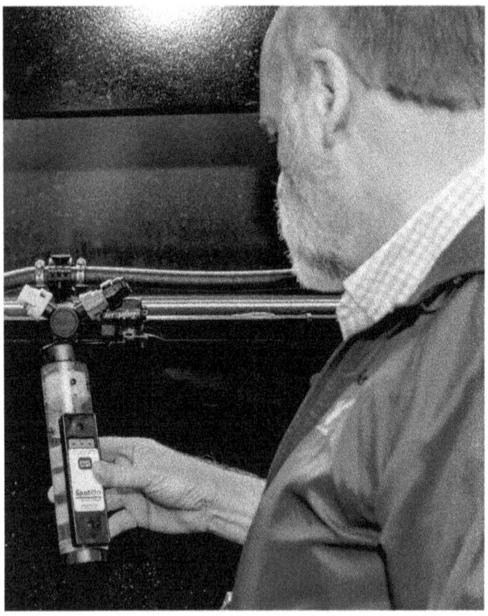

Figures 14.83–14.84. These photos show the flow rate of water being tested with a SpotOn Sprayer Calibrator (left). In the right photo, the SpotOn unit on the left displays the flow as ounces per minute (39), while the unit on the right displays gallons per minute (0.31).

Table 14.17.
Output of XR11004 Nozzle Tip Based on Pressure and Nozzle Angle

TIP PART NO. (STRAINER MESH SIZE)	PSI	DROP SIZE 80°	DROP SIZE 110°	CAPACITY ONE TIP IN GPM	CAPACITY ONE TIP IN OZ/MIN
XR8004 XR11004 (50)	15	M	M	0.24	31
	20	M	M	0.28	36
	30	M	M	0.35	45
	40	M	M	0.40	51
	50	F	F	0.45	58
	60	F	F	0.49	63

minute. Using the 10 percent threshold, the nozzle would be out of specification if the amount of water you collect is at least 0.04 more than (>0.44) or less than (<0.36) the 0.4 gallons per minute. The nozzle would need replacing.

A second method is to collect the spray output from each nozzle in a graduated cylinder for 30 or 60 seconds. For this method, remember the flow rate value of a new nozzle at a certain pressure. For example, in table 14.17, the XR11004 at 40 psi must apply 51 fluid ounces in 60 seconds, or 25.5 fluid ounces in 30 seconds. Add up the total amount of water you collect from the sprayer and divide that amount by the number of nozzles. This will give you the average output per nozzle.

For example, if the sprayer has four nozzles that applied 13.5, 16, 16, and 14.5 fluid ounces, then the average output is 15 fluid ounces per nozzle:

$$13.5 + 16 + 16 + 14.5 = 60$$
$$60 \div 4 = 15$$

Check the nozzle output. If any nozzle is more than 10 percent above or below the average output, then there may be a problem with the nozzle. If a nozzle has a lower-than-average output, it may be plugged or have a plugged screen. If a nozzle has a higher-than-average output, it is likely worn out and needs to be replaced. In the example above, the average output was 15 fluid ounces, so 10 percent is 1.5 fluid ounces. The acceptable range is then 13.5 to 16.5 fluid ounces. Because the output from each nozzle in our example is within this range, none of them needs to be replaced yet.

However, the problem with this approach is that you are comparing the nozzles to each other but not checking the flow rates of individual nozzles. This approach has the disadvantage that one nozzle may have a critical problem that can have a tremendous effect on the average flow rate. Thus, the first method described above is preferred.

Whether you use a SpotOn, a graduated cylinder, or another measuring device, you can only measure the real uniformity of a nozzle if the pressure is the same throughout the spray boom. There can be a difference in the flow rates of nozzles across the boom because of pressure differences from one end of the boom to the next.

A nozzle may be emitting 10 percent above the expected flow rate; however, that nozzle's output might be influenced by its location on the boom. Nozzles at the center of a boom might be at slightly elevated pressures, so you can expect more volume output. Nozzles near the ends of booms may have lower pressures, reducing output. Simply take those nozzles from the end of the booms and place them in the center to see if the location or wear is resulting in a reduced flow. You can also have a portable pressure gauge to use during the check.

You also can check for even pressure across the boom. The best method is to obtain two liquid-filled stainless steel gauges and put them on adjoining outlets. Compare the pressures, then switch the nozzles and the outlets, and again compare the pressures.

EMERGENCY REPAIR KIT IN-FIELD

Always have a few replacement nozzles, screens, gaskets, and even a couple of nozzle bodies handy in case you clip a fence post, or your booms get jammed in the dirt. No one wants to be in the field needing to clean a plugged, erratically behaving tip. Get the tip out and replace it with a spare new tip.

Clean the nozzles you remove at the shop or in the field at the end of the day after the spraying is done. Use nonabrasive cleaning devices (such as nylon bristles or compressed air) to clear obstructed nozzles without damaging the orifice shape.

Then write on one of the gauges how it compares to the other gauge. Then check pressures on the boom and check the pressure on the boom with the pressure gauge in the cab. Write on the pressure gauge in the cab how it compares to the pressures in the cab. The pressure at the boom is the correct operating pressure. If there are variations in different sections of the boom, then you should examine the plumbing and hoses to see if a more uniform application can be made.

A commonly question asked is: "Is it better to replace a few nozzles that are worn or replace the entire set?" The answer is that this is a personal judgment based on how much you have used the set. If you only have 2,000 acres on them, most will probably replace just the bad ones. If the nozzles have been used a lot over many years, replacing all the nozzles is the best choice.

While you could change only nozzles that are outside the 10 percent threshold as a way of saving money, it might not be the best solution. You will need to continually monitor nozzles for wear. It is much easier to put them on as a set and take them off as a set. This will ensure that the wear is consistent across the boom, but you must consider the cost of doing so.

"VINTAGE" PARTS KEPT FOR EMERGENCIES

It is common to find in many sprayer cabs or shops a box full of different-colored nozzles, different brands and designs, screens, tips, and plumbing pieces. Why? The answer is always, "I might need one in an emergency." But when was the last time you actually dug something out of the box and used it on the sprayer? It is better to discard these parts properly instead of saving them for just in case. They are no longer good, which is why they ended up in the junk box or drawer. If you leave them sitting around, you might be tempted to put them back on or confuse them with new nozzles. Did we mention that you should throw away the nozzles that are no longer being used?

THE UNSUNG HEROES OF LIQUID APPLICATIONS

With the many advances in pesticide application equipment, it seems that nozzles have been relegated to old technology—they are little more than pieces of plastic or metal attached to booms. However, all the bells and whistles that add precision to applications are of little value unless the simple nozzle performs in a desired manner.

A pesticide or fertilizer is only as good as the method with which it is applied. Ultimately, this comes down to the spray nozzle. All spray clouds may appear similar—they just create a spray mist, right? However, there are important parameters in that spray cloud. They may not be visible to the naked eye, but they can have an impact on spray pattern uniformity and pesticide efficacy.

Newer nozzle designs have definitely changed as the portfolio of pesticides and other products has steadily grown. With equipment and pesticides becoming more expensive and sophisticated, we must expect nozzles to hold up their end of the application process.

Since the 1940s nozzles, just like equipment and pesticides, have witnessed steady improvements in designs and applications:

- Nozzles designed for specific applications
- Consistent spray patterns and droplet size spectrums from one nozzle to the next
- Standards in nozzle manufacturing
- Technological advances in the nozzle body
- Better materials used in nozzle tip construction
- Techniques that can easily measure the distribution of specific droplet sizes
- Techniques that can control nozzle flow rate at variable speeds
- New spray angles and spray patterns that fit different targets
- Precise flow rate control
- Consistent nozzles with the same characteristics—flow rate, droplet size, spray angle
- Technologies that reduce off-target drift

Today's nozzles would not have made these rapid advances without modern equipment that could accurately measure the spectrum of droplets a nozzle produces. The ability to measure and quantify droplet sizes allowed the nozzle manufacturing industry to adapt their products to meet the needs of specific pesticide chemistries. The need for manufacturers to quantify and meet their customers' needs prompted continued research to find the droplet sizes, patterns, and coverages needed to make products work as intended.

HAVING CHOICES IS IDEAL

There is a reason to have all these different nozzle types in the market. Operators need to understand each application to select the right nozzle. You should ask:

- **What is your target, and where is it located?**
- **Which crop is being treated, and at which development stage will the application take place?**
- **What type of product is going to be applied? Is it a contact or a systemic product?**
- **What are the ideal weather conditions for applying this product?**
- **What is the correct droplet size for this product application?**
- **What is the necessity of drift control?**

A better understanding of how these and other factors interact with each other is no longer optional if pesticide product applications are to work as designed.

Today's analytical tools allow researchers to measure the performance of nozzles, including atomization mechanism, spray angle, overlap, droplet size, and drift potential. With all these advances, applicators should be aware that they should use different nozzles for different types of applications. The one-size-fits-all mentality no longer applies in the pesticide application industry.

Pesticide applications determine returns on investment of time and materials. It is not only what you put in the spray tank that determines the level of control your application achieves—how the product is applied influences the results too. It is more important today than in years past to know what a specific nozzle does and how it does it, because pest control is expensive.

Precision farming, postemergence application, minimum tillage, and precision planting have evolved and have become a much larger part of farming activities. Nozzles are an integral part of how modern farming is done today and will be done in the future. The nozzle is the last point at which any applicator has control over the application, so it is critical to get that right!

KEY TAKE-HOME MESSAGES

- Pesticides and nozzles are two wheels of the same cart; they must work simultaneously and in tandem to achieve a successful spray application.
- One nozzle doesn't fit every application. It is important to change nozzles to optimize pesticide applications.
- Plan ahead to get the right nozzle for the application.
- Don't overthink or overanalyze nozzle selection. It is a rather straightforward process, so let the science work for you.

15

Spray Drones

With contributions from Simerjeet Virk, Bryan Young, Steve Li, Alex Helms, Erdal Ozkan, Ashley Adair, Hunter Medenwald, Tommy Butts, Adam Shanks, and Kevin Leigh Smith

It's nothing short of amazing to see how the aviation industry has advanced since Orville and Wilbur Wright made their first flights on the sandy beaches of Kitty Hawk, North Carolina. Winning the flip of a coin over Orville, Wilbur made history on December 17, 1903, when his plane covered 852 feet at a speed of 34 miles per hour, becoming the first recorded flight.

In a little more than a century, planes have become essential for transporting people, mail, and cargo across continents and oceans. Advances in aviation technology have produced jets that can be undetected by radars, travel at supersonic speeds, and carry heavy payloads unimaginable to early aviation inventors and enthusiasts.

THE DAWN OF DRONES

The historic flight in North Carolina ushered in the dawn of aviation, which has a direct line to the development and deployment of unmanned aerial systems. While some people refer to these systems as *drones*, the term only refers to the aircraft itself. The phrase *unmanned aerial system*, or UAS, refers to the entire system required to operate the drone, which includes the remote controller and all the other components necessary to remotely operate the aircraft. A drone can be small enough to fit in the palm of a hand or as large as the units operated by the military.

Today drones are often considered new technology, but their military use dates as far back as the 1930s. The military invested billions of dollars to develop sophisticated hardware and interactive software programs for drones involved in surveillance operations, reconnaissance missions, and weapons deployment. A major advantage of drones over conventional military aircraft is that drones eliminate the need to account for the lives of pilots when planning military missions. This allowed decision-makers more objectivity when they considered the potential risks of operations.

Figure 15.1. The DJI Agras T30 is a spray drone designed for use by commercial applicators and farmers.

The Commercialization of Drones

In recent decades, the public release of military technologies (such as the internet and GPS satellites) has had important implications for commercial uses. Those who want to commercialize drone use quickly realized that the science and technology behind military drones provided a good starting point. They looked for ways to simplify the aircraft for public use and other specialized applications in different industries.

In the early days, one obvious disadvantage of military drones was the hefty price tag resulting from highly advanced capabilities such as detection, communication, and photographic technologies. Additionally, military drones were designed for specialized personnel trained in their operations and functions. For drones to become commercialized for the general public, manufacturers needed to develop a more turnkey solution so people of all skill levels could operate drones for hobbies or commercial businesses.

Additional disadvantages of adapting early military drones for public use included:

- They were large and bulky.
- Operations, features, and functions were too specialized for civilian uses.
- Systems were complex.
- Public operators could not use software that the military deemed confidential and restricted.
- Drones lacked important features for use in agriculture and other industries.
- Operators required a high level of competency atypical of the general public.
- A regulatory framework was lacking for flying and application.

Once drone technology was released to the public, drone manufacturers focused on the potential commercial uses of available technology. They developed new features that could be added or removed to improve the utility and value to end users. Much of the repurposing was developing, adapting, or modifying copyright-protected software that could be used by industries such as agriculture or the government.

This chapter describes the current state of drone technology for making liquid pesticide applications. Our goal is to inform and provide the current best management practices to those interested in using drones for applying pesticides. We want to ensure readers will be able to operate spray drones safely and effectively.

A DRONE BY ANY OTHER NAME IS STILL A DRONE

For many people, the term drone is associated with a tool used by the military for surveillance operations. As a result, many users have tried to rename drones to distinguish their nonmilitary uses from military uses. Some of these names include:

- Remotely operated aircraft (ROA)
- Remotely piloted aerial application systems (RPAAS)
- Remotely piloted vehicle (RPV)
- Remotely piloted aircraft (RPA)
- Small unmanned aerial system (sUAS)
- Uninhabited aerial vehicle (UAV)
- Unmanned aerial system (UAS)
- Unmanned aerial spray system (UASS)
- Unmanned aerial vehicles (UAV)
- Unoccupied aerial vehicle (UAV)

While some of these names are used, this chapter uses the more common drone throughout to refer to the aircraft, controller, and entire system needed to operate the remotely piloted unit.

BROADER USES FOR DRONES

For the reasons above, drone manufacturers had to make major modifications to convert them from military use, including redesigning aircraft, repurposing software, and changing onboard imaging equipment. One can only conclude that the wide acceptance of drones has been a function of these enterprises accomplishing their goal of producing a product with practical uses for the public.

Drones for Recreation

Remotely controlled, fixed-wing model airplanes have been around for decades. Many can fondly remember their parents or grandparents teaching them to fly model planes or attending events where pilots competed against one another for best of class. Newer models have more intuitive controls and more advanced features than the old models.

Drones for Imaging

One of the first and most popular applications of drones was to capture aerial images. These images may be traditional photographs or images from specialized cameras, such as thermal (heat) cameras or multispectral (specific light wavelengths) cameras. Some modern drones have the ability to hover and capture a 360-degree view, allowing them to closely inspect areas inaccessible by roads or vehicles, or potentially unsafe locations.

Figures 15.2–15.6. Components of an imaging drone.

(A) This quadcopter (four-rotor) drone hovers near the ground. Its camera can capture photographs and video. The camera is attached to a gimbal that can adjust the viewing angle.

(B) This controller can connect to a tablet, so operators can monitor the drone's movement. Many controllers have built-in displays.

(C) The single battery on this specific drone provides enough power for approximately 25 minutes before it needs to be recharged.

(D) This image shows the drone's USB port, which is one method for transferring data.

(E) The arrow points to the drone's obstacle sensor, which notifies the pilot about nearby objects and obstacles in the drone's path. Many newer drone models have forward, rear, side, top, and bottom obstacle sensors. Many of these features are shared across makes and models of drones in use today.

Industries and government agencies have taken advantage of the imaging capability of drones to:

- Fly under or along the sides of bridges and buildings to inspect for cracks and other defects
- Inspect utility structures in areas where ground equipment cannot reach
- Inspect sewer and storm drainage pipes
- Inspect the tops of silos for corrosion or other damage
- Monitor the movement of wildlife
- Examine pipes running through marshes for leaks
- Search for missing people and livestock
- Allow fire departments to discover remaining hot spots that could reignite wildfires or building fires and to assess the overall extent of fires during rescue operations
- Measure and document the impact of natural disasters such as floods, tornadoes, and wildfires
- Describe the extent of emergencies resulting from train derailments, oil spills, landslides, and collapsed roads and bridges

Drones are also useful for gathering immediate information about property damage. This drone information can direct people to specific areas of interest for more efficient inspections. Considering the dangers of inspecting a collapsed building, road, or bridge, the remote views drones offer can provide invaluable information while substantially reducing the risks to inspectors.

For these reasons, manufacturers worked with camera companies to equip drones with modern sensors to capture high-resolution images, record high-definition video, stream these images in real time to the pilot and crew, and document the images and videos for viewing and processing later.

In-Field Crop Scouting and Analyzing Field Conditions

Many drone companies thought that if they added better cameras or sensors, farmers would realize the benefits of seeing their farms, fields, or crops from different angles. However, farmers found it difficult to justify the expense of drones. In addition, many farmers may not have grown up with smartphones and related technology, so they may not be comfortable operating drones with seemingly little return on their investments.

On the other hand, advanced sensors provided decision-making value to consultants, ag retailers, agronomists, and precision ag companies. Drones became another tool integrated into the portfolio of services they offered to their clients and customers.

Some of the first drones flown in agricultural settings showed field drainage problems and detected crop nutrient deficiencies, crop weather damage, and the severity of pest damage. To some extent, satellite and traditional aircraft-based imagery was used in the past for these applications, but their resolution was not capable of capturing subtle differences that drones could detect. Many drones with modern sensors may

A CLOUD FULL OF DATA

Drones normally save data to an onboard memory card. Users exchange the memory cards when the cards are full, and it is most common for users to manually transfer data to a computer. Today there are software and service providers that offer automatic data transfer to a cloud-based database or that can even visualize data in real time if an adequate internet connection is available.

have a resolution of 0.25 square inches per pixel, depending on the flight height, while a satellite may have a resolution of several square feet per pixel.

Drone manufacturers were asked to equip drones to collect data. Manufacturers responded by equipping drones with built-in analytical capabilities to collect and send millions of data points from the aircraft to a receiver or computer. This data collection also allowed researchers to utilize a complex set of algorithms capable of assessing plant health status using the different wavelength bands the camera captured.

Demand for more sophisticated sensors grew as users learned about these new techniques available to monitor crop health. The new cameras needed to capture high-resolution images that collected many specific light wavelengths associated with crop stress. This led to multiple changes in sensors that are mounted on drones for data collection:

- **Upgrades to onboard RGB cameras.** Early digital cameras mounted on drones typically had poor resolution and no zoom features. These cameras could capture general patterns in a field. More advanced drones now have cameras that can capture high-resolution images (such as 48 megapixels) and feature mechanical zoom options. This means the camera can create pictures composed of some 48 million pixels, each pixel containing red, green, and blue color (RGB) wavelengths. These high-definition images can distinguish individual leaves and symptoms on leaves that show diseases, nutrient deficiencies, or insect pressures. Many of the cameras are comparable to professional-grade photography equipment.
- **Improvements in sensor resolution and angle of view.** The latest cameras have wide-angle views and higher resolution of individual sensors than cameras used on early drones.
- **Improvements in multilight spectral sensors.** Integrating multispectral sensors that can detect bands of light beyond RGB assisted individuals in making better field decisions. These sensors include near-infrared and red edge wavelength bands.

Figure 15.7. This DJI Mavic 3M has a multispectral camera, which can capture green, red, red edge, and near-infrared bands. The larger gray top is the RTK receiver.

- This additional information allows software to compute different indices such as the Normalized Difference Vegetation Index (NDVI) or Normalized Difference Red Edge (NDRE), which determines the "greenness" or stress of a crop beyond what can be seen with the naked eye. This spectral information can create color-coded maps where green means healthy, and yellow and red indicate stress. Users can then analyze data through multiple vegetative indices developed by university and industry researchers.
- **Improvements in thermal and infrared (heat) sensors.** Users added thermal and infrared sensors to measure the heat energy emanating from a crop, soil, or other objects. If objects are "hot," sensors will pick up more thermal energy, while cooler objects emit less heat on the surface. These thermal images convey information about varying levels of soil moisture across the field, detect crop canopy temperature differences, or identify plants that are water- or nutrient-deficient and stressed.

Figure 15.8. A thermal image of a field before any crop was planted: The lighter regions (yellow in the original) show tile lines that have higher soil temperatures resulting from water drainage. Darker colors indicate cooler soils, which are more likely to be wetter. Thermal imaging of bare ground can help determine elevation for ways to better manage water drainage.

- Since healthy plants with adequate water can cool themselves by evapotranspiration, the hot zones in a field may reflect poor crop growth or a reduced plant canopy. In addition to analyzing the crop, operators can use thermal imagery to inspect structures on the farm (such as barns and high tunnels) to ensure they are ventilated or insulated properly.

Applying Pesticides and Other Agricultural Inputs with Drone Sprayers

The next logical step in drone evolution in agriculture was to turn them into miniature aircraft equipped with spray systems to apply pesticides. Although some fixed-wing drones are equipped to apply pesticides, fertilizers, and seeds, multirotor drones seem to be better suited for agricultural applications in fields of varying sizes, shapes, and topographies. For these reasons, much of this chapter focuses specifically on rotary-wing drones, which are most common among drone applicators.

The idea of using drones to spray pesticides was put into practice first in Japan in 1997 with the Yamaha R-MAX drone. The drone looked like a small helicopter powered by an internal combustion engine. This high-powered drone had a single rotor with a diameter of about 10 feet, weighed 207 pounds, and had a spray tank capacity of slightly more than 4 gallons. With a full tank and spraying at 1 gallon per minute, the drone's tank was empty in about 4 minutes.

Ken Giles at the University of California, Davis, was an early pioneer with drones in North America. In 2010 he used an R-MAX drone that weighed 220 pounds when loaded to apply pesticides in vineyards, then on field crops and orchards. Giles was likely the first to get both federal and state authorization to use drones to apply pesticides in the United States.

Figure 15.9. Spray drones can apply pesticides. This one is applying fungicide to corn.

While this chapter focuses specifically on drones applying liquid products, many pilots have found them useful for applying dry materials such as fertilizers and cover crop seeds too. Most new drone sprayers can be equipped with optional kits to meter and spread these dry products.

BENEFITS OF SPRAY DRONES FOR PESTICIDE APPLICATIONS

The acceptance of spray drones in agriculture indicates a greater interest among growers in using and benefiting from them on their farms. We examine several documented benefits that spray drones bring to agriculture and industry on the following pages.

SPRAY DRONES FINDING WIDE UTILITY

One thing is for sure: Agriculture, industry, and government have quickly adapted to employing spray drones in ways no one could have imagined five years ago. Each year, spray drones continue to expand into new markets. Emerging uses for spray drone technology include managing issues in several settings:

- Mosquito control
- Ponds and lakes
- Utility rights-of-way
- Fruit trees
- Blueberries
- Grapes
- Agronomic crops
- Rangeland
- Turf
- Forests
- Pastures and hayfields
- Vegetables
- Greenhouse roof painting
- Cover crop seeding
- Fertilizer applications
- Spot-spraying invasive weeds

Drones will continually evolve to meet the ever-changing needs for each specific industry and agricultural sector.

Relatively Low Cost. Drones can be a lower-cost, more suitable piece of equipment for small- and medium-sized farms. A modern self-propelled sprayer can easily cost 15 times more than a current well-equipped spray drone. However, the cost of an enclosed trailer, mixing tanks and equipment, generators to charge batteries, and possibly a new truck can run to more than $100,000.

Spraying Challenging Fields and Difficult-to-Access Terrain. Many areas within fields or on farms may have limited access by ATVs, boats, wheeled sprayers, or trucks. This is especially true for applicators trying to control mosquitoes in the interior of remote areas or right-of-way companies treating brush around pipelines that run through marshes or under mountaintop power lines. Aerial and ground rig operators turn this type of work down because it is not profitable based on the low number of acres and the high costs of operating the equipment. There are situations where drones can spray fields when soil conditions are too wet for ground access. Using drones avoids wheel track compaction.

Furthermore, some manned aircraft or larger ground rig custom applicators require a minimum number of acres to be treated. Drones may offer farmers with smaller operations an opportunity to hire custom spray drone services that require fewer acres. Drones can make applications to small fields much easier than some equipment; however, efficiency and operator satisfaction increases for drones in big, square fields, similar to for conventional application equipment.

In many cases, steep hills can make pesticide applications with ground sprayers difficult and dangerous (think of equipment rollovers). A drone operator can fly hilly areas, which helps prevent fatigue and potential injury from people carrying backpacks up and down steep terrain and uneven ground.

Autonomous Presets and Obstacle Avoidance. Spray drones flying closer to the ground (10–15 feet) are typically equipped with supplementary radar sensors that detect obstacles in all directions—above, below, left, and right. These sensors allow drones to automatically follow the terrain, so they can maintain altitude for proper application and avoid obstacles. The 360-degree radar can essentially scan the surroundings continuously. The drones involved with applications require cameras that show front images (that is, field views of what are you about to encounter) to create an applicator log that can track where the drone has been.

Figure 15.10. Cameras on spray drones provide real-time visuals of the approaching terrain, crops, and other obstacles during applications.

Autonomous takeoff and landing flight options help overcome operator errors. These safety features are analogous to the parking sensors and backup cameras on modern passenger vehicles. The intent is to make piloting drones safer and less stressful for operators and to help avoid accidental crashes and midair collisions.

Applications Made Within Line of Sight. Drone operators know where they are applying, because they can view the flight paths and areas they are treating with onboard, real-time cameras during applications. In addition, pilots must maintain a line of sight with their drones throughout their flights to monitor any performance or safety issues.

Spray drone controllers provide operators real-time information about application rates, speed, flight height, and droplet sizes in most newer drones. All of these parameters are necessary to track applications. Although line of sight is a current Federal Aviation Administration (FAA) regulation, at the time of this chapter was written, petitions to remove these requirements for pesticide applications are awaiting FAA decisions.

Farmers Control Spray Timing. By owning their own drone sprayers, farmers can more easily control application timings than they can if they must rely on airplane services that have long lists of work to be done. This is more than just convenience—farmer-owned drones can enhance crop protection efforts with more timely pesticide applications.

Large Pool of Potential Pilots. Drone sprayers are easy to operate, and almost anyone can be trained to safely and effectively operate them in a relatively short time.

Reduced Environmental Footprint. Spray drones can reduce site disturbance, soil erosion, and site compaction compared to ground equipment.

Figure 15.11. Some fields (like this one) are easier for operators to maintain a line of sight than they could with rolling or uneven fields, especially when spray drones move farther away.

Avoids Physical Crop Damage. Drones are also an effective and safer option for crops with taller or thicker canopies where ground equipment can damage the crops, especially in the tire tracks or in the presence of viny weeds like bur cucumber and morning glory. Fungicide applications after corn tasseling have been one of the most common uses of spray drones in agronomic crops. Crop damage can also be reduced in smaller, irregular-shaped fields.

Combats the Spread of Invasive Weeds. Operators can map weeds that may be growing in certain parts of fields, pastures, rangeland, or wetlands and then use drones to spot-spray these areas at a much lower cost than broadcast applications. Because they are airborne, drones will not pick up invasive weeds, which helps to prevent their spread from field to field.

Reduced Applicator Exposure and Accidents. Using drones to spray can significantly reduce applicators' exposure to pesticides compared to making the same applications with certain types of ground-based equipment—particularly open-cab sprayers, backpacks, and ATVs. With spray drones, applicators are physically distant from the immediate area where the spraying is done. However, filling drone tanks, swapping and charging batteries, handling controllers, and taking sprayers back to trailers or trucks can create important sources of operator exposure. Spray drone operators still must follow the personal protective equipment (PPE) guidelines on pesticide labels to minimize this exposure potential.

Work Around Obstacles. Drones can spray around obstacles and tree lines much easier than airplanes and larger ground sprayers can.

Easier Repairs. Repairing spray drones after a forced landing or crash is fairly easy. Operators can fix most of the issues simply by replacing broken or malfunctioning parts. Additionally, if operators are intimidated by the thought of repairing their own drone, there are often trained technicians available to do the repairs.

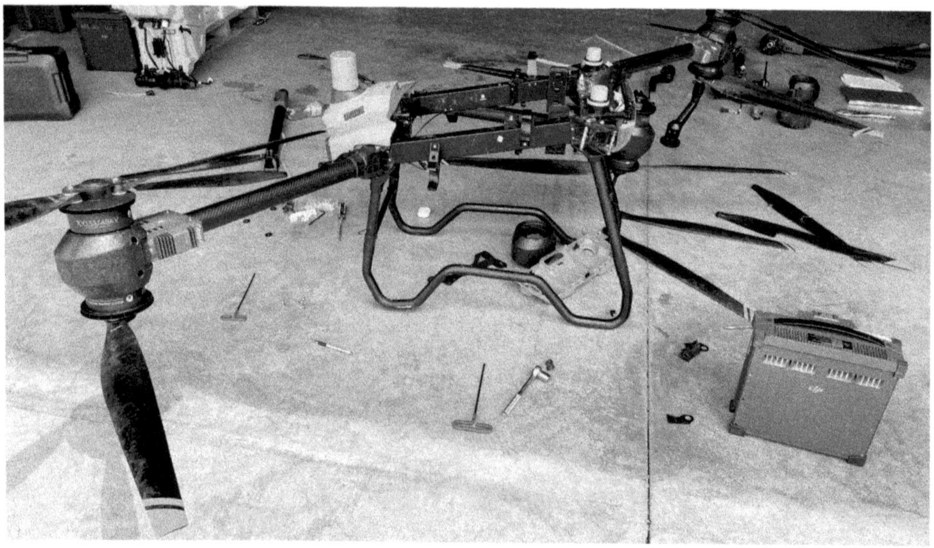

Figure 15.12. With a little experience, drone repairs are rather straightforward.

CURRENT CHALLENGES OF USING SPRAY DRONES

While there are many benefits associated with using spray drones, issues, questions, and concerns remain that users should consider before purchasing a drone and using it to apply pesticides. We describe some of these considerations on the following pages.

Volume Limitations. Drones cannot currently compete with ground rigs for applications that require high carrier volumes.

Pesticide Labels Lack Drone Information. In states where approved, spray drones could apply pesticides as long as the pesticide includes aerial applications on the product label. However, spray drone applications must still follow all of a pesticide label's aerial application requirements, which could be a challenge because these requirements were typically written for traditional fixed-wing aircraft or helicopters.

Some pesticide labels specify that if applicators cannot follow all aerial application requirements on the label, then applicators should not apply the product. This presents a challenge for spray drones with unique qualities that set them apart from traditional aerial applications, such as travel speeds, rates, and nozzle configurations. For example, many pesticides require a minimum spray volume of 5–10 gallons per acre, which spray drones may not be capable of providing. It is feasible for spray drones, but it just reduces the acreage completed per day and often increases the cost.

While both traditional aircraft and drones dispense pesticides from the air, there are likely many differences between traditional airplane and drone applications operating under standard procedures. The biggest one is the spray dynamics of each method. Aerial application sections of pesticide labels may list spray heights, carrier volumes, nozzle types, boom widths, and travel speeds specific to manned aircraft applications that drone sprayers cannot match.

It is unknown whether pesticide manufacturers will modify pesticide labels or provide supplemental labels specific to drone sprayer applications. Many manufacturers have indicated they may consider adding drones to new pesticide product labels but do not plan to modify existing product labels because of the expense involved.

Pesticide applicators are requesting label information to address pesticide safety for applicators and the environment, as well as product performance guidelines that are unique to spray drone applications. For some time, pesticide registrants have been working with the EPA on future drone labels, but dedicated drone labels are probably years away. The wide varieties of spray drone designs and operations create challenges to forming general application guidelines on labels that assure minimum adverse effects.

State-led agencies that regulate pesticides differ on how they interpret the aerial application instructions on labels. Some states allow drones to apply products under the aerial part of the label, while other states prohibit this unless the product labels specifically mention drone applications. In addition, operators often need to be licensed in each state where they will use their drones.

States also vary on what additional certification(s)—other than the ones required by the FAA—farmers and commercial applicators must have to use drones for applying pesticides. One state may require an aerial certification only, while another state may require additional certifications (such as aquatic or right-of-way) added to the aerial certification. Contact your state pesticide regulatory agency to discuss what certifications they require to use drones to apply pesticides.

Federal Regulations Require Two People. Current federal regulations require two people at each drone application site, which means you need to pay two people to spray with a drone rather than just one for a manned aircraft or ground rig—unless you receive a waiver from the FAA.

One is the dedicated pilot whose sole job is to pilot the aircraft. The second person is the visual observer whose sole purpose is to watch for obstacles or other hazards such as other drones or manned aircraft during the operation. Operators have filed requests for exemptions with the FAA regarding this regulation, so there may be a shift in this rule in the future.

Must Maintain Line of Sight. The visual observer must maintain a line of sight with a drone according to FAA regulations. This rule can make applications challenging because crop canopies can restrict visual lines of sight, especially in areas with rolling topography or areas behind tree lines.

What You Know Today May Change Tomorrow. Consider that boom sprayers and aerial sprayers have been used for 100 years. Billions of research dollars have been invested to determine the best spraying practices for ground sprayers that will lead to the most effective crop protection and to train applicators to best utilize these practices.

With commercial spray drone use still in its infancy, research is only beginning to be discussed at professional meetings, trade associations, and cooperative extension applicator training meetings. Experts expect that some of the information developed for ground sprayers and traditional airplanes will transfer to drones. However, new research undoubtedly will show significant differences between drone applications and other application forms.

Figures 15.13–15.14. Some spray drones are equipped with common ground equipment nozzles (such as the XR nozzle in the left image); however, these nozzles have much smaller orifice sizes than those commonly used with ground applications. More modern spray drones use rotary atomizer (spinning disc) nozzles, like that observed on the DJI Agras T40 (right).

Researchers are examining ways to better understand the spray deposition and efficacy of spray drones to make proper application recommendations. Agricultural researchers are investigating the similarities and differences between planes, drones, and ground rigs when applying the same product.

Researchers are investigating many questions that may change how drones apply pesticides in the future. The answers to these questions may also vary by drone manufacturer, model, and the overall size of the spray platform.

- What are optimum application heights?
- How will wind conditions, drone types, nozzle size/designs, or the types of pesticide being applied affect applications?
- Will spray heights vary by application type (spot versus broadcast)?
- What is the optimum carrier volume?
- Are there nozzles or droplet sizes that are better suited for drones?
- Are rotary nozzles better than traditional hydraulic nozzles?
- Will drones with linear booms perform spraying differently than drones with individual nozzles immediately below the rotors?
- For drones with linear booms, what are the optimum boom lengths in relation to the location of the propellers?
- For boomless drones, what nozzle configurations and how many nozzles under the drones will provide the best applications?
- What spray nozzle pressures on drones will provide better coverage, efficacy, and reduced off-target movement?
- How do drone speeds affect spray coverage and pesticide efficacy?
- Is there an optimal flight speed where the overall application is optimized to maximize pest control and minimize spray drift?
- What drone speeds are best for canopy penetration?
- How wide can drones spray effectively?
- What application parameters affect rotor downwash?
- What are the best flight patterns for effective pesticide applications with drones? Is it along, across, or at an angle with the crop rows or the wind direction?
- How does the lack of agitation in spray drone tanks affect pesticide applications and efficacy?

What You Use Today Might Not Be Used Tomorrow. Drones were initially designed for capturing imagery and collecting data, which largely remains true with today's aircraft. Just because they are equipped with a solution tank, pump, and spray nozzles does not change what they were originally designed to do.

We expect that the commercial use of spray drones will expand and be part of an evolution that will incorporate advances in spray system design, plumbing, nozzle placement, battery life, sensors, and pump specifications. There is no universal model, design, or software system that is currently considered an industry standard.

Figure 15.15. One of the many differences among drone models is how the spraying systems are equipped. This drone is equipped with conventional nozzles underneath the spray rotors.

Eventually, certain designs with the greatest efficiencies and value will come to the forefront and become the standard of how a drone should be designed for spraying. This process will be similar to the development and adoption of precision ag equipment currently used in crop production, where some level of compatibility and operation has been established.

Spray Drift Remains a Top Priority. Pesticides moving off-target during spraying is a major concern regardless of the equipment applicators use. It is nearly impossible to completely avoid spray drift. However, operators can minimize it by selecting the best application parameters, especially the nozzles, and operating them at optimal conditions (see chapter 14). One of the parameters that affects pesticide drift is the distance the spray droplets must travel from the nozzle to the soil or crop target.

For ground equipment the typical recommendation is to have the booms about 24 to 30 inches from the target canopy or soil. Even at that relatively close height, spray drift is among the biggest challenges with ground sprayers. Of course, the rotor downwash from spray drones may influence the spray cloud and serve as an air curtain to help drive spray droplets down to the target. However, what percentage of the spray cloud is in the downward rotor downwash or is pushed either laterally away from the drone or upward into the wind current is unknown.

Preliminary results from a small number of studies indicate a drone flying higher than 10 feet above the crop canopy is likely to cause higher levels of off-target spray drift. Drift potential from spray drones is expected to increase considerably as flying speed and flight altitude increase. Boom configuration in hydraulic systems can also change drift potential based on where booms are placed on the drones and nozzle locations in that particular configuration. As with ground and manned aerial applications, boom configurations for drones must be tested for effective swath before field applications to ensure best practices.

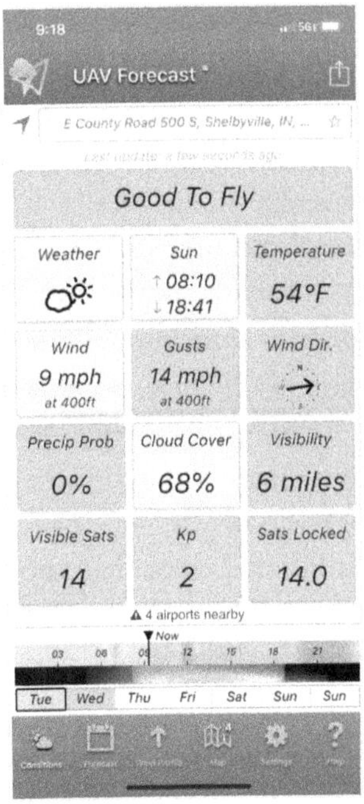

Figure 15.16. There are smartphone and tablet applications to determine if environmental conditions are suitable for application.

Figure 15.17. This photo shows the spray distribution and swath emitted from a DJI Agras T40 using rotary atomizer (spinning disc) nozzles.

Figure 15.18. Nozzle type, spray height, and carrier volume are some of the factors researchers are evaluating to determine optimal spray drone applications.

Regulations Pose a Large Hurdle. Individuals who want to operate drones have many regulatory layers to navigate, which takes time. For example, the FAA requires drone operators to comply with several requirements: For example, drones must be registered, pilots need to pass an exam, pilots must maintain a visual line of sight with their drones, and operators must obtain permission to fly in restricted airspace.

Drone manufacturers have provided the capability to swarm, meaning one remote can operate several drones at once. If the drone is less than 55 pounds, swarming has been approved for up to five drones; however, the vast majority of spray drones operate above the 55-pound limit, and in that case operators can only fly one drone at a time without getting a waiver from the FAA. This changed in 2024, when Hylio was granted FAA approval to fly three spray drones weighing more than 55 pounds with just one pilot and without an observer (docket ID FAA-2023-1833).

While swarming allows three drones to be flown with additional crew, that same crew of two now has three drones to refill at a time instead of just one, which would affect productivity. This has the potential to be a gamechanger for drone applications and to treat larger agricultural operations more quickly.

However, the exemption or waiver process may take several months. Applicants must provide proof that the extra precautions pilots have proposed to be exempt from regulation will provide a similar standard of safety as the rule provides. However, it is expected that the FAA will reduce conditions requiring exemptions and that the approval period for waivers will be expedited in the future.

Expansion May Depend on Battery Innovations. What started off as a drone carrying 2 gallons of water has now increased to 18 gallons—and that increases with each new model. However, each additional gallon of water makes drones heavier. This increased weight reduces the time batteries remain sufficiently charged.

Smaller drones used for photographs can get upward of 45 minutes from a single battery. For heavier spray drones, battery life has been reduced to approximately 7 to 12 minutes of flight. This means that each spray drone needs multiple spare batteries fully charged or actively being charged to keep the drone operating continuously over the course of a day's operations. We expect that battery research fueled by the electric vehicle industry will lead to transferable technology that will power drone batteries in the future.

Another issue associated with batteries is how long they require to recharge. Battery recharging can take anywhere from 10 to 30 minutes depending on battery size, the type of charging power available (110 vs. 240 volts), and how hot the batteries are. Generators that are large enough to produce sufficient wattage to charge multiple batteries at once can be difficult to find and are extremely bulky; their price and quality range widely.

Newer drone models allow for a fast charge at 240 volts and 50 amps. This can reduce charging time if the power supply (generator) is rated to provide the full current the battery charger needs to operate most efficiently. Efficient drone operators will typically have several batteries so they can use one battery while two or more others are being charged (or already charged) and ready to be swapped when drones land to refill.

Protecting Your Investment Requires Insurance. With today's drones costing $30,000 or more, crashing and destroying one without insurance coverage is risky. Flying without insurance is like buying a new truck and driving off the lot without coverage. When you have that much money invested in a drone, you don't want to take a chance on an accident or being involved in a lawsuit. Some pesticide regulatory agencies require proof of liability insurance before they will issue the certifications or licenses necessary to operate a spray drone.

Two generally accepted types of insurance are hull and liability. Hull insurance protects the drone, while liability protects against whatever damage the drone causes. Currently, there are very limited options for insuring spray drones, but this, too, is rapidly changing. Limited availability often means insurance coverage for a drone can be expensive, especially for liability insurance. Some drone operators are paying $4,000 to $6,000 annually per unit, but this cost includes spray drift and other liability coverages. Since drones can be repaired relatively easily, some operators choose to get only liability coverage to reduce their premiums.

If you are interested in hiring a custom spray drone application, be sure to ask if the applicator is fully licensed specifically for spray drone applications, because the insurance is not the same as what is issued for ground applications. Accidents or off-target incidents that occur from custom spray drone applications not carrying insurance could result in legal ramifications for all of the parties involved—the farmers, customers, and applicators.

Figures 15.19–15.21. A spray drone contacted a transmission line and irrigation pivot, knocking out the power line. Beyond the damaged equipment, what other liabilities could the operator be responsible for?

LEARNING TO FLY WITH COMPUTERS

Each drone manufacturer has their own specific software for planning flights and controlling drone functions. Like much of agriculture's new precision technology, drone technology requires skills to effectively operate and navigate computers. For some, this may be as simple as the interface and navigation systems. Most drone technology is not much different from other common software systems or computers and displays used in modern agricultural equipment.

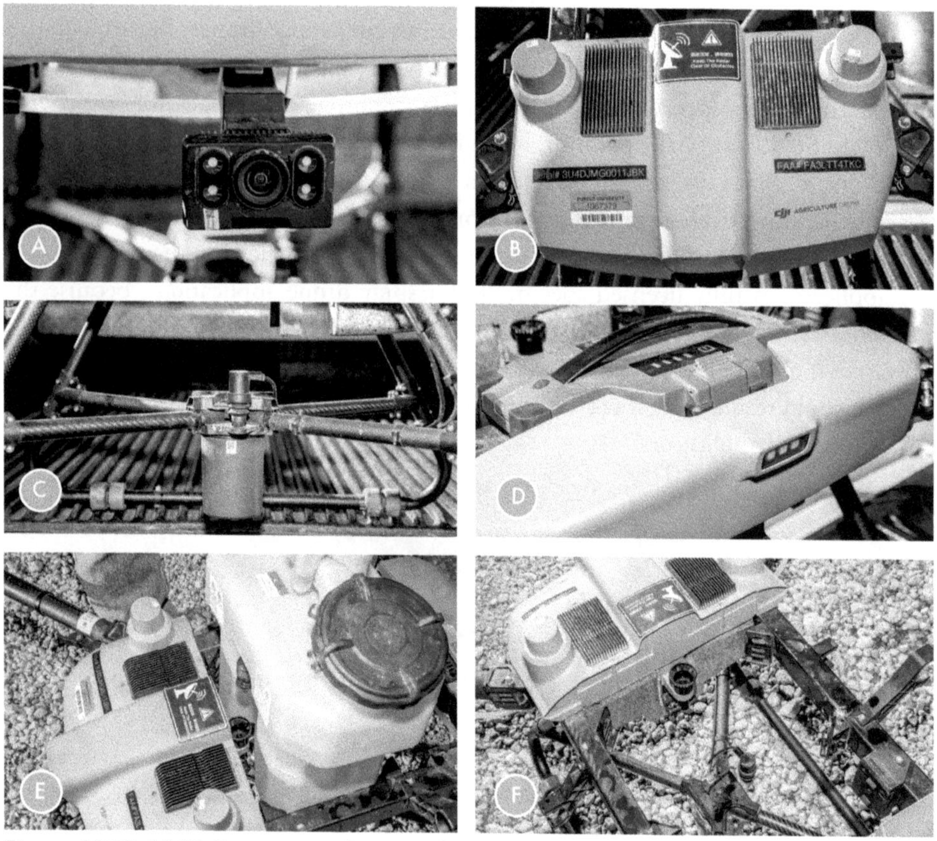

Figures 15.22–15.27. Components of a spray drone.

(A) This DJI Agras T10 has a camera in the front and is flanked on either side by an obstacle avoidance sensor.

(B) The drone receives a signal from the controller via its two antennae. The two black panels on the side of the receiver allow air flow that helps keep the hardware in the drone from overheating.

(C) This device is called an omnidirectional digital radar system. It measures the altitude that is recorded on the controller.

(D) This battery snaps into place. The four light panels on the battery indicate how much battery life remains.

(E) This drone's tank holds 2.5 gallons.

(F) The drone with the spray tank removed.

Figure 15.28. This drone has four nozzles. When it is flying autonomously, only the nozzles in front of the travel direction turn on for spraying. In other flight mode types and at higher application rates, the drone can turn on additional nozzles.

However, for those who are not as tech-savvy, adopting these technologies can be daunting. Some interfaces can be difficult to manage, and some operations may be difficult to troubleshoot in the field. These challenges are compounded by a lack of adequate product and software support in the industry.

In addition to the steep learning curves with the technology, employees need to thoroughly understand the federal and state laws that govern drone use for agricultural applications. The FAA, Environmental Protection Agency (EPA), state regulatory offices, and local governments can all be involved with creating and enforcing the rules that surround the use of spray drones. Some of those rules can be confusing, conflicting, or generally unclear because the rules and laws are continuously being updated to account for this novel technology. The good news is it should be easier to find eager employees who have both drone and computer experiences as the technology becomes more popular and widespread.

CALIBRATING SPRAY DRONES FOR LIQUID APPLICATIONS

You might assume that manufacturers would perform a standardized calibration test on their spray drones before shipping them out of the factory. It appears to be a pretty straightforward process: Open the box, assemble the drone, enter the flight parameters into the controller, and mix and pour the product into the tank. You are now up and running, ready to start spraying ... right? If only it was this easy!

This "ready to spray out of the box" assumption can lead to some major application problems because every spray drone must be calibrated properly to ensure it applies products at the correct rates and across the desired swaths. This should not come as any great surprise because the same holds true for new ground sprayers or any application equipment.

Regardless of the type of sprayer used, sprayers can be effective, efficient, and safe only if they are properly checked and calibrated before they are taken to the field. Accurate applications also depend on periodically checking and calibrating drones during the spraying season. This is especially important when operators significantly change any of the components in the setup (such as new nozzles, nozzle spacings, and so on).

The primary goal of calibrating spray drones is to determine the actual application rate over a given area for a given time. Operators should adjust their equipment if there is more than a 5 percent difference between the actual application rate and the intended application rate.

Selecting Nozzle Type and Size

Although nozzles are some of the least expensive components of spray drones, they significantly influence sprayer performance. Nozzles meter the amount of liquid sprayed per unit area, control the application rate, and affect the uniformity of spray within the swath. Nozzles also influence droplet sizes, which affects both target coverage and spray drift risk (see chapter 14 for more).

We don't have as many nozzle options for drones as we do for ground equipment because spray drones have smaller pumps and limited output. The electric pumps on spray drones may be limited to less than 5 gallons per minute (gpm) depending on the drone's size (these pump systems are similar to ATV spot sprayers). This is why most drones use nozzles with low flow rates (0.1 or 0.15 gpm) instead of the higher flow volumes fitted on ground sprayer booms.

Using nozzles with higher flow rate capacities will likely create problems because drone pumps will not be able to generate enough flow to operate the larger nozzles effectively. It should be noted that our nozzle selection and calibration discussion here does not include the rotary atomizer nozzle types used in newer spray drones. For a more thorough review of these spray drones and nozzles, see "Care and Calibration of Rotary Disc Atomizers."

To determine the size (flow rate in gpm) of a hydraulic nozzle that will adequately satisfy your intended application rate in gallons per acre, follow these steps:

Figure 15.29. Drones are often equipped with flat fan nozzles, but the spacing and orientation can be very different from that on traditional application equipment.

Step 1. Select the desired application rate in gallons per acre (gpa).

Step 2. Select a practical and safe flying speed in miles per hour (mph).

Step 3. Determine the effective spray swath for the drone in inches (W). Do not use the maximum swath manufacturers list in drone manuals. See "Determining the Effective Spray Swath Width" for the differences between maximum and effective spray widths. This step differs from traditional spray calibrations that use the width of the nozzle spacing for the calculations.

Step 4. Determine the flow rate (gpm) the drone requires for the application:

$$\text{gallons per minute (gpm)} = \frac{(\text{gpa} \times \text{mph} \times \text{W})}{5{,}940}$$

(By the way, the 5,940 number used in this equation is a constant to convert gpa, mph, and inches to gpm.)

Let's use an example that shows the calculations using a T30 that has 16 nozzles but only turns on 12 when spraying. You want to spray a fungicide at 2 gpa, at a speed of 15 mph. There are 12 nozzles on the drone, and you would like to maintain a swath width of 25 feet (300 inches). Based on the inputs, what flow rate per minute must the nozzle need? Enter the values given above into the equation:

$$\text{gpm} = \frac{(2 \times 15 \times 300)}{5{,}940} = 1.515 \text{ gpm for entire drone}$$

Step 5. Divide the output for the entire spray drone by the number of nozzles activated during a spray pass.

$$\frac{\text{gpm for entire drone}}{\text{number of working nozzles}} = \text{gpm per nozzle}$$

$$\frac{1.515}{12} = 0.126 \text{ gpm per nozzle}$$

Please note that most drones come equipped with nozzles from the factory or have rotary atomizers. These calculations verify that the nozzle on the drone meets our application requirements.

The conventional hydraulic nozzles on spray drones have small orifices (01 and 02), which tend to clog. Spray drones with rotary atomizers are relatively less prone to clogging.

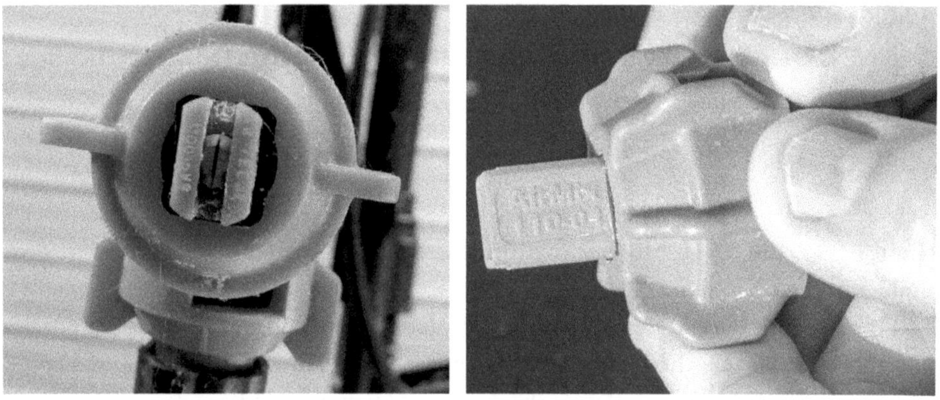

Figures 15.30–15.31. These two nozzles have a 110° spray width. The left nozzle has a flow rate of 0.15 gallon per minute (gpm) at 40 psi, while the right nozzle has a flow rate of 0.1 gpm at 40 psi. The first nozzle would more closely align with what we expect to get out of the drone.

Select a nozzle size from the manufacturer's catalog (or from their website) that will provide the flow rate (gpm) that you determined in Step 5 when the nozzle is operated within the recommended pressure range. For drones with nozzles, you are basically limited to two sizes—0.1 and 0.15—because drones cannot generate enough flow and pressure to operate larger-orifice nozzles properly. So you should select a nozzle that will deliver 0.126 gpm of spray output.

Take care when using spray width (W) in the equation above to determine the nozzle flow rate (gpm). Several things affect the actual effective swath width that you should use in the equation, including the flight altitude (see the following section).

Determining the Effective Spray Swath Width

One of the variables in the equation used to determine the flow rate (gpm) is the spray swath. It is extremely important that when determining a value for W in the equation, you should use the effective swath width, not the maximum swath width the drone user manual may provide. The effective swath width considers the need for some spray overlap between passes to achieve a uniform application across the field.

Spray drone manufacturers often advertise their maximum swath widths. The values they provide represent how far away from the drone some droplets may deposit on the spray target. However, the deposition pattern of droplets across the spray swath will not be uniform. A relatively small number of droplets reach the outer edges of the spray pattern while there are greater deposition levels toward the center of the drone.

A general rule of thumb from drone users, the scientific literature, and practitioners is that the effective spray width is between 65 and 75 percent of the advertised spray swath width. For example, if a drone has a maximum spray width of 30 feet, it would have an effective spray width between 20 and 22 feet. This can vary, so you should verify the effective spray width by performing a swath test. Researchers should come up with methods to easily find and refine effective spray widths in the near future as new findings are reported.

For that reason, the drone operator using the controller screen shown in figure 15.32 has entered 25 feet for Route Spacing instead of the 30-foot maximum swath width suggested by the drone manufacturer. One other factor that plays a key role in determining the actual spray swath is flight altitude. The lower the flight altitude (height from the ground or the crop canopy), the narrower the spray swath will get. You also should consider altitude when you select nozzle sizes.

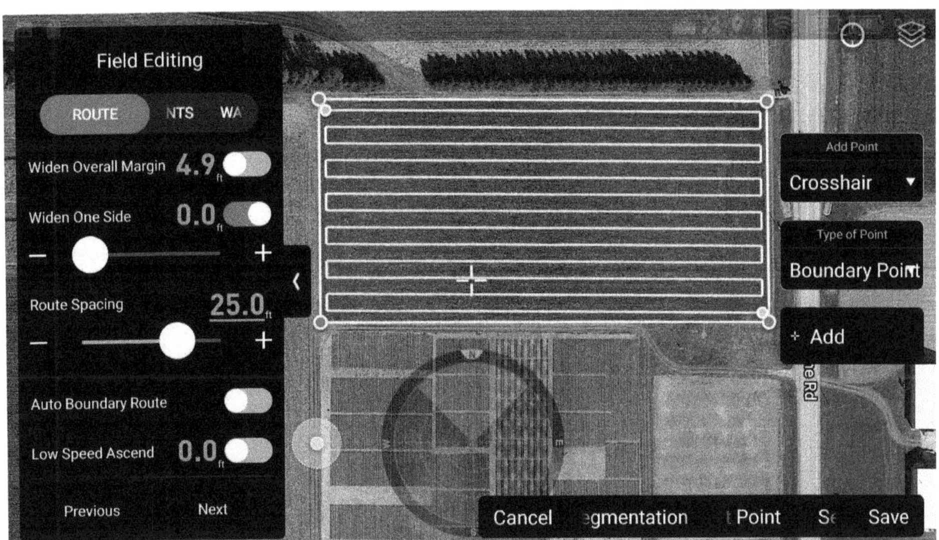

Figure 15.32. This drone is using 25 feet as the route spacing, to use an effective spray width instead of the maximum spray width. Using a reduced route spacing allows some overlap between passes to achieve a more uniform application. Sprayers101.com provides an excellent explanation of the difference between maximum and effective swath widths when using spray drones.

NOTES FROM THE FIELD

Crosswinds can cause swath displacement, which drone operators should account for when establishing spray passes. When flying in wind, spray speeds and pressures will likely change whether flying against a headwind or with a tailwind. This can also affect how the spray swath lays down. This reemphasizes the need to swath-test in multiple different environmental scenarios.

Considering the variations in nozzle sizes, spray pattern angles, wind direction, wind speed, and flight altitudes, there is no theoretical mathematical solution to determine the effective swath width. Ideally, operators can check the spray swaths of their drones using methods and instruments that can quantify the spray solution volume or the amount of spray droplets being deposited across the swath. However, this may be limited to university researchers and spray application consultants, who commonly use these methods.

The best way to determine the spray deposition from multiple passes is to visually observe the variations in the spray deposits across the sprayed area. For visual observation, the most effective approach is to lay a strip of white paper tape or water-sensitive paper (WSP) under the drone's flight path. Add some food-grade coloring or foam marker dye in the spray tank, and then fly the drone over the tape or paper and observe the spray pattern and droplet size spectrum, as shown in figure 15.33. Depending on what you observe, you may have to change the flight altitude and/or the overlapped distance between each flight pass. You will need to replicate several adjacent spray passes until you are satisfied with the uniformity and deposition of the spray across the test area.

Another approach researchers have used for decades to determine spray deposition uniformity with ground equipment is to use water-sensitive spray cards (figures 15.34–15.35). Users place spray cards perpendicular to the sprayer's travel route. The cards turn blue anywhere water drops hit them. In general,

Figure 15.33. The most practical way to find out if a spray deposition is uniform is to add a very small amount of dye to the sprayer tank and then make several passes over a tape laid on the ground just like you would be spraying the area. The dye used for sprayer foam marker systems is commonly used to visualize spray deposits. It also is safe and inexpensive for calibration runs.

Next, observe the deposition of spray droplets on the tape. Heavy dye concentration on parts of the tape is a good indication of too much overlapping of adjacent passes. No dye on some sections of the tape indicates not enough overlap. This may require you to change the flight altitude and/or distance between passes. Individual applicators can use several pattern evaluation systems, including Swath Gobbler, DropFlight, and AccuPath.

Figures 15.34–15.35. An inexpensive and practical way to determine effective spray width, spray uniformity, and spray drift is to use water-sensitive paper (WSP). Here the applicator fastened WSP to boards lying on the ground perpendicular to the drone flight path. Operators can use the information from the WSP to adjust the settings (height, speed, etc.) that determine the effective spray swath.

the more water that hits the cards, the more blue droplet stains appear on the cards, and the better the spray deposition or coverage.

As you would expect, water-sensitive cards placed toward the end of the swath show less coverage compared to the center. Researchers study the distribution of spray droplets and use analytical methods to better quantify the intensity of those spray droplets. There are some phone applications (like DropLeaf) that can quantify the differences in the spray cards. Just take a photo of the card, and the app will generate spray droplet data.

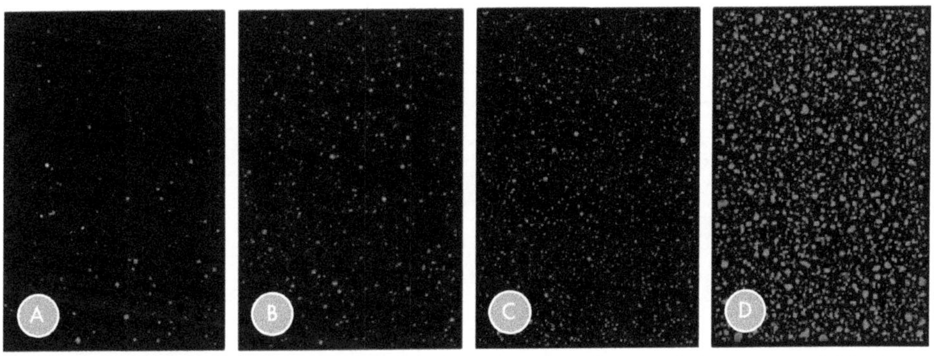

Figures 15.36–15.39. Using a fluorescent dye along with Kromekote cards shows spray distribution when applicators use (A) 1 gpa (gallons per acre); (B) 2 gpa; (C) 3 gpa; and (D) 15 gpa. (The dots are purple in the original.)

Figures 15.40–15.47. Height influences spray pattern.

A drone sprays **2 feet** above the pavement.

Spraying **5 feet** above the pavement is not high enough to deliver an acceptable and uniform pattern.

A drone spraying **10 feet** above the pavement.

A spray width testing card showing deposition after being sprayed by a drone.

Calibrating the Spray Drone

Regardless of the type of sprayer you use, you can only achieve your goal of protecting crops against pests if you properly check and calibrate the sprayer before you take it to the field (and recalibrate and recheck periodically during the spraying season). Like for ground applications, the various combinations of pesticides, adjuvants, and other chemicals can alter the physical properties of the spray solution. In turn, those properties influence sprayer calibration, swath width, and spray deposition.

Arguably, the chemical inputs in spray solutions for drone applications may have an even greater impact because they are 5 to 10 times more concentrated than the more diluted solutions used for ground applications. The primary goal of calibrating a spray drone is to determine the actual rate of application in gallons per acre and then to make the necessary adjustments if there is a difference greater than 5 percent between the actual application rate and the intended application rate. As is true for all sprayers, travel speed, nozzle type, flow rate, and spray pressure (which influences both the flow rate and the droplet size) all affect application accuracy. The three rules of application accuracy apply to all spraying equipment, including spray drones:

1. Maintain a uniform flight speed. This is essential to keeping the application rate relatively uniform throughout the operation. Faster flying speeds will result in lower application rates, while slower flying speeds will produce the opposite—higher application rates. Most spray drones are equipped with rate control capabilities that provide consistency in application when speeds change; however, application speed should remain the same throughout the field.

2. Maintain an optimum spraying distance between the nozzle and the target. Drones have radar altimeters to take care of height. Operators cannot control height during autonomous flights with most drones. This often requires operators to make hard choices about productivity and coverage.

3. Maintain a uniform spraying pressure. This is essential for keeping both the application rate and droplet size consistent during the application. Lower spray pressure leads to lower application rates and larger droplets. Conversely, high spray pressure leads to higher application rates and a greater number of smaller, drift-prone droplets. Both extremes waste pesticides and provide poor pest control.

However, directly setting and adjusting spray pressure may not be possible because spray drones do not have pressure gauges or pressure controls. Alternatively, operators could adjust travel speeds to reach desired gpa outputs. Similarly, keeping spray height uniform could be difficult if a drone is not equipped with a terrain sensor. Luckily, most, if not all, spray drones come with terrain sensors.

All spray drones come with controllers that allow drone operators to set their spray parameters, such as height, swath width, gallons per acre, and travel speed. Controllers also allow users to plan autonomous routes for applications. These are good starting points for accurate spraying. However, operators can attain proper application rates only if sprayers are in proper working order and correctly calibrated. Sprayers that

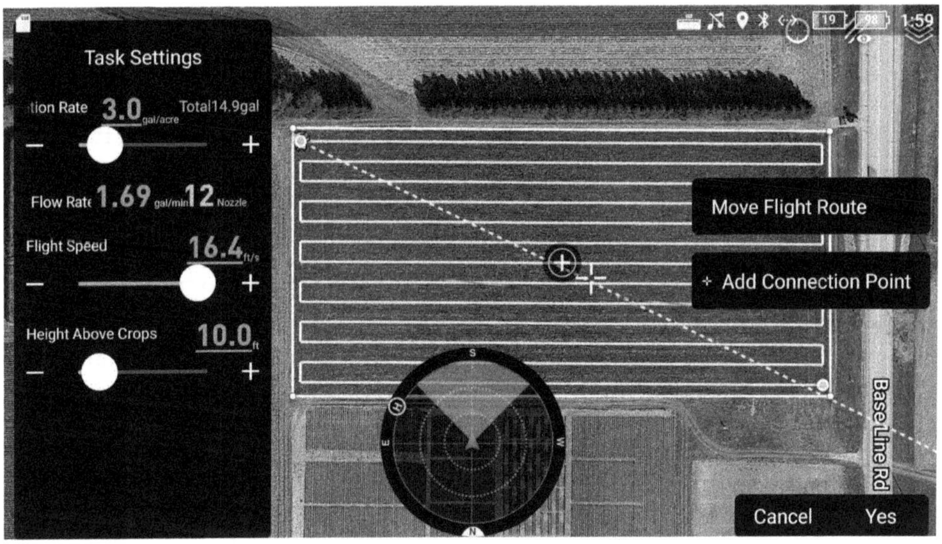

Figure 15.48. This controller says the drone is applying 3.0 gallons per acre. The only way to know if this is accurate is to measure the flow rate in gallons per minute and convert it into an application rate in gallons per acre.

are not accurately calibrated may apply too much or too little chemical, resulting in unsatisfactory efficacy and wasted pesticides. Poor calibration can also increase the risk of injuring the crop, fail to control the pest, or increase drift risk.

In several ways, calibrating drone sprayers is somewhat similar to calibrating ground sprayers. Drone remote flight controllers are really acting like in-cab display controllers on ground sprayers. Both systems require initial calibration checks before hitting the field for the first time after purchase or at the beginning of each spray season.

For ground sprayers, you enter calibration mode and run water through the sprayer for a known amount of time, then measure the volume you catch from the nozzles during that period. You record the volume you caught and enter that into the in-cab spray controller, so the computer can correct the built-in calibration settings. This essentially improves the accuracy of the flowmeter.

WHEN 1 GALLON IS NOT EQUAL TO 1 GALLON

Several drones may use imperial gallons instead of US gallons. An imperial gallon equals 1.2 US gallons. In the calibration section, 2 gallons per acre using imperial gallons would require 2.4 US gallons. Review the owner's manual or contact the sales representative to see which gallon unit your drone uses.

Spray drone remote-control modules do not have ways for users to enter any calibration data. Drone control modules use internal application formulas to control the output through the pump and flowmeter. You cannot directly influence the flow rate of a nozzle by turning a knob or entering a value in a pop-up window that alters the pump and flowmeter. We also don't know the spray pressure, because this is not displayed on the controller.

There are several ways to calibrate spray drones to work around this problem. Use the calibration method that you are accustomed to if it happens to be different from the procedure outlined here. Regardless of the method you use to calibrate sprayers, you will have to take certain measurements. You should follow the important operational principles mentioned in this chapter to achieve maximum efficacy from the chemical applied.

There are a few things you must do before starting the calibration process. Needless to say, you should always wear gloves and goggles for safety reasons when mixing, spraying, and calibrating sprayers. After you have your proper PPE, clean spray systems to remove any pesticide residues to allow for safe calibration.

Next, run the sprayer on the ground, inspect it for leaks, and make sure that all vital parts function properly and that the sprayer has a good set of nozzles. Check the flow rate of each nozzle for a set period (minimum 30 seconds) to identify nozzles that may be clogged or worn out (see chapter 14). If it is difficult to run the spray process with the drone on the ground, you could remove the nozzles from the drone and put them on a standard sprayer to check the flow rate.

Spray drone nozzles are more prone to clogging because the nozzles designed for drones have smaller orifices to accommodate the lower carrier volumes drones apply. Compounding the problem is that most drone nozzles do not have screens in the nozzle body assembly. Furthermore, drone sprayers apply more concentrated spray solutions, which have a greater risk for chemical incompatibility and for particulates to build up in the tank. For these reasons, you must frequently check the flow rates of spray drone nozzles. Even partially clogged nozzles will cause underapplication and nonuniform spray deposit on the ground.

Be sure to clean all clogged nozzles before you calibrate sprayers, and clean nozzles daily (especially after spraying mixtures that contain dry products). Never use sharp objects (such as wires or knives) to unclog nozzles. They will damage nozzle orifices, which will affect the spray pattern. You should also not hold any nozzle close to your mouth and blow air into it—you don't want any chemical or debris in the nozzle splashing on your face or in your eyes. Always use a soft brush to unclog nozzles.

If the flow rate of an existing working nozzle is 10 percent (or more) higher or lower than the flow rate of a brand-new nozzle of the same type and size (at the same pressure setting), then consider the nozzle on the drone worn out and replace it with a new one of the same type and size.

A word of caution: Larger spray drones usually have two sets of pumps and flowmeters. Each set controls one set of nozzles. So when you check flow rates on such spray drones, pay attention to which pump and flowmeter the nozzle is connected to. If there is a noticeable difference between the flow rates of the pumps, the discrepancy is likely caused by a malfunctioning pump or flowmeter.

Once you complete these precalibration checks and have a good set of nozzles on the drone, you are ready to fine-tune your spray operation by going through a set of steps we describe as calibration. Although the concepts presented here for calibrating drone sprayers apply to most drone makes and models, we encourage you to check if your drone's manufacturer provides specific procedures because drones may come with different nozzle configurations.

There are eight steps to calibrate a spray drone, which apply to most drone types:

Step 1. Measure and mark off an area equal to at least 1 acre (make it larger for more accuracy).

Step 2. On level ground, completely fill the sprayer tank with water, or at least add more than half the tank volume, and mark the water level on the tank. Prime the sprayer, and note the volume remaining in the tank.

Step 3. Spray the marked area from Step 1.

Step 4. Return the drone to its launch point and land it.

Step 5a. Fill the tank to the same level you marked before spraying the marked test area. Record the volume of water you added to refill the tank to the marked level. (Note: Weighing the drone before and after application can be an alternative if you have a reliable scale available where you calibrate.)

Step 5b. Subtract the volume of water you added to refill the tank from the volume of water you originally put in the tank. This determines how much liquid the drone used. This can work if the tank has sight markings on it and they are accurate enough to make a reasonable estimate.

Step 6. The refill volume is equal to the application rate per marked test area.

Step 7. Convert the number of gallons used to treat the test area in Step 6 to gallons per acre.

Step 8. Compare the measured (actual) application rate with the recommendation on the pesticide label (the intended rate). If the difference between the recommended rate and the measured rate is greater than 5 percent of the recommended rate, adjust the drone sprayer to reduce the application error so that it deviates less than 5 percent from the intended rate.

AN ALTERNATIVE APPROACH

An alternative method for calibrating drone sprayers is to record the time it takes to spray the marked area and then collect and measure the spray from each nozzle for the same time period. The total spray you collect is the number of gallons applied in your test area (as in Step 6). Then follow Step 7 above to convert this to gallons per acre.

How to Reduce Application Error

Currently, there is no way for drone operators to adjust the flow rate except by entering a slightly lower or higher value in the controller. With ground sprayers, when applicators need to adjust the application rate, they usually do one of three things:

1. Change the travel speed
2. Change the spray pressure
3. Change the nozzle (if the first two methods are impractical)

Unfortunately, there is no ability to adjust the flow rate in drones by changing spray pressure because only the internal control system adjusts the pressure.

Working Through a Calibration Example

Let's say you want to spray a product using a drone that has 10 nozzles at a rate of 2 gallons per acre (based on label recommendations) while flying at a speed of 10 mph.

To determine the actual application rate, you marked off an area equal to 1.5 acres. From start to stop, it took 4 minutes to spray this area. Then you landed the drone and ran the pump in the same way as described in the calculation but with each nozzle directed into a bucket for 4 minutes (see figures 15.49–15.50). You measured the amount of liquid in each bucket: 315 ounces. This would be the total for all nozzles in the sprayer, not from an individual nozzle.

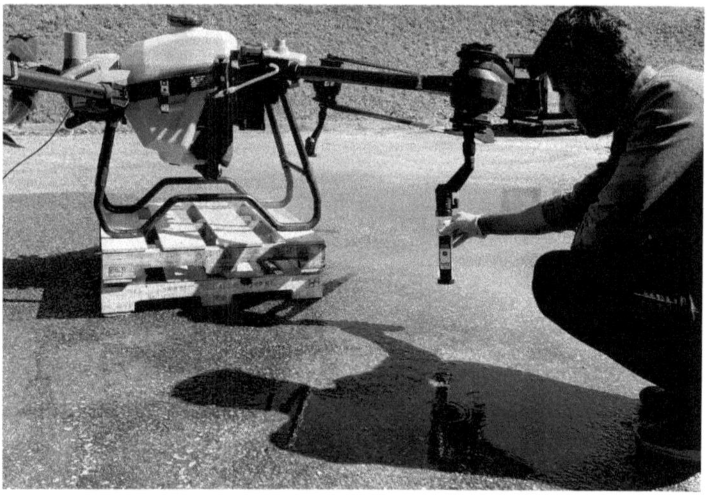

Figures 15.49–15.50. As with traditional sprayer calibration methods, operators can use a calibration catch to determine the accuracy of the spray drone's output. Although the sprayer was cleaned, it is always a good idea to wear protective chemical gloves.

What Is the Application Rate in Gallons per Acre?

First, convert 315 ounces to gallons (a gallon is 128 ounces).

$$\frac{315}{128} = 2.46 \text{ gallons}$$

Next, determine the application rate per acre. You applied 2.46 gallons to a test plot of 1.5 acres.

$$2.46 \text{ gallons} \div 1.5 \text{ acres} = 1.64 \text{ gallons per acre}$$

What Is the Percentage Application Error?

Your intended rate is 2 gallons per acre, but the actual rate was 1.64 gallons per acre. Subtract the actual rate from the intended rate and divide that by 2 gallons, then multiply by 100 to get your application rate as a percentage.

$$\text{application error} = \frac{(2 - 1.64)}{2} \times 100 = 18\%$$

This 18 percent application error (underapplication in this case) exceeds the 5 percent acceptable margin of error. Therefore, you must make some changes to bring the application error as close to 0 percent as possible—but at least less than 5 percent.

One way to do that is to enter a higher value in the controller (2.5 gpa) and then catch the spray and see how close you are to the target rate of 2 gpa. Another approach is to tell the drone to spray 2.36 gpa if it is underapplying by 18% (2 gallons × 0.18 = 0.36 gallons; 2 gallons + 0.36 gallons = 2.36 gallons).

Care and Calibration of Rotary Disc Atomizers

Some new spray drone models come with a set of rotary disc atomizers instead of hydraulic flat fan nozzles to atomize the spray solution. An advantage of rotary atomizers is that operators can use their tablet/controller to select from a range of droplet sizes—very fine to very coarse—without having to change nozzles. However, there is limited research to verify whether the rotary atomizers actually achieve those classifications as designated through ASABE S572.

The calibration process for drones with rotary discs is generally the same as that for drones equipped with hydraulic nozzles. You start with a set of rotary discs that come with the drone when you purchase it. You do not need to worry about the type or size of nozzle you should have on the drone, as you would if you had hydraulic nozzles.

To determine if you are getting the intended gallons per acre application rate from drones with rotary discs, use the same procedure explained earlier for spray drones with hydraulic nozzles. With a rotary atomizer, you cannot use a single-catch container under each nozzle, because the rotary atomizer sprays liquid in all directions. Instead, use a bag that can guide all the spray into it. Although it is a bit more difficult to do, you still need to check the flow rate of the disc atomizer and calibrate the drone as described earlier.

Clogging is not a major issue with these atomizers. However, other issues may surface after extended use of the rotating disc assembly. Some have suggested that the increased pump sizes on newer drones have resulted in rotary atomizers being flooded. This has led to improper droplet formation and a drooling-type effect. In such cases, the drones produce larger droplets and deposit a high concentration directly beneath the nozzles, leading to streaking in fields.

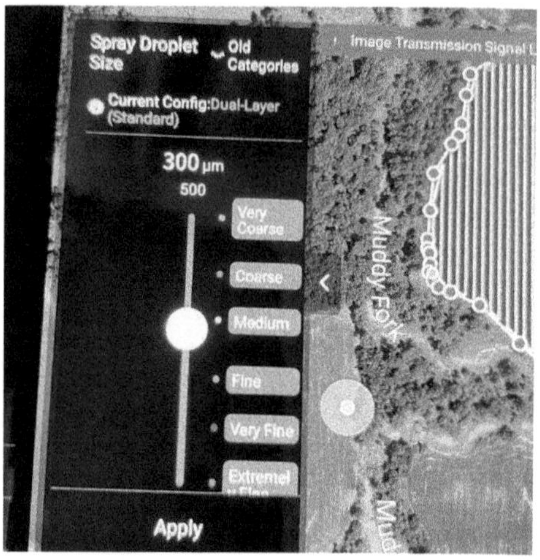

Figure 15.51. Rotary atomizers, seen on many new drones, allow operators to select the desired droplet size in the remote controller. More research is needed to verify how accurate these droplet sizes are and whether they meet label requirements for droplet size specification.

Another potential issue is the change in rotation speed of the discs. The rotation speed of the discs influences spray droplet size. Higher rotation speeds tend to produce smaller droplets, while lower rotation speeds tend to produce larger droplets. Unfortunately, it is not easy to determine the rotation speed of these discs. It requires a digital tachometer, which many newer models now have. If you have one, stick a tape on the side of the disc, and while the disc is rotating, aim the digital tachometer at the tape. This will tell you the speed of the rotating disc in revolutions per minute (rpm).

You need to compare your measurement with the motor's top rpm. Some manufacturers provide recommendations on when operators should replace these discs. Droplet formation will also be affected, to a smaller degree, by the cleanliness of the serrations (grooves) on the disc. So cleaning the discs (and other parts of the drone) becomes even more important if spray mixtures contain solid particles.

Figure 15.52. The rotary atomizer (spinning disc) nozzles (like the ones shown here) can create a very uniform droplet size distribution, but more research is needed to fully validate their capabilities on spray drones.

Figure 15.53. The small grooves in the spinning discs act as tiny channels that distribute liquid while it is dispersed. The liquid eventually forms droplets as it leaves the spinning disc. The grooves wear with more abrasive materials. Always carry spare discs to quickly replace them in the field.

Figure 15.54. Grass lanes within fields provide excellent places to set up drone operations.

OPERATIONAL SETUP IN THE FIELD

Arguably, spraying with a drone takes more effort to prepare and complete than with a ground rig in the same field. While drones appear simple, setup and operation can be quite involved. Before entering a field, you must consider the following:

- What equipment, materials, and supplies you must bring to each site
- Where to park a trailer or truck
- Where to take off and land to protect the pilot and equipment from spray drift, dust, and debris
- Where to position the pilot and visual observer so they can always have a visual line of sight during flight while avoiding pesticide exposure
- Where to place the generator and batteries so they do not overheat but are easily accessible for frequent charging (charging efficiency is better in the shade)
- Where chemicals will be mixed
- Where to place a nurse tank and/or where the nearest fill station is located
- Where the drone should be rinsed and cleaned
- What size the field is, which will help determine whether a single flight or multiple flights will be conducted
- What the topography of the field is and what the flight direction will be
- Where in-field irrigation equipment is located
- Where sensitive crops, sensitive areas, water bodies, highways, residential areas, and nearby commercial properties are located
- What the local wind speeds and the potential for temperature inversions are
- Where any potential overhead hazards, such as power lines and tree limbs that can hinder safe takeoffs and landings, are

Figure 15.55. Because of a drone's relatively small tank size and limited battery life, all of the items it needs to continue operating should be strategically positioned at the application site to improve the efficiency of the operation.

Scan the Application Area for Flight Obstacles

Investigating the lay of the land remains a top priority when using drones to apply pesticides. Take a few minutes to scope out the property to save hours or days dealing with a problem that should have been spotted before the start of the mission. A collision with an obstacle due to hasty flight planning can shut down operations for the day, or even longer, depending on the severity of the drone damage and part availability.

A crashed drone might mean it has flown its last mission. That means an outlay of cash for a new drone, a significant amount of time to receive it, and time to file the proper paperwork to legally operate it. The good news is that if you have the parts, most repairs are rather straightforward. What we see as obstacles can change how we fly the field to work around them. Scout the property for power lines, irrigation pivots, radio towers, high-tension lines, guy wires, fence lines, trees, and overextending branches in and around a field.

In larger row crop fields, operators can use smaller, less costly mapping or recreational drones to look for obstacles, keeping the hours off the spray drone. Operators can also use binoculars to help look for nearby obstacles. However, the quickest first step in reconnaissance could simply be viewing the area on various platforms that provide free satellite imagery, such as Google Earth.

Exercise caution when using satellite imagery because it may not be up to date—new infrastructure could have been added, or certain obstacles could have been removed. Satellite image resolution is also low, which prevents you from seeing obstacles well.

Drones have onboard sensors that are programmed to sense and avoid obstacles. The sensors can stop the drone in midair several feet from the object, but some drones can begin to detect obstacles up to 30 feet or more away. Unfortunately, avoidance systems can fail to detect power lines, guy wires, and smaller branches along tree lines (especially when the branches have lost their leaves or when you're operating drones at maximum speeds).

It's also important to map obstacles beforehand and mark them in the mission because most drones are operated at full speeds for application. While obstacle avoidance systems still work at top speed, drones require longer distances to slow down. Some applicators operate their drones at top speeds by disabling the obstacle avoidance. So,

NOTES FROM THE FIELD

A drone operator indicated:

> We mitigate obstacle hazards with a scouting flight before we start each field. This allows us to observe the field for any unexpected hazards that would not show up on satellite imagery. It also allows us to verify that our boundaries are far enough away from hazards, by observing the relationship between the boundary on our screen, the drone icon on our screen, and the drone's location relative to the hazard in the real world.
>
> For example, if we fly the drone to the minimum safe distance away from the power lines in the field before application, and then observe the relationship between the drone's position on our screen to the boundary of our application area, we can determine if our boundary needs to be moved closer or farther away from the power lines to optimize our application.

when radar detects the obstacle, it might be too late. Experienced drone operators concede after the fact that it doesn't take much of a limb sticking into a field or a misjudged power line to bring a drone down.

Obstacle avoidance systems give you room to manually maneuver around or above detected obstacles. Stopping in midflight to avoid a detected obstacle is an important safety measure, but realize that hovering can quickly drain batteries, which could lead to forced landings. Fortunately, once drones detect an obstacle, they will shut down pumps, thus stopping product application until the obstacle can safely be avoided. Afterward, safe application can resume.

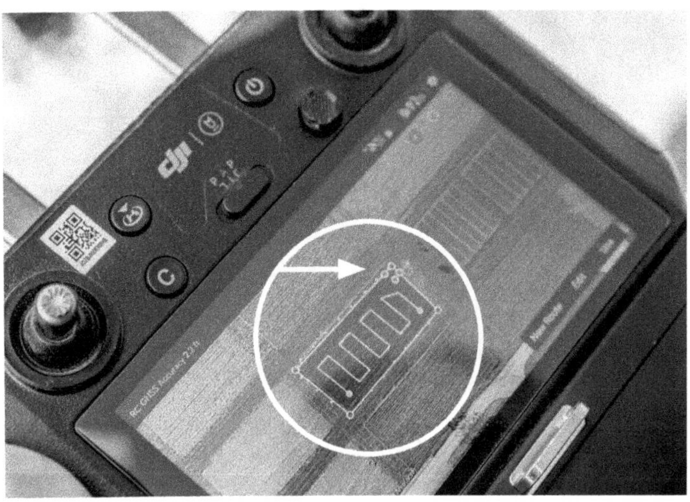

Figure 15.56. Today's controllers can mark out areas that should remain untreated.

More than 90 percent of spray drone crashes were caused by operators flying drones with obstacle avoidance radar turned off or by operators failing to map obstacles and wires before flight. We strongly recommend you map obstacles before flight to avoid the drone continuously sensing and avoiding them in flight.

Crossing waterways can cause obstacle avoidance sensors to stall a drone after it lowers into the depression of the waterway. The drone may identify the abrupt change back to the tall corn on the other side of the waterway as an obstacle. A related problem is flying too low. A tall pigweed or random plant can stall a drone. Make sure that power lines, trees, and guy wires are not part of the flight pattern as it takes off and lands and are not a concern in the return-to-home route.

Setting the Field Boundaries and Inputting Variables into the Controller

There are many ways to map the boundaries of a field to help keep your drone on a set spray pattern. One method is to drive around a field using an ATV. Every 30 to 50 feet, enter a waypoint (boundary point) into the controller or tablet. If there are obstacles the drone needs to fly around, then operators need to manually add more waypoints. Most modern spray drones allow users to mark around the obstacles with different waypoints. Certain brands also allow users to map no-spray waypoints so drones will fly over these areas without spraying any product.

A second option is to set the boundary by using the spray drone software and background imagery to mark waypoints when the field has too many obstacles or is inaccessible to ATVs for mapping boundaries. Some models and brands do not have these capabilities.

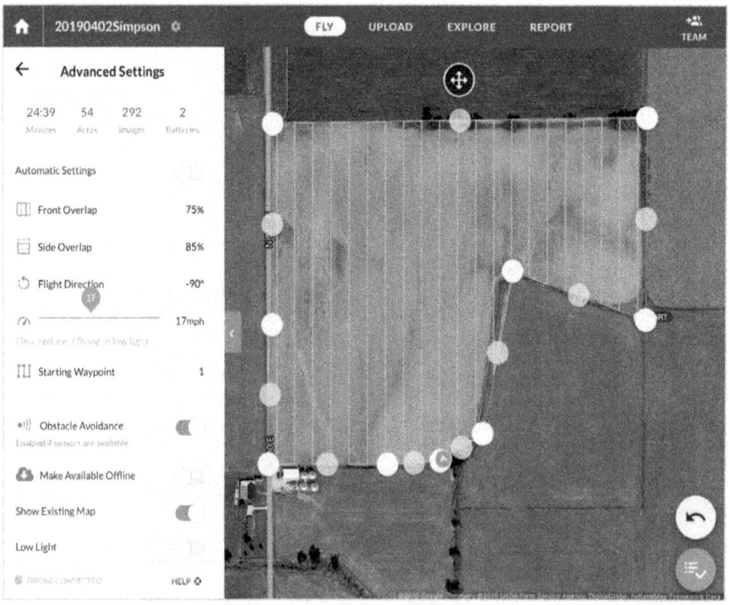

Figure 15.57. Several programs are available for marking field boundaries.

Figure 15.58. This is an example of the software available with many drones. This image shows an application in progress after the operator developed a detailed flight plan. The light-colored vertical lines indicate the drone's flight paths. Each dot is a waypoint that indicates a change in the application boundary, which signal the drone to change direction.

In many cases, field edges run alongside roads and steep ditches, making it impossible to drive ATVs in those areas. A field might be adjacent to a tree line or marsh, which also limits ATV access. Of course, one of the benefits of using spray drones for pesticide applications is their ability to spray fields with irregular shapes. Thus, mapping a field's boundaries by tracking the entire perimeter may be necessary instead of simply marking the four corners of a field.

A third option to mark field boundaries is to use mapping programs from Google Earth, Google Maps, DroneDeploy, ArcGIS, and precision ag software. You can export the boundaries as shapefiles and upload them to the drone remote controller. These mapping applications provide acceptable resolution at the time the images were made. If the field remains unchanged, then these satellite images work well for drawing the boundaries. However, make sure to do a site verification before the actual flight.

Remember that satellite images cannot spot small obstacles, and telephone poles are hard to see. If you use a precision ag software platform, you can make accurate field boundaries from planting, application, or harvest maps that most farmers currently use.

Don't forget that with satellite imagery the data displayed might be a few years old. This is only a problem if, for example, the field has added or removed a building, parts of the property were sold for homes, power lines or cell towers were added, or trees were removed to make larger fields. Sometimes the satellite data will document the image dates; however, in-person scouting is always required to confirm that the satellite imagery is correct, and the operation can be safely completed. When using these images, it is best to scout the field to verify the boundaries and any obstacles the imagery could have missed.

A fourth option is to map the field with an RGB drone, download the images from the drone to a computer, and then stitch images into a precision map (technically called an *orthomosaic map*). This lets you easily mark obstacles and field boundaries on the orthomosaic map and then export those to the spray drone as shapefiles. This is also the most time-consuming method, but the precision level is the highest among the options mentioned.

Operators may input certain application variables in the planning phase or at the edges of fields. Operators enter spray variables, such as the height above the canopy, drone flying speed, spray volume (gallons per acre), and application swath width. On newer drones with rotary atomizers, there is also an option for selecting desired droplet size. Operators can enter other parameters (such as distance offset from the boundary) in the field. Operators also need to determine flight directions that result in more efficient operations with the fewest turns. These input variables can change from field to field depending on the crop being treated, the product being used, and field conditions.

The Pilot's Eyes Focused on the Drone

Under federal law, the pilot and visual observer are required to have a visual line of sight with the drone, so the operator can maintain full control of the unit and safely ground the aircraft if unforeseen circumstances occur. It is important to find an area where you have a continuous and clear line of sight to watch the drone as it makes its application across the field, pond, or county right-of-way. That position is no longer limited to the entrance of the application site like ground equipment, because now you can select a location that gives you a clear view of the area being treated.

A useful tactic to maintain the line of sight is to find the highest elevation where you can watch the drone make the application across the field. This spot becomes the staging area where the drone will land and take off. For fields with low-canopy crops, you can also walk along field boundaries to have a line of sight. In some cases, you may need to relocate the staging area to have a better line of sight. This is not always a fixed spot but can change as needed to maintain line of sight and to attain greater efficiency with drone applications.

Another advantage of using an optimal field observation site has to do with keeping the drone and controller in constant communication. The antennae on the remote controller use a radio signal to remain connected to the drone. Drone and controller communicate best when no major obstacles interrupt the signal. If communication is lost, the drone may initiate the return-to-home function. Furthermore, finding a central staging location can help make efficient use of the batteries to limit empty payload flight time when returning to refill.

CAN YOU SEE THE DRONE?

Some horticultural crops like orchards, vineyards, and fruit and bush crops may make it very difficult to maintain a visual line of sight. It is not easy and will require a lot of thought and changes in practice to accomplish the legal requirement for visual line of sight.

Figure 15.59. Always pick the highest point possible for a staging area, which will give you a clear view of the field and obstacles.

Figure 15.60. This is an example of a difficult field to fly because of the rolling hills and potential to lose contact and visual line of sight with the drone. The operator may have to fly half of the field from this end of the field and find another vantage point to cover the rest of the field.

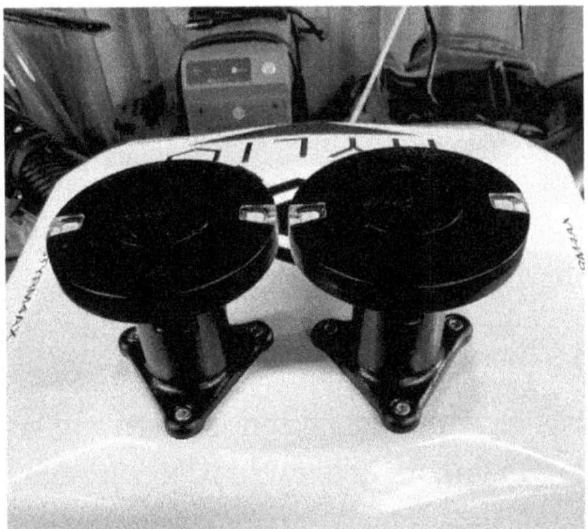

Figures 15.61 – 15.62. (Left) The antennae on the controller. (Right) Antennae on top of the drone that receive the signal from the controller.

Figure 15.63. This pilot placed a ladder on a flatbed trailer to improve line of sight. Safe ladder use now must be addressed.

When you are standing on the edge of a tall crop (like tasseling corn) or rolling hills, the controller and the drone can drop the signal with one another. This can result in the pilot losing the ability to direct the drone in flight. Flying from higher ground to lower ground helps maintain a strong signal and provide the pilot full operational control.

A flight can be autonomous (preprogrammed flight) when the drone is programmed to fly only over a certain area. Autonomous flights are typical of most applications because it can be difficult to achieve optimal and constant speed, height, and field coverage during manual flight.

With an autonomous flight, the operator creates the flight by uploading data from the controller to the drone. If the drone and controller lose connection during an autonomous flight, the drone can continue its programmed mission until the radio signal is reestablished with the controller. However, you will not be able to control the drone until the signal between drone and controller is reestablished. Be aware that current rules require that if the drone loses communication with the remote controller for a longer period, then it must initiate its return-to-home function so that the drone will return to the takeoff location and communication can be regained.

An important consideration with autonomous flights is battery life. If the return-to-home function is set to initiate when battery life is lower than what is required to return to the home base, the aircraft will be forced to land elsewhere. This can damage the drone if the landing area is not ideal for the drone, such as in the middle of a cornfield.

The return-to-home function also poses a risk when the signal connection is lost. When a drone is returning to home by itself and you cannot control it, it will fly a

Figure 15.64. Drone operators have found many ways to equip vehicles to operate drones in the field.

straight line back to home. This means the drone can run into obstacles. In such cases, it will either crash into or stall in front of the obstacle. If it stalls in front of the obstacle and connection cannot be reestablished before the battery runs out, the drone will do a forced landing and crash into the crop or whatever is on the ground, which could be an irrigation pond.

Those operating drones are developing innovative solutions for seeing over tall crops and other obstacles. For instance, spraying fungicides on tall corn presents unique challenges to keeping lines of sight on drones. One recommendation is for the pilots to stand on double-deck trailers to see above the top of corn. Operators should also use signal-enhancing devices to extend signal range.

Taking Off and Landing from Hard Surfaces

Spray drones produce strong downdrafts when taking off and landing. The downdrafts throw dirt, dust, and small pebbles into the air. In dusty areas, operators may need to clean drone motors, connectors, and electronics more often due to debris finding its way inside the drones. Dust storms during takeoffs and landings frequently clog flat fan nozzles, which is yet another concern.

This makes it important that drones take off and land on grassy sites or landing areas that do not have much debris. Many commercial operators purchase or build trailers customized for drones where they can store the supplies, aircraft, and essentials. These trailers can also have higher decks where drones can land and allow pilots and observers to have good vantage points on the operations. Custom trailers can add significant expense but may be necessary for efficient, legal, and safe operations.

Figures 15.65–15.66. (Above) A landing platform can provide a solid base from which to operate a small drone. (Right) Application drones are much larger and require larger landing pads.

Equipping a Trailer for Your Spray Drone

Tanks and product mixing typically will be like larger ground rig operations, except on a smaller scale, since drones apply a lower volume of water in a given area. However, drone operators will need to add power generators and charging stations that ground sprayer operators do not normally need.

Figure 15.67. Highly specialized trailers can support spray mixing, loading, battery charging, and elevated observing. In addition, they can provide adequate room for safely landing and walking around drones for inspections and repairs between flight missions.

A Clean Water Supply

Under most circumstances, application sites do not have accessible clean water for mixing pesticides, rinsing or flushing tanks, washing off residue from drone surfaces, or cleaning hands. The amount of clean water operators bring to application sites varies by the size of the operations. Consider the number of acres you plan to apply in a day and what volume you plan to apply to estimate the amount of water you will need.

For example, if a drone pilot plans to fly 200 acres at 2 gallons per acre, the operator will need at least 400 gallons of water, plus additional water for rinsing, cleaning, and hand washing. If the application site is near a farm, the operator can refill smaller tanks. But some areas may be far away from a water source, so it is always best to bring all the water you need for that day.

It is vital that drone operators carry separate freshwater supplies with them. Not only is this needed for operator safety, but operators need fresh water to rinse out drones and clean equipment. Use pressure-regulated pumps and garden hoses to ensure plenty of fresh water.

THINK PERSONAL SAFETY FIRST

Not all spray drone operators have applied pesticides with ground equipment or even worked around pesticides. Regardless of the mixing method, make sure operators follow all safety precautions for handling and mixing pesticides. At a minimum, all operators should wear gloves, which are fundamental to personal safety. We also suggest operators wear at least one layer of clothing (long sleeves and pants). After a day's application, be sure to wash these clothes separately from household laundry.

Mixing Operations

There are two options for mixing pesticides at the job site:

1. Add the pesticide product(s) directly to the drone solution tank for mixing.
2. Fill the drone from a premixed solution of the pesticide product.

You should avoid the first option as much as possible for several reasons. Products may be incompatible, there may be a lack of agitation during mixing, it could compromise worker safety, and it can increase the risk of spills. Filling drone tanks by hand also can be very inefficient.

We recommend using premixed pesticide solutions in large mixing tanks. Newer mixing trailers/trucks even have enclosed mixing systems that can directly draw chemicals from original containers into mixing tanks, then pump the premix from the mixing tanks into drone tanks without exposing chemical handlers. Still, you should wear gloves and any other necessary personal protective equipment (PPE) that pesticide labels require when mixing and loading spray mixes.

Figures 15.68–15.69. (Top) This operator is filling a drone solution tank. (Bottom) Mixing tanks offer enclosed systems to directly transfer mixed pesticide solutions to drone tanks.

Incompatibility Issues

Since drone applications use carrier volumes of 2 to 3 gallons per acre, which are dramatically lower than for ground applications, the pesticides mixed for drones are in highly concentrated forms. With so little water and no agitation in the spray tank, some products that work well in more dilute concentrations using 10 to 20 gallons per acre may be incompatible at the higher concentrations in drone tanks. Such mixtures can clog nozzles, spray lines, flowmeters, and pumps. Clogged nozzles require extra time to clean and flush the spray system.

PREVENTING LIQUIDS FROM GELLING

To optimize tank mixes, first add half to two-thirds of the required water to the tank, then add the pesticide, and finally top up the tank with the rest of the required water. Never mix concentrated pesticides together first. You should always add each pesticide individually to the tank with water to reduce incompatibility.

We highly recommend you conduct a jar test if you are mixing more than one pesticide in a drone. A detailed explanation of how to conduct a jar test is in chapter 11.

Poor Mixing due to Lack of Agitation

Most drones lack the agitation systems that are common in ground application equipment. We should keep in mind that the pumps and spray configurations on drones are more similar to ATV spot-spray systems than to sophisticated commercial ground application systems. Agitating the spray tank on a drone requires two people on both sides of the drone to pick it up and shake the tank to mix the product.

Anything other than pesticides that easily dissolve in water will have difficulty becoming suspended due to the lack of agitation. It is important to start spraying as soon as possible to avoid chemicals settling to the bottom of the tank and clogging the flowmeter, pumps, nozzles, and filters. Also, at the end of the day, do not let leftover spray solution sit in the tank after the drone gets back. The best option is always to spray any remaining solution out in the field and come back empty.

Mixing Pesticides in Drone Tanks Is Inefficient

Given the relatively small solution tanks on drones, operators may need to load and mix a fresh spray solution every 2 to 5 acres. If you decide to mix pesticides in the drone tank, you will be devoting considerable time to both mixing the product and agitating the product once it is in the tank. This method also introduces more opportunities for mixing errors, which can reduce pest control or severely injure crops.

Premixing Pesticide Solutions in Large Supply Tanks

The general consensus of experienced drone operators is that operations are more efficient when they premix pesticides in larger tanks such as shuttles or minibulks. Make sure to mix only what you need. This process would be like many commercial applicators that mix "hot loads" in larger supply tanks at mixing facilities and then transfer them to sprayers in the field.

Remember, the pump and spray configurations on drones resemble ATV spot-spray systems more than sophisticated commercial ground systems. However, most new drones are equipped with rate and droplet size control capabilities that are comparable to those found on ground rigs.

RACE CAR PIT STOP MENTALITY

Watch a pit stop when a race car driver enters for fuel and tires. In a matter of seconds, the driver is back out on the racetrack. Some drone operators have a similar philosophy: One minute on the ground, and the drone needs to be back in the air.

These larger tanks should be able to agitate mixtures and to pump chemicals directly into a drone's spray tank. Remember, since these mixtures are highly concentrated, some form of agitation in the shuttle or minibulk may be required to ensure that operators achieve the correct dosage and compatibility from their first fill to the last.

Figure 15.70. When this drone returns to base, the operator swaps the battery, quickly fills the tank from the larger one (sitting in a utility vehicle), and sends the drone back out to spray with minimal loss of time. Many commercial operations are set up with larger-volume tanks to be more efficient when filling up drone sprayers. But the extra weight might mean operators need to upgrade trailers and trucks.

One operator offers this tip:

> We commonly use IBC totes because they are readily available and often free. They are a challenge to properly agitate and clean out because of the flat bottom. We highly recommend using agitation and some kind of round tank with a sump for complete, thorough cleanout.

Battery Management at the Job Site

Batteries are generally the heaviest components on empty spray drones. Battery capacity determines flight time and the effective payload. However, there is no firm answer to the question, "How long will a battery last?"

It all depends on the drone's load, battery size, battery age, drone weight, field layout, route design, and local weather conditions. An empty drone may be able to fly for 15 minutes, while the same drone fully loaded may only fly 6 to 8 minutes. This is why it is important to plan your operation strategically. For example, you do not want a fully loaded drone to fly to the other end of the field (and use up half of its battery life) before it begins dispensing product.

It is true that larger payloads allow drones to treat more acres per tank, but the extra weight from the larger spray tanks also reduces battery life. Heavier drones (such as those with larger payloads) will discharge their batteries quicker to keep them aloft and spraying. Operators must replace batteries on spray drones more often than on drones used solely for imaging and surveillance.

How fast a battery discharges can also depend on the effects the ambient air has on a drone—wind speed and direction, air density, humidity, and temperature all affect battery life. Battery life can also be linked to the mission itself. Manual flying requires more battery power as the pilot adjusts the drone's course, speed, and altitude. Autonomous flying can help conserve battery life and improve efficiency.

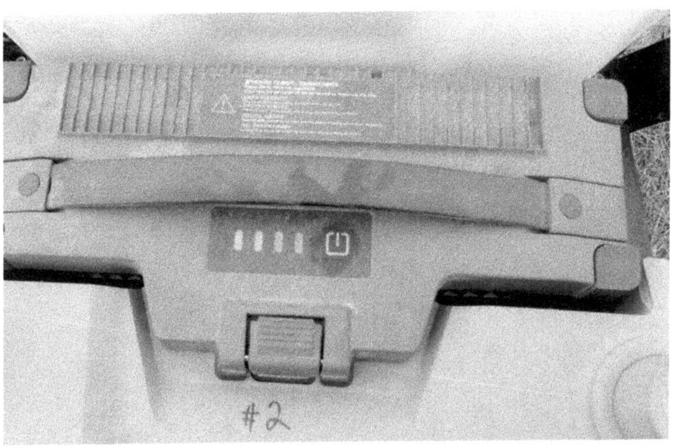

Figure 15.71. Most batteries have lights that indicate how much battery life remains.

KEEP BATTERIES CHARGED

It may be a good idea to replace your spray drone's battery each time you refill the tank—even if the battery is partially charged. You don't want to run out of battery life before you need to refill the tank. When its battery gets low, a drone is programmed to stop the mission and go back to the staging area even if it has material in the tank.

All drone spray operations vary by site. Missions that have many turns or short rows will use more battery power. Turning also makes the operation less efficient because the drone must slow down as it approaches the end row, then speed back up once it begins its next swath. When planning a mission, minimize turns as much as possible because they slow down the drone. Flight plans with longer rows will get longer battery life than flight plans that contain many turns.

Battery life is also significantly reduced when the drone spends time hovering and operating at slower speeds. If possible, fly in long rows without losing the signal, and then fly at higher speeds with moderate swaths. This increases the drone's efficiency by maximizing battery life. Be careful when flying on very sunny summer days with temperatures greater than 95°F. Batteries can overheat and cause the drone to make a forced landing.

Always arrive at a job site with your full complement of charged batteries. The number of batteries you need will vary depending on the drone, battery charger, and the generator you use to power the charger. Some chargers charge batteries as quickly as drones can discharge them, while other chargers are much slower. If you have slower chargers, you will need many batteries to support continuous flight.

Depending on the type of drone you purchase, you may need a minimum of 3 batteries or as many as 12 for continuous flying. Some batteries have a history of overheating and may require cooling stations to keep them within specified operating temperatures. Overheated batteries can cause aircraft to unexpectedly lose power mid-flight, potentially resulting in crashes.

It is always good to speak with people who have experience with the specific drone you are interested in purchasing. They will have insight into how many batteries and chargers you will need to keep your drone in the air.

Generators for Charging Batteries

It is never wise to purchase a generator for your drone without knowing the generator specifications or your battery charger requirements. A generator will be the workhorse behind the scenes while the drone is operating. Generators work constantly to recharge batteries while operators continually swap them in and out during the drone's operation.

Some charging systems run on a 110V supply, but you can achieve a faster charge rate if converted to a 220V and 50A system that may be custom designed. Electronic load management for the generator is always a nice feature. This feature can minimize fuel consumption and noise during periods of reduced generator demand.

Before you purchase a generator, check the requirements for your charger. You want to be sure any generator can maximize the efficiency of your charger. Additionally, you might consider how loud the generator might be—there are low-noise options if the pilot and visual observer will be near the generator all day long.

Generators differ in the number of outlets they have and in their output ratings for the circuit. Sometimes the generators require special wiring to achieve what you need. To provide faster charging for current spray drone batteries, operators require a 240V 50A plug. This means generator output needs to be greater than 9500V when running. Remember, generators may also be running agitation pumps, cooling fans, or other electronics at an application site so make sure your generator is properly sized and has the number of outlets you will need.

Figure 15.72. Generators are necessary for operations that include multiple batteries that require charging. Do not charge batteries in full sunlight because they can overheat very quickly. In addition, do not place generators inside sealed trailers because the exhaust is a health hazard, and oxygen depletion will eventually compromise a generator's engine combustion. Generators that are too small or that do not have well-regulated voltage output can damage drone battery chargers.

TIP FROM THE FIELD

Some advice from a drone operator:

Most air-cooled, gas-powered generators are not designed to operate at maximum capacity 12 hours per day in 80° to 90° heat in July.

Many operators rely on these small engines and get by in the short term, but in the long term, a better option might be a liquid-cooled generator. We utilize commercial-grade liquid-cooled diesel generators that are enclosed in noise-deadening housings. These will provide service for years to come and ensure we deliver proper voltage to our chargers because we are only operating between 50 and 70 percent of generator capacity.

We do not require our operators to wear hearing protection because of a screaming small engine all day. The drawbacks of our commercial units are cost, size, and weight, although we feel in the long run that cost will be comparable to constantly replacing consumer-grade generators.

Cargo-Carrying Trailers Are Popular Additions

As an alternative to trucks, operators often use open and enclosed-type trailers to carry drones and other items they need for spraying, mixing chemicals, and charging batteries. These trailers are comparable to farmers outfitting trailers with minibulks, tanks, water, pesticides, adjuvants, pumps, and metering systems to keep their operations functioning.

Enclosed trailers offer additional advantages. They can protect equipment and supplies against dust and poor weather conditions, and they can deter theft because people cannot see what is inside. For additional information about equipment theft, see volume 2.

Figure 15.73. Landing a drone on a flatbed truck.

Operators can place generators in enclosed trailers if the trailers are ventilated to the outside. Be aware that carbon dioxide can build up in trailers and can choke out combustion for the generators. Keeping generators in ventilated trailers can keep operators from having to load and unload some supplies at every job site.

An advantage of enclosed trailers is that operators can park them overnight by simply disconnecting them from their trucks. Getting to the field is as simple as hooking up the trailers and getting to the first job that day, or simply keeping trailers constantly connected during the busiest times of the year. One major benefit of enclosed trailers is that you can outfit the tops with landing surfaces or observation decks that provide a higher elevation to maintain line of sight with the drone during flight.

Other options for transporting necessary equipment to the field for drone applications exist. Light- to medium-duty trucks with flatbeds may already be available, and you can convert them to spray service on a seasonal basis. This same style of trucks may have an enclosed bed, known as a *box truck*. Some examples on farms would be retired moving trucks, such as U-Haul vehicles, or retired food service trucks used to supply commercial passenger aircraft. Box trucks can provide a secure storage area for equipment and supplies, plus allow for a possible landing and observation deck on top of the box.

TIPS FROM THE FIELD

Here is some insight from a drone operator:

> **Open versus enclosed trailers come with trade-offs.**
>
> **We prefer open trailers, because that way, we always have a landing zone on the back of our trailer. When moving from field to field, the drone lands in its transportation spot, and we do not have to carry the contaminated drone into our enclosed trailer. Furthermore, by landing on our open trailer, we reduce the amount of dust and debris that is kicked up into the air on takeoff and landing. Finally, we land the drone on the same level that our operators stand, so there is no up and down ladder or stairs for our operators to climb to refill the drone and swap batteries.**
>
> **We can park inside overnight, and thus theft protection is not a concern for us. Also, many enclosed trailers are not built heavy-duty enough to carry the weight required for all the water and spray solution we need to operate all day.**

TIPS FROM THE FIELD

Here is some insight from a drone operator:

> **Many times, trailers need some special modifications to securely hold down all this equipment for transport.**
>
> **It is important to carefully consider equipment placement and adding the necessary secure points to make sure the equipment stays in working order and does not get jostled. This is especially important when traveling on rough terrain to reach a site/ recharge point.**

On-Site Parts and Repairs

Just like other mechanical devices, drone parts break down and need to be replaced. Keeping spare parts on hand can reduce downtime and get you up and running within a day or sometimes within several hours. The main supporting frames of most drones are constructed of carbon fiber, a material that is light but durable. However, drones are also constructed of plastic materials to reduce their weight, so it is easy to see how external parts can degrade from sunlight.

Common operational factors can also cause wear and tear on drones. Rotors get chipped and broken when they hit obstacles or make rough landings. Motors can fail if dirt and debris infiltrate the internal parts.

Having spare parts on hand can save a lot of time, considering it might take days to get your shipment anywhere from within the United States. Your spare parts inventory could include nozzles, inline valves, pumps, flowmeters, batteries, electronic speed controllers, motors, and a set of rotor blades. How many of these parts you carry in your inventory will depend on how quickly you can access parts from dealers or manufacturers. If you have a reliable dealer who can ship parts quickly, you may not have to keep as many items in stock.

Quick access to spare parts could mean a job comes to a standstill for only a few minutes rather than days. Some argue that keeping replacement parts in inventory is a lot of money just sitting there. While that is true, ask, "How much money are you losing while the drone is inoperable?" A $10 part can cost you hundreds, if not thousands, of dollars in downtime. It can also upset customers while they wait for their fields to be treated.

Downtime is money lost for you and potential crop losses for the farmers you serve. Furthermore, replacing drone components can be relatively simple. Circuit boards and motors are often connected with a few screws and a plug that you can switch quickly. While some diagnostics are too difficult for the average person to do, some dealers can help diagnose an issue over the phone or with a simple video call. As drones become more popular, increased customer and technical support will become available too.

Personal Safety like for Any Other Application

With drones, apply the best management practices for handling and mixing pesticides that you normally follow for any other application method. First, operators must wear whatever pesticide product labels list as the minimum personal protective equipment (PPE).

What makes drone applications different is the higher pesticide concentration of the product solution compared with application methods that use higher carrier volumes and more dilute solutions. While both methods apply the same amount of pesticide active ingredient per acre, the pesticide solution in the tank can be more concentrated at the extremely lower carrier volumes that drones use.

Figure 15.74. A good example of maintaining a line of sight even when spraying a tall crop like corn.

Figure 15.75. Wear chemical-resistant gloves when handling spray drones because the aircraft exterior will have pesticide residues.

A DIFFERENCE IN CONCENTRATION

Keep in mind that drone tank mixtures are low-volume applications rather than what you may be accustomed to dealing with in other equipment. With ground rigs, some pesticides are often diluted in 10 to 20 gallons of water per acre.

However, drones may be applying the same pesticide rate per acre in 2 gallons of water. You are mixing, applying, and handling a more concentrated spray solution. That requires you to follow additional safety precautions when spraying with drones. At a minimum, you need to use all label-recommended PPE.

Make sure that no one is working or playing in application areas. It goes without saying that you should not operate drones when people are fishing on a pond, highway workers are repairing the shoulder of a road, workers are weeding in a tomato field, or students are detasseling corn.

An issue you need to address early is the wind direction relative to the staging area. If you would be uncomfortable spraying with a ground rig because of the potential for spray drift, then there is no reason to apply pesticides with a spray drone. Always follow label directions regarding wind speeds and/or specific state requirements.

While the tops of hills provide good lines of sight as required by the FAA, pilots and visual observers should be aware they could be positioned downwind of applications. Even if they are not positioned downwind of spray passes, droplets may still drift toward staging areas during the first couple of passes made. You can always relocate yourself and your crew temporarily until you complete those passes to avoid exposure to spray drift.

COMMON QUESTIONS REGARDING SPRAY DRONE APPLICATIONS

In this section we answer frequently asked questions about using drones for pesticide applications.

Q: Should I rinse the tank if I'm planning to use the same product tomorrow?

If you plan to use the same product the next day, the best practice is to flush out the system so that the product does not remain overnight in hoses. Remember, the higher pesticide concentration in drone tanks presents greater risks to drone parts over time. As products sit in tanks and lines for extended periods, there are more opportunities for pesticide solutions and formulation solvents to seep into lines, hoses, and poly tanks. They could even allow solid residue to form on surfaces of solution tanks. Studies with blue dye show some products have the potential to remain in lines for weeks. Rinsing out the system reduces the likelihood that materials will solidify, which could corrode pumps, seals, and rubber parts.

The problem with spray drones is that there are no rinse tanks with clean water onboard to flush lines like there are with ground rigs or manned spray planes. What remains in partially filled tanks must be sprayed over the edges of fields or back over recently sprayed fields. After that, rinse drone tanks with clean water and spray it out, repeating the process several times.

Q: Should I use a tank cleaner when I switch products?

After you rinse a drone, you should use a tank cleaner when you switch from one pesticide to another or if you switch to another crop that may be sensitive to a chemical you have previously used. Just like for traditional ground sprayers, when you use tank cleaners with proper technique, they remove and break down any remaining ingredients and make them inactive.

Figure 15.76. Many drone solution tanks have removable strainers that allow for easier cleaning.

Make sure to remove and flush out any internal screens your drones may be equipped with. Sometimes you can remove drone spray tanks, which allows you to add tank cleaner solutions to tanks and manually shake them.

Q: How do I agitate the mixture in the tank?

Currently, there are no agitation devices on spray drones. Mixing pesticides directly in the tank requires two people to pick the unit up and shake it. If you use anything other than a water-soluble formulation, it is best to make a slurry of water and the pesticide mixture in a 5-gallon bucket or a 2.5-gallon jug to get the products well suspended throughout the mixture. Another option is to use a separate mixing tank that has a built-in agitation pump to improve mixing before transferring the mixture into the drone.

Q: Should I rinse off a drone before leaving an application site?

When drones are spraying liquid pesticide solutions, some of the spray mixture will undoubtedly get on drone exteriors. Many mistakenly believe that rinsing drones damages electronic components. In reality, spray drones are largely watertight, which prevents damage to the onboard electronics when you rinse their exteriors.

Figure 15.77. Due to the wind turbulence drones produce, pesticides are often deposited on drones, requiring thorough cleaning. Use a mild detergent solution from a squirt bottle and then rinse after a few minutes.

Getting into a habit of rinsing drones helps remove solvents that could gum up moving parts and remove grime from sensors. Rinsing drones not only protects them but also keeps unwanted residues out of the trailers or truck beds when you store the drones.

We do not recommend using any type of power washing equipment to rinse drone surfaces—water can get into electronic modules with high pressures. Periodically scrubbing drone surfaces is required to remove oily residues. In addition, removing those residues can help prevent carrying unwanted, sometimes herbicide-resistant weed seed from one field to another.

Although the concentration of chemicals in rinse water is extremely low, it is a good idea to rinse tanks in areas far away from ditches or other areas (such as grass waterways) that present a risk of the rinse water reaching nearby water resources. Consider using large confinement trays in your trailers or truck beds to place drones on after you rinse them. Confinement trays keep any water from dripping onto trailer and truck floors. The trays also could contain possible spills if drones still have pesticide in their tanks as you travel from site to site.

DUMPING PRODUCT NEVER A GOOD THING

One option for leftover solution in spray tanks you should never consider is dumping a tank's contents on the ground. Dumping is illegal and can lead to serious consequences with federal, state, and local regulatory authorities. It is one of the easiest ways to tarnish your reputation and image.

Q: How should I handle leftover pesticides in the tank that are no longer needed?

In a pesticide application, the guiding philosophy is that you should leave minimal product in spray tanks. Consider this principle when you are mixing pesticide solutions. Consider the size of the field to be treated or the overall job. Assume 1 gallon left in the spray tank.

Some think it would be much easier to empty a tank's contents back into the premix tank it was originally pulled from. However, there is always a risk that the leftover mixture is contaminated and that you would be combining contaminated mixture with uncontaminated premix. While the likelihood of that might be slim, why take the chance?

Draining tanks directly on the ground is not a good option either. There is a high risk that the person opening the drain cap under the tank will get tank mix all over their hands, arms, and legs, creating excessive chemical exposure. Then there are always the regulatory repercussions of dumping chemicals on the ground.

The only real option is to dilute leftover mixture and spray it on the field you just sprayed or around the field edges. Normally, you will need to apply less than a gallon or two. Putting herbicide products to use on field edges where weeds are normally thicker is a good idea, while insecticides and fungicides can go back over the crops. As long as the spray from the previous application has dried, going back over it a second time with drones will not wash off any previously applied products. Diluting the remaining chemical mixture avoids overapplying any chemicals beyond label limits.

Q: How do I manage wind and streaking?

Streaking is a function of improper swath management. You can help prevent streaking by swath testing and adjusting application parameters (height, speed, or droplet size) to determine effective swath widths. Swath displacement caused by excessive winds can also result in streaking because product deposition will be highly concentrated in certain portions within the swath.

Q: Do I have to post and/or leave information at the field?

Following any pesticide label's use directions includes meeting the requirements of the Agricultural Use Requirements box. This box provides further instructions about posting, reentry times, and information exchange when the area treated is a farm, forest, nursery, or enclosed space production (for example, a greenhouse). If the area being treated is not one of the above areas (for example, aquatic), consult the box called Non-Agricultural Use Requirements, which is shaded and placed directly below the Agricultural Use Requirements.

> **Following a pesticide label's use directions includes meeting the requirements in the Agricultural Use Requirements box. This box provides instructions for postings, reentry times, and information exchanges when treated areas are farms, forests, nurseries, or greenhouses.**

DRONE MAINTENANCE AND WINTERIZATION

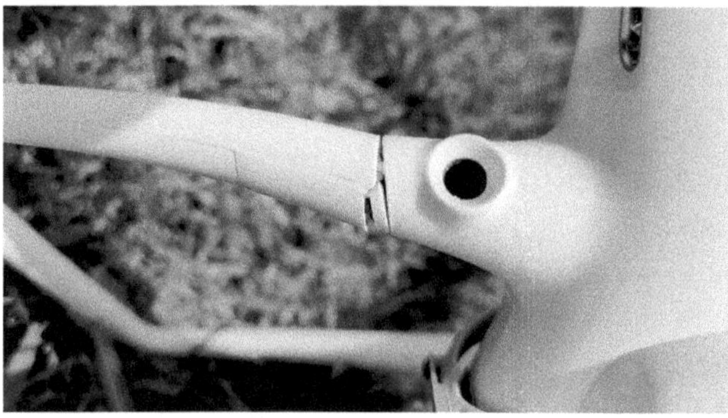

Figure 15.78. Observe and inspect drones before and after operations to detect any damage that may have occurred.

End-of-season maintenance on drones ensures quick start-ups in the spring. Winterizing drones is straightforward, since most of the time they will be stored in temperature-controlled rooms like heated shops. Perform a visual inspection that focuses on looking for cracking and splits in the frame, because there is a lot of torque when drones take off and impact force when they land. Check all the parts to see if anything needs tightening or replacing. Make sure to replenish your spare parts inventory so you are not surprised the next time you need that part in the field.

There is really no inspection you can perform with the drone's software. The software is designed to check the functionality of the electronic components. Drones will go through self-check diagnoses when they are first powered on. The propellers move a little, the cameras will move, and lights will flash. If any part fails the self-diagnosis, the controller will tell you there is a problem. A message will appear on the controller saying the drone can't take off because the following issue x, y, or z needs to be addressed.

It is important to ensure that you are using current firmware for your drone's software. Manufacturers may send software updates to fix bugs or other issues. Also, ensure drones and battery firmware are compatible; otherwise, a pilot may show up to the field and not be able to fly until the software is updated with current software patches. If you have not flown for a while, power up the drone and controller at the shop where you have a consistent internet connection to help avoid any untimely firmware updates at jobsites.

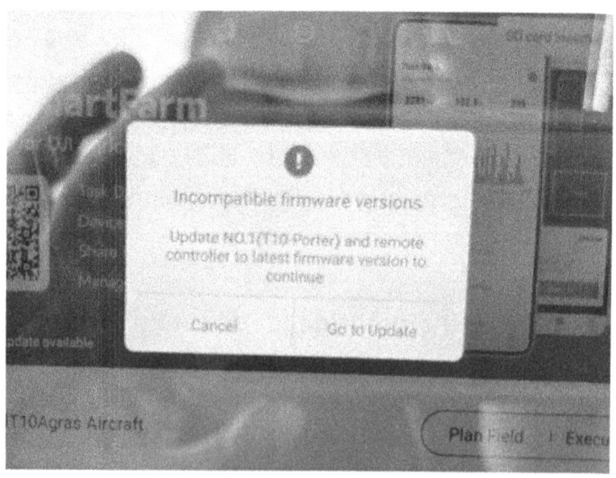

Figure 15.79. Software errors are often displayed when starting up a drone. Keep software up to date to ensure best performance.

WHAT IS MY DADGUM PASSWORD?

Make sure you have carefully noted your username and password to log into drone software and keep it where you have easy access to it when you want to use the drone next season.

Make sure to keep a battery in place to help protect the copper connection when you wash a drone and even when not in use. Use a paper towel or cloth to wipe any water from the copper connectors. Do not take out the battery because this exposes the terminals and can cause a short if water is present.

Before storing batteries, examine each battery for signs of swelling on the top and sides of the unit. It is always a good practice to label your batteries 1, 2, 3, and so on. The labels help ensure that you use each battery regularly instead of charging and using the same battery over and over again.

The owner's manual should provide directions for storing batteries. Triple-rinsing drone tanks, flushing out lines and pumps with clean water, and using tank cleaners overnight are good practices to help remove any unwanted chemical residues. Replace hoses that appear to be checkered from sunlight damage or appear to be stained or colored from absorbing chemicals.

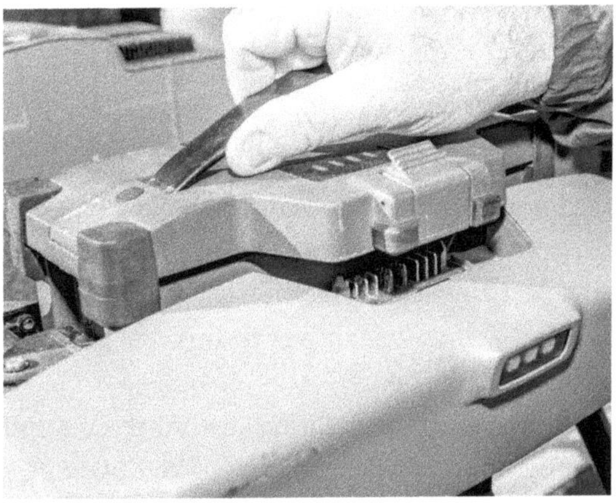

Figure 15.80. When batteries are not in place, electrical components are exposed and are at risk for damage.

Figure 15.81. Batteries can swell if they overheat, which in turn decreases performance or causes permanent damage.

When cleaning drones, do not pressure-wash them or remove any components (such as batteries) to avoid getting water into the drones.

If your drones will be subjected to freezing temperatures, then you should flush their systems with RV antifreeze after cleaning them. This helps prevent any remaining water from freezing. Remove as much water as possible by emptying tanks, open any valves to release any trapped water, and blow air through the systems. Fill drone tanks with enough RV antifreeze to flush the entire drone.

At first, the color of the antifreeze coming out of the nozzles will be lighter, because it has been diluted with any remaining water left in the drone. Continue to flush until the antifreeze coming out of the nozzle is the same color as the antifreeze you put in.

PESTICIDE LABELS: A QUANDARY FOR CLEAR GUIDANCE

Using drones for spraying pesticides requires a high degree of interpretation of label instructions. Currently, specific information about the carrier volumes, application heights, nozzle types, and droplet sizes that are typical of drone applications is not incorporated in the aerial application sections on product labels. However, we expect that information for drone use will begin to appear on pesticide labels as soon as reputable sources adequately research the issue and regulatory agencies review further application requirements or considerations that need to be provided.

Can I Spray Any Pesticide Using a Drone?

To answer the general question of whether you can apply a pesticide using a spray drone, you should ask yourself three specific questions.

Question 1: Does the Pesticide Label Prohibit Aerial Equipment?

Many pesticides work well when sprayed with drones. No one could have foreseen that drones would be applying products on golf courses, greenhouses, ponds, tree plantations, and nurseries. Regardless, if a pesticide label has language that reads "Do not apply this product using aerial equipment," then the answer to this one question is clear: You can't legally apply this product with a drone.

If a product label prohibits drone applications, then drone operators take full responsibility if it fails or causes problems, whether on- or off-site. Manufacturers have no liability when applicators use their products if labels prohibit that application method. There could also be repercussions from regulatory agencies for using the product off-label.

Instructions on product labels may be different from one product to the next even though the chemicals contained in those products are similar or almost identical. For example, some generic 2,4-D formulations typically used in burndown herbicide applications are labeled for aerial application. However, some 2,4-D formulations that have been developed to spray postemergence in 2,4-D-resistant soybeans are not approved for aerial equipment. Although both pesticides are 2,4-D, they cannot be treated the same with a drone.

WARNING: REGULATIONS AND INTERPRETATIONS ARE CHANGING

In this chapter we have provided what we believe is true as we were writing it. It is always important to check with FAA websites and your state pesticide regulatory agency for up-to-date regulations and policies.

You must read and follow label instructions for the specific pesticide you are using. Do not assume it is OK to use drones to spray this one pesticide because you previously used a similar product by different manufacturer with the same active ingredient that allowed using drones.

Question 2: Does the State Regulatory Agency Allow Drones to Follow Aerial Labels?

The US EPA currently allows each state regulatory agency to determine whether drones can operate under instructions for aerial applications. You will need to ask your state pesticide regulatory agency if they allow drones to apply pesticides using the instructions provided on product labels for aerial applications. You also should ask what state-specific pesticide certifications they require for you to use drones to apply pesticides. These certifications will vary from state to state. Not all states observe reciprocity, so applying across state lines may require multiple examinations and certifications.

Question 3: Can the Drone Follow Aerial Application Instructions?

If a pesticide product has an aerial application section, then a drone (if your state allows) can apply it by following all requirements listed under the label's aerial section. You are required to follow the minimum or recommended application rates per acre, carrier volume, and other criteria listed on the product label. If you cannot follow some of the aerial application label requirements because of unique application parameters used with drones, then you cannot legally apply the pesticide with a drone.

Following only part of the label and ignoring requirements in other sections can create efficacy problems, increase the chance of pesticide resistance evolving, and allow illegal residues to remain on harvested commodities. It also violates one of the key statements on all labels: "It is a violation of federal law to use a product inconsistent with its labeling."

How you interpret a label depends on how the instructions are worded. A label's recommended carrier volume is often a make-or-break deal for legally using spray drones. Remember, drones use significantly less carrier volume of water per acre than ground rig applications. Some pesticide labels may require 10 to 20 gallons of volume per acre for ground rigs and 5 gallons per acre aerially. Spray carrier volume rates exceeding 2 or 5 gallons per acre become difficult for some drones to accomplish because these rates severely limit drone efficiency. At that point applications can become cost prohibitive.

How would you interpret the following label direction about the application rate? "Do not use less than 10 gallons of spray solution per acre." There is no flexibility in this language. The label clearly states that you must spray this solution at a rate of 10 gallons or more per acre, which means you would be off label if you use a drone (or any other application method) to apply less.

Other labels are written with some flexibility. For example, two labels provide the following information:

1. Use the recommended rates in 3 to 25 gallons and
2. Minimum spray volume per acre is 2 gallons

Such language gets you into the lower range of volumes, which drones can take advantage of. It also means that the manufacturer has done research to support that the amount of carrier they recommend on their label will work and that doing so adheres to other regulatory safety conditions, such as crop residues.

You can increase drone spray coverage by optimizing swath speeds, flight heights, adjuvants, spray tips, and spray volumes.

Regardless of what spray equipment they use, sprayer operators seek to adequately cover the target with the pesticide while minimizing spray drift. There are two common ways to increase coverage on the target:

1. Increase the gallons per acre application rate—perhaps by using nozzles that have higher flow rate capacities (gallons per minute).
2. Reduce the droplet size.

Relatively smaller spray droplets are highly susceptible to spray drift. To minimize this risk, pesticide manufacturers have been including droplet size recommendations on labels (for example, ASABE Droplet Category). Some labels may even recommend not using fine spray droplets to avoid spray drift.

Unfortunately, to achieve maximum coverage with drones' low application rates, most drone operators apply their products using nozzles that produce smaller droplets. However, some current research indicates that larger droplet sizes do not have the same coverage penalty from drone applications as they do from ground-based applications, so larger droplet sizes could be used for both drift mitigation and coverage. Remember, the label is the law. Therefore, you should read the label carefully to determine if there are specific requirements related to droplet sizes, nozzle types, and spray pressures when using specific nozzles.

Manufacturers commonly equip spray drones with standard nozzle bodies, so applicators can select which spray tips fit their specific needs for their intended applications and the pesticide requirements. Alternatively, spray drones may come with rotary atomizer nozzles that applicators can adjust to deliver a range of spray droplet sizes (figure 15.52). These "spinner" nozzles are similar to the controlled droplet applicator (CDA) nozzles that were popular for ground applications in the 1980s, which lost favor due to the complexity of the system, which was driven by hydraulic motors. Current atomizer nozzles on drones use electric motors.

Rotary atomizers provide some value because applicators may not have to change spray tips to adjust the spray droplet sizes. However, the additional electric motors add weight to drones, which can drain batteries faster.

Rotary atomizers also have the advantage that they don't clog like hydraulic nozzles can. You can spray a thick tank mix with rotary atomizers, which often is not possible with standard hydraulic nozzles fitted for drones. Rotary atomizers are not without challenges, because the spray quality can be compromised if the gallons per acre and travel speed are set higher than the intended nozzle capacities. Furthermore, adding spray adjuvants may alter the performance of rotary atomizers and impact droplet sizes so that they don't match what is indicated on controller displays.

FEDERAL AVIATION ADMINISTRATION REGULATIONS

The US Congress assigned the federal regulatory oversight of drones to the Federal Aviation Administration (FAA). With commercial use of spray drones being a recent phenomenon in the United States, FAA regulations and policies outlining who can pilot drones and under what conditions they may operate them are still in flux. Current drone regulations will surely change as policies evolve.

Therefore, we advise you to check the FAA website (https://www.faa.gov) often for the most current regulations concerning drone use. Additionally, when drone operation requirements conflict with FAA rules for which no authorization or waiver authority is available, the secretary of transportation has authority to grant conditional exemptions based on petitioners' requests.

The FAA regulations drone users must follow depend largely on three factors:

1. Whether the drone is flown for recreational purposes or commercial uses (all spray drones will be classified as commercial use)
2. How much the drone weighs
3. What kind of commercial use the drone is used for—that is, applying agricultural products versus capturing images

Recreational Versus Commercial Drones

Although the FAA has a clear definition of recreational, or hobby, drones, the interpretation of the rules can encompass purposes that are not well defined, especially in the world of agriculture. If you buy a drone to fly over your farm for enjoyment, then you would assume you are flying for recreational uses. But what if you see something in your crop field and you take a closer look and find out that you need to control some weeds? Or what if you see a nutrient deficiency that you go out and remedy? Do these flights still count as recreational since you made business decisions?

Our interpretation is that recreational purposes include flying drones for fun, using them in competitive races, or taking pretty photographs. Conversely, drones that provide any value to a business or farm are designated as commercial.

For instance, a drone flying a field sends back images that suggest a pesticide applicator in a ground rig had not cleaned out the booms properly before spraying the field. The drone provided the farm a commercial benefit. Taking photographs with your drone that you use on advertising brochures or websites would also be considered commercial use. As with most cases, if you are unsure if the use is recreational or commercial, it is best to play it safe and follow the rules of a commercial drone pilot.

Drone Weight: Three Categories

Some FAA drone regulations apply to all drones regardless of their size, while other regulations are targeted to certain weight classes. The FAA currently divides drones into three weight categories:

1. Less than or equal to 0.55 pounds
2. Greater than 0.55 and less than 55 pounds
3. Equal to or greater than 55 pounds

Figures 15.82–15.83. Using a drone to (A) measure heat loss or (B) determine if the evidence collaborates a drift complaint would be considered commercial use. Operators are using both photographs to make decisions that affect the revenue or profits for a business or farm and would be classified as commercial uses.

Figure 15.84. This drone weighs less than 0.55 pounds and is exempt from many FAA regulations.

Figure 15.85. This drone weighs around 11 pounds.

A drone's total weight is calculated by adding the weight of the platform, all onboard accessories, and the payload it carries.

Drones that weigh less than 0.55 pounds have limited commercial uses. These lightweight drones do not require federal registration and can be flown without being a certified pilot if they are used for recreational purposes.

All drones that weigh more than 0.55 pounds and are used commercially must comply with a set of baseline FAA regulations that deal with pilot certification and drone registration as well as product applications.

If you are unsure whether you are using your drone for recreational or commercial purposes, it is best to get a commercial license and follow the commercial drone operation rules. If you are using it for agriculture, your drone is most likely considered commercial.

Remote Pilot Certification for Commercial Drones

The FAA requires certification for pilots to operate aircraft remotely (operators not in cockpits of planes) for commercial purposes. The multiple-choice exam, called the Airman Knowledge Test (or Part 107), is taken at an FAA-approved testing site under review of a proctor. You will need to visit the Integrated Airman Certification and Rating Application (IACRA) website to create a personal identification number. This number, known as the FAA tracking number (FTN), is unique to you and is used to complete your airman certification application. In addition, you will need to provide proof of identity and current address at the time of testing.

Once you receive notice that you have passed the test, you will need to complete FAA Form 8710-13. Bring the completed 8710-13, a photo ID, and your online course completion certificate to an official site to validate your identity (see Step 5 on the FAA's page "Become a Certificated Drone Pilot": https://www.faa.gov/uas/commercial_operators/become_a_drone_pilot).

Alternatively, you can also fill out this form online, and you will receive your certificate in the mail. At that time, you will receive a temporary airman certificate. In several weeks, the FAA will send you a permanent remote pilot certificate. Remote pilot certifications are valid for two years. At the end of each two-year period, you can renew your remote pilot certification by retesting or by completing a free online continuing education course known as FAA Drone Pilot Recurrent Training for Part 107 Small Unmanned Aircraft Systems Course.

After 24 months, you will need to have a recurrent certificate of achievement with you in addition to your remote pilot certificate whenever you fly. You can complete this additional training online using a computer at your business or home. Continuing education is an easier way of keeping apprised of changing FAA regulations, polices, procedures, and interpretations than the retesting option.

PESTICIDE CERTIFICATION THROUGH PESTICIDE REGULATORY AGENCIES

All states require individuals applying pesticides through spray drones to be certified. Some states may require a basic core test and possibly additional tests. This includes both commercial businesses and private applicators (farmers). The National Pesticide Information Center provides a list of state pesticide regulatory agencies and contact information: http://npic.orst.edu/reg/state_agencies.html.

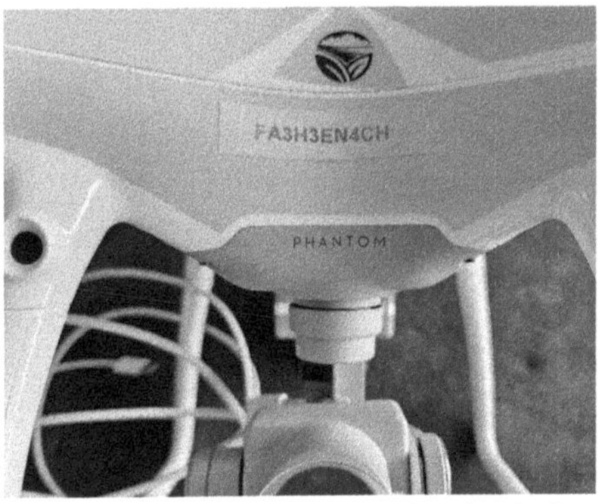

Figures 15.86–15.87. Can you identify which one of these two drones weighs more than 55 pounds? Clue: Check the registration number.

FAA Drone Registration

All drones greater than 0.55 pounds must be registered with the FAA. When you register a drone, the FAA will assign a unique number specific to that aircraft. This is like a vehicle registration. The FAA also collects information about the drone owner and their address. The FAA number must be attached and outwardly visible on the drone. If a drone goes down or someone gets hurt, the number on the drone can help identify the pilot or owner of the aircraft.

The current FAA registration process depends on the weight of the drone and its intended use. Drones weighing less than 55 pounds can be registered online in a few minutes. A drone that weighs 55 pounds or more requires a notary to validate the purchase of the drone and that the purchaser is the one registering the aircraft.

The letter designation that precedes the FAA-assigned registration number differs for drones that weigh more and less than 55 pounds. A drone that weighs 55 pounds or more will be designated with the letter *N*, which is similar to manned aircraft registration. Drones that weigh less than 55 pounds will designated with the letters *FA*.

Commercially registered drones that weigh less than 55 pounds must be renewed every three years at a cost of $5. Drones that weigh 55 pounds or greater must be renewed every seven years—like manned aircraft. The number assigned to a drone remains the same throughout repeated renewals.

FAA RemoteID Requirements

As of 2022, all drone manufacturers are required to have a built-in RemoteID that broadcasts the ID of the drone with its location and pilot's location. For drones manufactured before September 16, 2022, without RemoteID hardware, the drone must be retrofitted with a RemoteID module to fly in an FAA Recognized Identification Area (FRIA).

The RemoteID signal can be accessed by the public using applications like Dronetag or Drone Scanner. However, the drone owner's name and contact information can only be obtained by the FAA or law enforcement, as a means of protecting the pilot's privacy.

Altitude Limits and Airspace Restrictions

A drone may not fly more than 400 feet above the ground without an altitude waiver. Most spray drones fly up to a maximum of 50 feet. They usually operate at 15 to 20 feet during application and may reach 40 to 50 feet during takeoff and landing. With crewed aircraft (the pilot is onboard), flying no lower than 500 feet is a requirement. This ideally provides a 100-foot buffer between drones (remotely operated) and other aircraft with the pilots onboard.

Part 107 allows drones to fly up to an altitude of 400 feet, but spray drone operations are limited to 200 feet. However, spray drones routinely stay within 10–30 feet of the ground or crop canopy.

There are limitations when working in and around an airport in an area known as controlled airspace. Air traffic control manages controlled airspace. Flying in controlled airspace is more restrictive and requires clearance before flying in. Rarely are drones allowed to fly higher than 400 feet in controlled airspace without additional authorization.

Within each airport-controlled airspace at surface level, the FAA has established a set of grids to describe the maximum altitude a drone may be flown with prior authorization. It starts at 400 feet when entering the airport-controlled airspace, but the maximum altitude decreases as the drone nears the airport and runways.

The FAA requires pilots flying within controlled airspace to get authorization for operating within that airspace. Pilots flying under Part 107 rules may request immediate authorization to fly in controlled airspace from a Low Altitude Authorization and Notification Capability (LAANC) service supplier. More information about LAANC is available from the FAA: https://www.faa.gov/uas/programs_partnerships/data_exchange.

When a smartphone places a marker pin for the pilot's location in a LAANC application, the FAA reviewer will know the pilot's exact location. Some drone software is integrated with LAANC as well. This service can be used to quickly preapprove your flight plan from your controller or smartphone without any advance notice in airspace that may have low-altitude flight restrictions. If the operation is in a zero-altitude grid or a higher altitude is needed than provided in the grid, you may request an airspace authorization through the FAA DroneZone website: https://faadronezone-access.faa.gov.

Using the altitude assigned for your location within that grid, the reviewer will check the flight altitude provided to them on the waiver application. If you fly at or below the maximum allowed altitude within that grid, it generally means they will quickly approve your flight plan. The approval is sent to you via text or email and serves as documentation that the flight was preapproved.

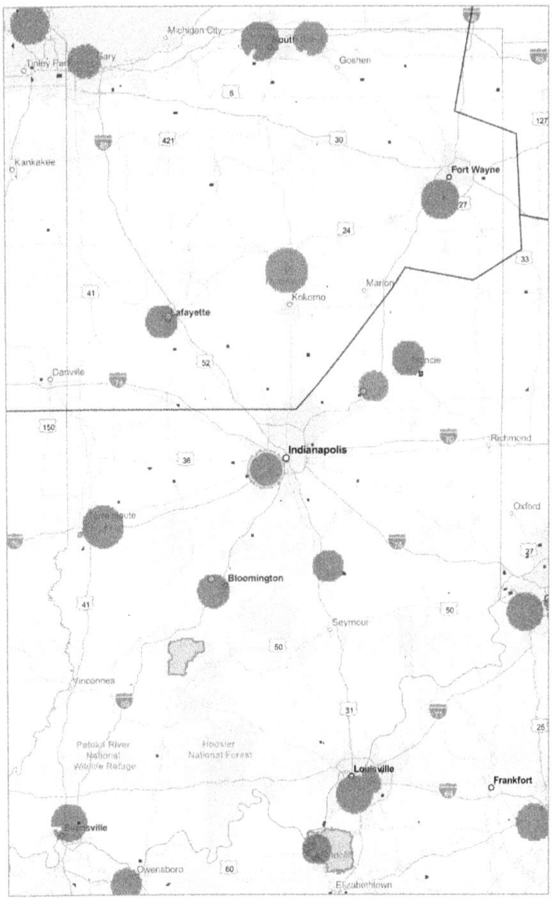

Figure 15.88. Some of the airport-controlled airspace in Indiana from the FAA website https://faa.maps.arcgis.com. On the website the dots are in different colors indicating how to obtain permission to fly there.

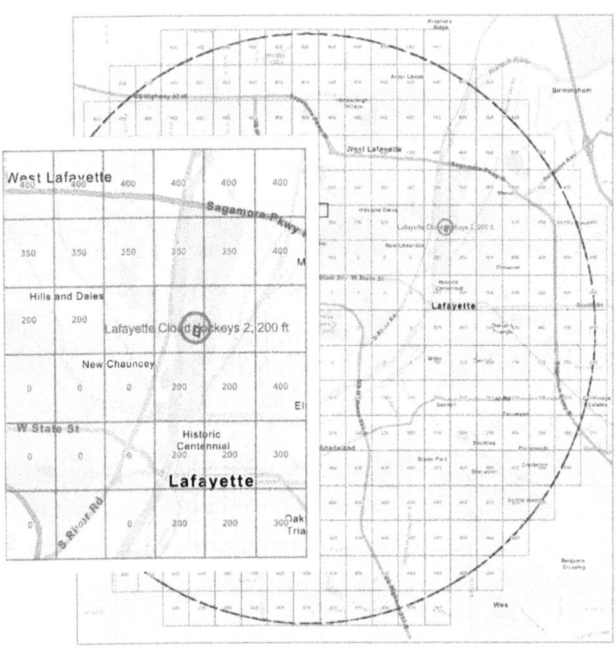

Figure 15.89. If you click on the airport you will be working around, the FAA website (https://faa.maps.arcgis.com) provides this level of detail. This is the controlled airspace for the Purdue University Airport in West Lafayette, Indiana.

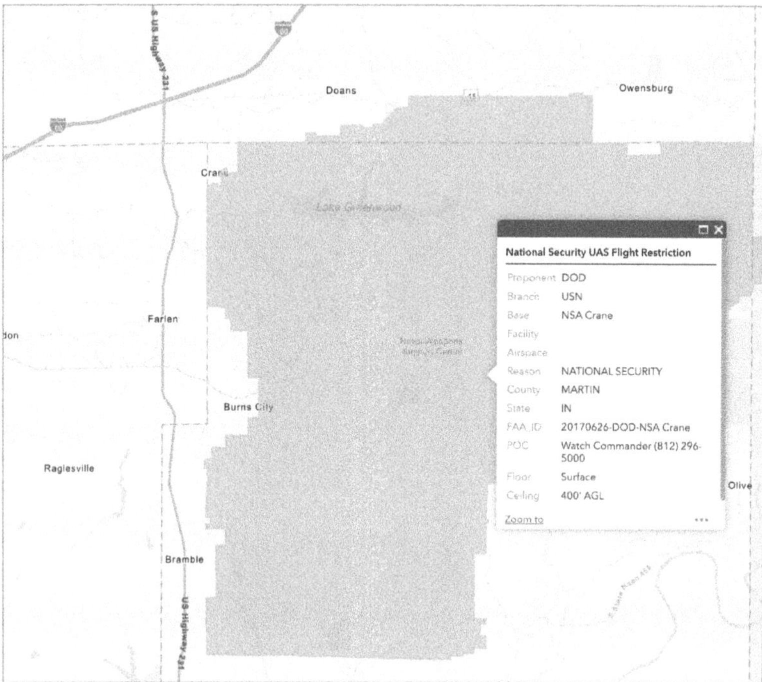

Figure 15.90. The Crane Naval Base in southern Indiana is listed as prohibited airspace on the Federal Aviation Administration website (https://faa.maps.arcgis.com). Prohibited airspaces do not have grids, which indicates drones and other aircraft are not allowed within the shaded borders.

The airspace around most military installations is either restricted or prohibited. Restricted airspace includes areas where drones may operate but are subject to restrictions when determined necessary to confine or segregate activities considered hazardous to nonparticipating aircraft. Prohibited means that a drone cannot fly within the designated area—for example, members of the public are prohibited from making flights around the White House.

Knowing whether you are near or in a restricted or prohibited airspace is important before you fly a drone. You can use an FAA application called B4UFLY (https://www.faa.gov/uas/getting_started/b4ufly) to check which airspace you are currently in and whether you need any waiver or are not allowed to fly at all. Failure to comply with airspace restrictions can result in significant regulatory enforcement actions by FAA investigators.

Visibility Requirements: 3 Miles

Under Part 107, a drone can be flown if aerial visibility is at least 3 statute miles, 500 feet below clouds, and 2,000 feet laterally. This ensures that drone pilots can easily see manned aircraft and take evasive actions in case aerial applicators are in proximity to drone operators. Waivers or exemptions are required when pilots want to operate drones when visibility is less than 3 statute miles. Drone pilots should always yield to crewed aircraft operating in the same vicinity.

Night Flying Allowed with Anticollision Lights

Pilots can fly their drones at night if the aircraft are equipped with anticollision lights that can be seen from 3 statute miles away, and they flash at a sufficient rate to prevent collision. No waiver is required. However, we do not recommend applying pesticides at night regardless because of concerns about temperature inversions and, often, reduced pesticide activity.

No Hazardous Materials

Agricultural products are considered hazardous materials by FAA definitions. At the time this chapter was written, if the operator wants to use a drone to apply pesticides, fertilizers, or other ag products deemed hazardous by FAA definitions, they must petition the FAA for exemptions from certain rules governing the use of drones.

This includes carrying hazardous materials in spray tanks and operating a drone that weighs more than 55 pounds. The rules of Part 107 on the following pages are not a comprehensive list but only a small subset of rules that are most applicable to using spray drones. Visit the FAA's website for a full list and description of the Part 107 rules you must follow when operating a commercial drone.

Application of Agricultural Products from a Drone

Any drone that weighs less than 55 pounds that is used to dispense dry or liquid pesticides (referred to as *economic poisons* in FAA statutes), plant nutrients, and other products to agricultural, horticultural, and forest areas are subject to rules under CFR 14 Part 137 and Part 107. To legally operate under Part 137, an organization must first obtain a Part 137 operating certificate. Part 137 provides FAA clear authority to provide oversight of aerially applied products. Remember that pesticides applied with drones are regulated by both the EPA and FAA.

WAIVER, AUTHORIZATION, AND EXEMPTION ARE DIFFERENT TERMS

When you are flying under Part 107 rules, certain actions are not allowed but may be granted a waiver. Pilots or operators may apply for waivers and explain what their operations are, what safety precautions they are taking, and what kind of training will be required for remote crew members. An application for authorization is similar to a waiver but involves access to controlled airspace that the LAANC does not grant immediately. The FAA reviews the proposed wavier or authorization and will accept or deny the application within 30 to 90 days. An exemption, on the other hand, is required to perform an operation that is strictly prohibited or "unwaiverable" under the current FAA rules.

Since drones that weigh less than 55 pounds are operated under Part 107 rules, and agricultural operations operate under Part 137, pilots or operators must comply with both 107 and 137. However, as previously stated, there are certain rules that spray drones cannot meet, such as the rule outlined under Part 107.36 Carriage of Hazardous Material. Since spray drones and spray drone pilots cannot abide by all the rules outlined in Part 107 and Part 137, pilots or organizations must ask the FAA to exempt them from certain rules.

The first step to obtain a Part 137 operating certificate is to petition for exemption from certain rules that cannot be followed. The rules that the organization or operator needs to request an exemption from for drones under 55 pounds include:

- Part 107.36 Carriage of Hazardous Material
- Part 137.19(c) Certification Requirements, Commercial Operator—Pilots
- Part 137.19(d) Certification Requirements—Aircraft
- Part 137.19(e)(2)(ii) Certification Requirements; Knowledge and Skill Tests; Skills; Approaches to the Working Area
- Part 137.19(e)(2)(iii) Certification Requirements; Knowledge and Skill Tests; Skills; Flare-outs
- Part 137.19(e)(2)(v) Certification Requirements; Knowledge and Skill Tests; Skills; Pull-ups and Turnarounds
- Part 137.31(a) Aircraft Requirements; Certification Requirements
- Part 137.31(b) Shoulder Harnesses
- Part 137.33(a) Carrying of Certificate; Certificate Carried on the Aircraft
- Part 137.33(b) Registration and Airworthiness Certificates Available
- Part 137.41(c) Personnel; Pilot in Command; Demonstration of Knowledge and Skills; Commercial Certificate
- Part 137.42 Fastening of Safety Belts and Shoulder Harnesses

For the FAA to approve an exemption, the petitioners must explain how they will maintain a comparable level of safety to what the rule creates. The exemption process can be challenging and confusing. Some organizations and operators defer to a legal authority to write the exemption for them, or, in some cases, the drone manufacturer or sales company may assist operators in submitting exemption requests.

WHAT REGULATIONS AM I FOLLOWING?

Part 137 regulations cover all spray drones, regardless of weight. Part 107 regulations govern all drones less than 55 pounds. Part 61 and 91 regulations apply to all drones that exceed 55 pounds.

Any pilot who flies a spray drone (regardless of weight) must have a Part 107 pilot's license. Instead of the manned pilot license required for Part 61, FAA regulations allow the Part 107 pilot's license to serve as a substitute.

NOTES FROM THE FIELD

A drone operator indicated, "There are many lawyers who help with exemption filings. It is a very common practice and has become quite streamlined. What used to take six months is now flowing through in six weeks."

Part 107 applies only to drones that weigh less than 55 pounds, so spray drones that weigh 55 pounds or more require exemption from 14 CFR Parts 61, 91, and 137. Parts 61 and 91 take the place of Part 107, since drones that weigh 55 pounds or more are not "small" drones, so they are covered under rules more like those for manned aircraft (for example, aerial applicators).

The rules that organizations must ask to be exempt from when flying drones that weigh 55 pounds or more include:

- Part 61.3(a)(1)(i) Requirement for Certificates, Ratings, and Authorizations
- Part 91.7(a) Civil Aircraft Airworthiness
- Part 91.119(c) Minimum Safe Altitudes: General
- Part 91.121 Altimeter Settings
- Part 91.151(b) Fuel Requirements for Flight in VFR Conditions
- Part 91.405(a) Maintenance Required
- Part 91.407(a)(1) Operation after Maintenance, Preventive Maintenance, Rebuilding, or Alteration
- Part 91.409(a)(1) Inspections
- Part 91.409(a)(2) Inspections
- Part 91.417(a) Maintenance Records
- Part 91.417(b) Maintenance Records
- Part 137.19(c) Certification Requirements, Commercial Operator—Pilots
- Part 137.19(d) Certification Requirements—Aircraft
- Part 137.19(e)(2)(ii) Certification Requirements; Knowledge and Skill Tests; Skills; Approaches to the Working Area
- Part 137.19(e)(2)(iii) Certification Requirements; Knowledge and Skill Tests; Skills; Flare-outs
- Part 137.19(e)(2)(v) Certification Requirements; Knowledge and Skill Tests; Skills; Pull-ups and Turnarounds
- Part 137.31(a) Aircraft Requirements; Certification Requirements
- Part 137.31(b) Shoulder Harnesses
- Part 137.33(a) Carrying of Certificate; Certificate Carried on the Aircraft
- Part 137.33(b) Registration and Airworthiness Certificates Available
- Part 137.41(c) Personnel; Pilot in Command; Demonstration of Knowledge and Skills; Commercial Certificate
- Part 137.42 Fastening of Safety Belts and Shoulder Harnesses

To see an example of an exemption that was written to allow Purdue University drone pilots to operate spray drones, visit Regulations.gov (http://www.regulations.gov), and type "Purdue spray drone" in the search bar. You can also search specific drone manufacturers and drone model numbers or company names to find exemptions written for each particular case.

Once you submit the exemption, the FAA will approve your exemption petition, decline your petition request, or ask for more information. This process may take many months to complete and can be the most time-consuming part of obtaining a Part 137 operating certificate.

If the FAA approves the exemption, they will send a document that outlines the conditions and limitations under which you must operate to comply with the waiver. The document also will explain when the exemption will expire and when it will have to be amended to remain current.

Once you secure the proper exemptions, you may apply for your operating certificate by filling out form 8710-3 and sending it and the exemption to the FAA authority over 137 drone certifications (UA137Certificates@faa.gov). Once the FAA processes the request, a knowledge and skills test can be self-administered whereas previously a group from your FAA Flight Standards District Office (FSDO) would have administered the exam. This streamlined application process has greatly increased the speed with which operating certificates are processed and distributed.

Once you have satisfactorily completed the certification requirements, you will receive an operating certificate that designates the chief supervisor of the operation and any exemptions that the organization is operating under. The chief supervisor is the person who self-administered the exam and will be in charge of the Part 137 operations for your organization. Once you identify a chief supervisor, that individual will oversee the pilots who operate within your organization and will have to ensure compliance with FAA rules and regulations.

CONSIDERATIONS BEFORE PURCHASING A DRONE

Drones, particularly spray drones, are not cheap, especially when you add the expenses of generators, trailers, and supporting equipment. Individuals who are considering purchasing a spray drone should spend time thinking about how they will use the drone and their return on investment in the form of saving time, generating profits, or solving particular application problems. While the marketplace is not very competitive now, different companies manufacture a number of spray drone models.

Timely Applications Can Justify the Cost

Some drone owners upgrade their drones every few years because newer models may provide distinct advantages, such as being able to fly at a higher speed. A farmer might be able to justify the cost of a drone based on the need to make timely applications, particularly to high-value cash crops like watermelons, tomatoes, sweet corn, or seed corn. In some seasons, manned applications may be delayed because of high demand during critical time windows or subpar weather and field conditions. It's not that there is a problem with fixed-wing aircraft or helicopters, but your need for a timely spray application might not align with their availability or with weather conditions.

Even if growers see drones as worthwhile purchases and good fits for the farm, they still must consider who will make the applications and whether they will have time to do their own spraying. Most important, growers must carefully consider who will be responsible for obtaining pilot certification and keeping up with the changing technologies and regulations.

It is time-consuming to meet the regulatory challenges (for example, obtaining pilot certifications) and obtain any waivers that are needed to operate drones for spraying pesticides and other products. For any drone to pay for itself, growers must understand how the drone operates and work out how they will find the time and labor during the busy season to make those applications.

Most growers are so busy with tillage, planting, and side-dressing that they have little time left for spraying with drones. This is why many farmers currently prefer to hire commercial applicators. Time challenges do not change just because you have a drone. Using a drone might even be more difficult in busy times because the visual observer requirement currently in place ties up two people. This alone may be impractical if you do not have more than one person working on a regular basis.

A couple of points are worth considering before buying a spray drone. The first question is: Am I really going to use it? Without a lot of use, it would be difficult to justify the cost of a drone because of the accelerated electronic depreciation of most drones. Also, keep in mind that the drone you buy today could become obsolete in the near future. Spray drones are electronic consumables, just like televisions, computers, phones, and tablets. They do not hold value over time, even if they are still brand-new in their boxes and never opened.

MORE THAN AN ALTERNATIVE

Spray drones are more than just an alternative to traditional aircraft. Spray drones may be the only option available to apply products to fields with unique shapes, small sizes, or flight obstacles that traditional ground equipment and aircraft cannot treat. Many ground-based or manned aerial applicators do not want to treat smaller fields because it is not profitable for them. Application companies increasingly refer the work they do not want to perform to drone companies.

The second question to ask is: Does the farm have enough acres to help spread the cost of the drone over those acres? This question is one you would also ask for ground equipment. One farmer may have a land base large enough to spread the cost of a $650,000 self-propelled ground sprayer, while another farmer with less land would not be able to pencil it out. Before you buy a drone, you should calculate the costs of buying and operating it, and then compare that with how much you are charged for custom applications.

Considerations for Using Drones Commercially

The number of commercial businesses offering drone services to their customers has been increasing steadily. These commercial businesses can gear up by employing multiple drones to make applications.

Questions about doing commercial work with drones are similar to questions you might ask about commercial ground, plane, or helicopter applications. How many acres are best applied with a drone versus other sprayer types that are already available? What will my labor and equipment costs be? What is my cost per acre for the application? What is my expected return on investment? These and many other questions are all part of the equation of making sure the spray drones you are considering will provide enough value to your farm operation.

As you explore purchasing a drone for commercial use, consider these questions:

- Will this be my full-time business or just seasonal?
- What crops am I going to target?
- What licensing and certification will I need from the state pesticide regulatory agency and FAA?
- Does my state allow drone pesticide applications if the pesticide label does not provide specific instructions about drone sprayer use?
- Can I find reasonably priced commercial insurance? What will it cost?
- What market(s) can I fit into? Some operators focus on row crops (like spraying fungicides on corn and soybeans). Others use their drones to treat specialty crops like blueberries, grapes, or other fruits. Outside of crop production, drone applications can be used for forests, ponds, highways, and livestock operations.
- Do I plan to use a drone under a government contract? Make sure that performing work for a government agency does not require using a drone that is different from what you have or what you are thinking about purchasing. The US Department of Agriculture and the US Department of Defense have concerns about digital security breaches, so they don't allow drones that are manufactured by certain companies to operate on state and federal property.
- How many acres can I cover with a single drone? You will want to ensure your fleet is big enough to cover enough acres to make your investment worthwhile.

- What is the backup plan if a drone goes down?
- Will I rely on someone else to keep track of the regulations? Either you or someone on your team will need to be comfortable with understanding the ever-changing state and federal rules, regulations, and policies. Rules can be complex and require more than just filling out paperwork. Putting those requirements into practice will require forethought and patience.
- Can I find qualified labor? Justifying the cost of a drone may not be a problem, but finding people to work will be tough when there are labor shortages or when you need to find workers who can put in the long hours often necessary in agriculture. Many potential employees are unwilling to work the long hours outdoors, including evenings and weekends, that are required in agriculture.
- Are the products your customers commonly use labeled for aerial applications?
- Will you or the customer provide the chemicals for applications?
- What size spray tank will you need? Larger fields require tanks with larger capacities. But larger drones bring more stringent FAA regulations. Depending on the market, a drone with a smaller tank capacity might work just as well.
- Does the drone manufacturer provide customer service? Are they able to supply parts quickly to repair drones?
- Where will I get the drone repaired?
- How will I map the field for boundaries and obstacles with the highest efficiency?

Figure 15.91. Several spray tanks have labels that indicate different volumes, which can be helpful for accurate mixing. This 38-liter tank is equivalent to 10 gallons.

- Can I be competitive with manned aerial and ground applicators?
- What will be my travel limit? Some applicators have been known to travel as many as 12 hours to do a job, while others will not take on work outside of a 2-hour radius.
- How will I promote my business and gain new business outside of my area, since spray drone trailers can easily travel on highways across the state?

- What software comes with the drone package? Every drone comes with a software package. At this time, you cannot add third-party vendor packages to spray drones as you commonly can with ground equipment. Even if a vendor says they have better technology and software, drone manufacturers will not allow access to their drone's software.
- How comfortable are you using the software and flying the drone? Is it worth investing your time to understand the software by accessing websites and attending trade shows? Make personal contact with manufacturing representatives or with extension educators who can demonstrate their drones at your place of business.
- What other supplies or equipment will you need to purchase? As you have read in this chapter, the drone is just one component of the operation. Think of your total cost by adding expenses such as an additional truck, specially designed trailers, generators, tanks, spare parts, and other supplies. These all raise the total investment required for a drone spraying operation. This may be especially challenging for start-up businesses.
- Will you be doing only single-rate applications? Or will you also consider variable-rate applications? Not all drones can do variable-rate applications so you will need to think ahead.
- Do you plan to also provide application services for dry products like urea or cover crop seeds? If so, you need to think about purchasing a dry box that can handle different dry product rates.
- What will be the equipment depreciation?

INSURANCE FOR PROTECTING YOUR INVESTMENT AND DEFENDING AGAINST LAWSUITS

Protecting your investment in your drones and shielding yourself against liability claims are two reasons to consider insurance as part of your protection plan. Insurance is not just about replacing a drone after it crashes in a field. You also should consider having coverage in the event your drone hits a vehicle on the highway, crashes into a building, or produces spray drift that lands on an organic farm or commercial apiary. Even more dangerous is if your drone hits a person who is unaware of the ongoing application.

Some drone owners inadequately insure their drones to save on their insurance premiums, but some state regulatory agencies may require operators to carry minimal liability insurance to cover drift or other accidents. Some drone operators also wrongly assume that their current vehicle insurance policy will automatically cover them for drone replacement and liability lawsuits. Without proper insurance coverage, everything you own could be up for grabs by the person who alleges your drone physically hurt them or damaged their crop or property.

As a general rule, insurance companies typically exclude coverage for all aerial applications, including those made by drones. When speaking with an insurance company, make sure they understand if you are seeking coverage for a drone that will be used for monitoring or if you are seeking coverage for a drone that will be applying pesticides. It might change their answer about whether they provide coverage and for how much.

Sometimes a policy might provide limited coverage for a drone used for monitoring and surveillance only. It is almost guaranteed that a drone that sprays pesticides will not be covered under most insurance policies. More than likely, your insurance company has contacts with other providers that deal with aerial applications and can provide coverage. Three important points deserve to be restated:

1. Don't assume that the liability insurance for your property, vehicle, or trailer will also cover your drone operation.
2. Don't assume that the insurance you have for ground application equipment will also cover your drone.
3. Just because you have a million-dollar liability policy, that does not mean the drone is covered for that much. Read the policy carefully because you may only have a limited amount of coverage for replacing a wrecked drone or when facing a lawsuit.

Know What Is in Your Policy

There are three main areas regarding spray drone insurance. Each area can be purchased or excluded, which can influence the premium:

1. General liability property casualty
2. Chemical liability/crop spraying liability
3. Hull coverage

Let's use an example to illustrate how these might fit together: Let's say your drone crashes into a car traveling on the highway. The crash ruptures the tank, which releases a fungicide into a nearby creek and causes a fish kill. The general liability coverage covers damage to the car. It does not cover the fish kill nor damage to the drone. The chemical liability covers the fish kill. The hull coverage takes care of damages to your drone.

The main point is that the general property casualty insurance excludes any liability resulting from the contents of the spray tank. That can include a drift issue, or environmental hazard, or crop injury to a crop that was sprayed. People often think that a $1 million general liability limit will cover all exposures from flying a drone. But unless you carry chemical liability insurance, that is not true. Some insurance agents suggest that hull coverage is the least valuable of all, as its premium-to-coverage ratio is poor. If an operator is looking to save on insurance, hull coverage is the place to make a cut, not the general or chemical liability.

In any case, it is better to visit your insurance agent to make sure you fully understand what your policy will and will not cover. Only a handful of companies in the United States offer special spray drone insurance with liability coverage. Without competition, a spray drone insurance premium can be $4,000 to $5,000 versus $35,000 for a new drone. But you must evaluate insurance costs against what you might lose in the event of an accident where someone is hurt. It is a rare event, but always possible.

CONCLUSION

The drone application industry is evolving so quickly that new models are becoming available as quickly as smartphones and other consumer electronics—and, like phones, they are becoming obsolete after a few years. This makes buying drones different from buying ground sprayers, which you know can be used for decades to come.

Still, drones present new opportunities for making pesticide applications that we have not had before. Drones are viable and have been effective in other countries for a long time: Farmers in Japan use them to spray rice fields. Using drones for pesticide applications is mostly positive because it checks a lot of boxes that we can't with traditional ground and aerial equipment. Drones have become very versatile tools used across many different application sites beyond traditional agriculture.

Figure 15.92. Attending classroom lectures and field demonstrations are good ways to find the newest information. Keeping current is important because drone technology changes so quickly.

Figure 15.93. Spray drones are unique pieces of application equipment compared to ground boom equipment. Researchers are evaluating drones to find optimum use practices.

We have seen spray drones used for forestry, mosquito control programs, container nurseries, vegetable crops, invasive weeds, roadside spraying, greenhouses, fly control in dairy operations, and organic production. However, drones require specific calibration procedures, application considerations, and insurance—commercial operations, in particular, require these things to use drone technology legally, safely, and effectively.

While spraying with drones has many positive attributes, it is unlikely that they will replace ground rigs, spray planes, or helicopters in the foreseeable future. However, drones complement these other application methods precisely because there are situations where drones are a better fit than other methods. Traditional aircraft or ground sprayers can generally cover more acres in a day because they travel faster, have wider spray widths, and can carry greater payloads. When weather conditions are ideal and operators must cover many acres, ground rigs, airplanes, and helicopters typically are more effective.

However, we should not undervalue drones because they likely will evolve into another critical tool to augment ground, water, or aerial applications. In some cases, using drones to spray pesticides could be the only viable and safe option we have.

A 21ST-CENTURY CAREER OPPORTUNITY

Does the next generation of the workforce have experience with handheld controllers, joysticks, and software systems? We could never be more confident of an emphatic yes. Drones will attract a new generation of employees who are computer and technology savvy. Ag technology companies and government agencies will use their skills to provide valuable insights on using this mode of application to its fullest.

Figure 15.94. Drones have become an important tool in making pesticide applications.

NOTES FROM THE FIELD

A drone operator indicated, "I encourage people to not be threatened by this new technology. Rather, think through how drones can supplement or improve what we are currently doing. Also do not look only at what it is now, but what it could be in the future."

There are many unknowns the research community will need to address for drone applications of pesticides and other inputs to achieve the successes everyone is hoping for. Drones are a different type of aircraft, designed differently, and have significant differences in how they spray compared to traditional fixed-wing agricultural spray planes and helicopters. Ongoing research at many research institutions around the world is currently identifying those differences.

Technological advances and improvements are made through more research and testing. Designs, spray characteristics, regulations, and business models are also subject to considerable change as more research and testing pave the way for technological advances and improvements.

This new research will undoubtedly help create new science-based guidelines that will guide spray drone operators. The good news is that (as application scientists mention) drone platforms have more in common with each other than they have differences.

If we look back only a few years, we see that drones could only do small fields or certain high-value applications. Today a pilot can spray a couple hundred acres or more when conditions are right. Spray drone technology will continue to evolve as the customer base increases and demand increases for more features and practical applications. This means that current pilots (with or without much experience) will be required to keep current with changing technology.

The benefits of using spray drones with today's technologies outweigh the many concerns just raised. We are just seeing the tip of the iceberg when it comes to what we can do with drones. Keep your nose up and stay tuned.

■ ■ ■

16

Removing Herbicide Residues

With contributions from Dennis Nowaskie, Bryan Young, Kirk Foreman, Ples Spradley, Thia Walker, Joe Becovitz, John Obermeyer, Daniel Reynolds, Bill Johnson, and Kevin Leigh Smith

All of us say that quality is important, but it is easy to cut corners when we're pressed for time. Consider the following example, as recalled by one applicator:

> Everyone at our agricultural retail store was under pressure to spray a bunch of backlogged acres. Two weeks of rain kept us from spraying, and it really put us behind the eight ball. We started early in the morning and worked late into the night seven days a week until we caught up. We were spraying fields like there was no tomorrow. Late one day, I sprayed my last field of corn and was eager to start spraying soybeans.
>
> I knew I needed to rinse the tank because the chemicals I'd been spraying on corn could damage soybeans. But I thought I could save a little time, so I made an executive decision to quickly flush the sprayer at the edge of the field instead of completely rinsing it.
>
> I quickly flushed the tank and booms with 100 gallons of clean water out of the reservoir. I decided not to clean the screens, figuring I'd save an hour by skipping that step. With the tank rinsed and the booms flushed, I filled with water, added the soybean herbicide, and headed out to spray the field.
>
> Keeping the spray rig rolling instead of wasting time cleaning it seemed like a good idea at the time. But as it turned out, it was not one of my better decisions.
>
> A couple days later, a farmer called in. He was furious about the application I made to his soybean field. Before meeting with the farmer, my manager asked me if anything unusual happened before or during the application. My manager couldn't understand how the herbicide damaged the soybeans. I told him that I skipped cleaning the screens, assuming that running 100 gallons of water through the booms had cleaned out any remaining residue. All he could do was shake his head and ask why I had not followed the proper sprayer cleaning procedures when changing to a different crop.

"I was trying to save time" was all I could answer. When we got to the field, all we could do was stare and gasp at the acres of damaged soybeans. I could hear the farmer tell my manager and an agronomist, who had been called to explain the damage, that he had paid our company to kill the weeds, not his soybeans.

Then I heard the farmer ask my manager, "Who is going to pay for the damage to my soybeans?"

Unfortunately, and more to the point, he threatened to take his business to one of our competitors. This meant a loss of sales not only for future applications but for fertilizer, fuel, chemicals, and seed. I could see from the expression on my manager's face that losing the income from a 5,000-acre farm would put a serious strain on year-end profits.

About that time, the farmer turned to me and shouted, "What were you thinking?"

I knew the answer but decided it would be more prudent not to say anything. The conversation shifted back to the manager, which gave me time to relive the week before I had pulled into the soybean field. I recalled that we had gotten enough rain during the two weeks before that we had to stop making applications until the soils dried out. The sprayers were sidelined.

If only I could go back and undo that decision to cut corners. But that was no longer an option. My manager told the farmer he would immediately contact the insurance company and start the process of getting the crop damage complaint resolved. Thankfully, it seemed to be enough to pacify the farmer.

TIME CONSTRAINTS WITH WEATHER DELAYS

Many of today's herbicides are active at very low rates. If you cut corners when cleaning application equipment, it can cost thousands of dollars in damage claims even when only small amounts of herbicide are left in the spray system. "We don't have time during the busy season" should never be the excuse for not cleaning out sprayers. As the old adage says, "You can pay me now or pay me later."

Figure 16.1. The lighter areas of soybeans were damaged by a spray application.

Now all we could do was wait until the insurance company settled with the farmer to see if he would remain one of our customers.

Obviously, this situation could have been prevented if the applicator had taken the time to clean and flush the entire sprayer system—tank, hoses, screens, booms, and nozzles. However, bad things happen when we rush or take shortcuts to save time. Using contaminated tanks seldom ends happily. Saving a few minutes by taking shortcuts during the cleanout process can almost guarantee that you will spend more time in the long run resolving contamination issues. It is much smarter to spend the time to do it right the first time.

This chapter discusses and describes the proper procedures to remove residues from pesticide application equipment. These cleaning procedures can help prevent applicators from unintentionally introducing damaging herbicide residues to sensitive or nonlabeled crops.

FACTORS THAT LEAD TO HERBICIDE INJURY

Herbicides can damage crops in many ways. Applying the wrong herbicide can damage susceptible crops, as can herbicides that drift from nearby applications, as well as excessive water runoff that contains herbicides. Herbicides also can injure crops when applicators don't follow the application directions on product labels or when they spray under abnormal environmental conditions.

And, of course, herbicides can damage susceptible crops when applicators do not thoroughly clean sprayer equipment in between spraying different crops. Even

Figure 16.2. An ag retailer sprayed this non-GMO corn, thinking it was Roundup Ready® corn. The retailer acknowledged the mistake and determined the total financial damage from the misapplication. The farmer asked the seed dealer where the mysterious corn plants that survived came from.

Figures 16.3–16.4. These watermelons were injured by 2,4-D drift. The watermelon is without leaf cover, which can lead to sunburned melons.

very small amounts of herbicide residue in tanks, hoses, screens, fittings, or booms can seriously damage crops.

If you leave herbicide residues in the spraying system, you will not only pay up front to settle the claim but may face other costly long-term repercussions. For example, if you damage a crop, the farmer might decide to take their business elsewhere. That same farmer might also tell everyone about your mistake, which could turn away potential customers. Neither of those outcomes is good for business. The following sections describe some reasons that unwanted herbicide residues continue to find their way onto the wrong crops.

Herbicides Are Active at Low Concentrations

Some herbicide products still call for applying pints or quarts of herbicide per acre, but many herbicides now have application rates measured in ounces per acre. Residues from products with these lower application rates can affect crops even when only trace amounts are left in the application equipment.

Postemergence Herbicides Are More Common

By definition, postemergence herbicides are applied over an existing crop. So if there are any unwanted herbicide residues in the equipment, future applications using the same equipment on a different crop can damage or kill it.

Dry Formulations Are Popular

Herbicides formulated as dry products have become quite popular, because they are easy to handle, ship, and store. Most dry formulations create suspensions of the herbicide (similar to mixing clay in water), so they do not dissolve completely into the spray mixture. So unlike most liquid formulations, dry formulations require constant agitation to evenly suspend the product.

When applicators fail to allow enough mixing time, larger particles of the dry product can get trapped in a series of screens. The particles may remain in the sprayer system until applicators run enough water through the screens, or another product solubilizes them into smaller pieces—which can then pass through the screens and out the nozzles.

Adjuvant Use Is Growing

Adjuvants, such as surfactants and crop oil concentrates, make foliar-applied herbicides work better. Many herbicide products are formulated with adjuvants to improve their effectiveness. But adjuvants also may dislodge old herbicide residues embedded in tank walls or hoses, or they may help break down particles in screens. When they do, the adjuvants may cause an old, unwanted herbicide residue to become part of the spray liquid (for more information on adjuvants, see chapter 13).

FARMERS HAVE CHOICES; RETAILERS DON'T

Everyone with an interest in a crop takes pride in watching it grow during the season. No one likes to see a crop damaged by herbicides. When farmers are pressed for time, they may decide that some crop damage can't be avoided—there are thousands of acres to treat, and waiting is not an option. Farmers who are behind schedule know that if they don't apply the herbicide before it rains, taller weeds will become more difficult to control and reduce crop yields. Yields also may be affected if sprayer tanks are not thoroughly rinsed or not all of the screens get cleaned. Farmers must decide whether to have a weedy field that will yield less or a field with some herbicide residue injury that reduces yield.

These are tough choices farmers have to make. Every hour is critical. Farmers can decide to quickly flush their sprayer systems instead of spending the time to break down the system for a complete cleaning. They can do this because they are spraying their own fields. **Retailers never have the flexibility or choice of not cleaning the sprayer system—they are spraying someone else's fields.**

Tank Mixes of Herbicides Are Common

Glyphosate-tolerant crops (Roundup Ready® soybeans, corn, and cotton) simplified weed control by allowing one active ingredient to control a broad spectrum of weed species without killing the crop. As glyphosate-resistant weed populations have grown, however, it has become common to use multiple herbicides to control resistant weeds. It is important, of course, to learn which herbicides can be tank mixed to control these resistant weeds.

Products Can Be Incompatible

Problems with product hiding or building up in application equipment can be traced back to how different products get along. If you do not mix agricultural chemicals in the correct order, they can settle out of solution or form a paste. If that happens, there are going to be problems. It is important to pay attention to compatibility issues, especially when field mixing. If you're receiving premixed "hot loads" from a mixing facility, always tell the people at the facility if you observe potential compatibility issues. Incompatible mixtures often leave difficult-to-remove residues inside spray equipment (see chapter 11).

Application Timing Is More Crucial than Ever

Optimum performance of postemergence herbicides often depends on the size of the weeds at the time of the application. As weeds get larger, the herbicide's effectiveness is greatly reduced, which is why herbicide labels provide maximum weed size restrictions. When weather delays early-season herbicide applications, controlling weeds becomes more challenging. With little time to spare, applicators may gamble and only partially clean sprayers in order to save time and cover more acres.

Plumbing Is Complicated and Interconnected

Modern sprayers have many places where herbicide residues can remain trapped, even after flushing hundreds or thousands of gallons of water through the system. Not opening a single valve might leave quarts or gallons of herbicide in a hose. When the applicator opens the valve to that hose during the next application, the residue will then be mixed with the new product and pushed out the boom.

WHERE DO RESIDUES HIDE?

Self-propelled sprayers are becoming more popular than pull-type sprayers with farmers, and they have become standard for agricultural retailers. In crops, sprayers are sophisticated machines outfitted with the latest technology, including rate controllers, chemical injectors, long booms (up to 120 feet), 1,800-gallon tanks, boom section controls, boom levelers, state-of-the-art nozzles, pulsating nozzles, advanced

Figure 16.5. There are many places that a pesticide can remain in a sprayer, which is why spending time to effectively remove the residue is important.

tires, computer recordkeeping, global positioning with light bars, and auto-steering. These advances enable applicators to get more work completed more efficiently and effectively throughout the season.

Underneath these self-propelled sprayers, you will see a sophisticated system of interconnected valves, pumps, screens, hoses, nozzles, and boom sections. Although flushing the system with clean water will remove most remaining herbicide from the system, clean water will not remove it all. *Most* and *all* definitely have different meanings when we are talking about potentially damaging herbicide residues. Rinsing with clean water will not remove larger, undissolved clumps of herbicide particles or areas of heavy residue buildup. Those particles could remain trapped in the boom's end caps or in a screen, no matter how many thousands of gallons of water have already run through the system.

Think of rinsing sprayer equipment like using just water to wash grease from your hands. If you just run your hands under water, the grease isn't going to come off; nor will the residue come out of spray equipment if you just flush a few hundred gallons of water through it. Furthermore, to get the system completely clean, you must remove the filters and nozzles to remove the particles or paste from the screens at the ends of the booms. There are a number of locations where residues can remain trapped within the sprayer.

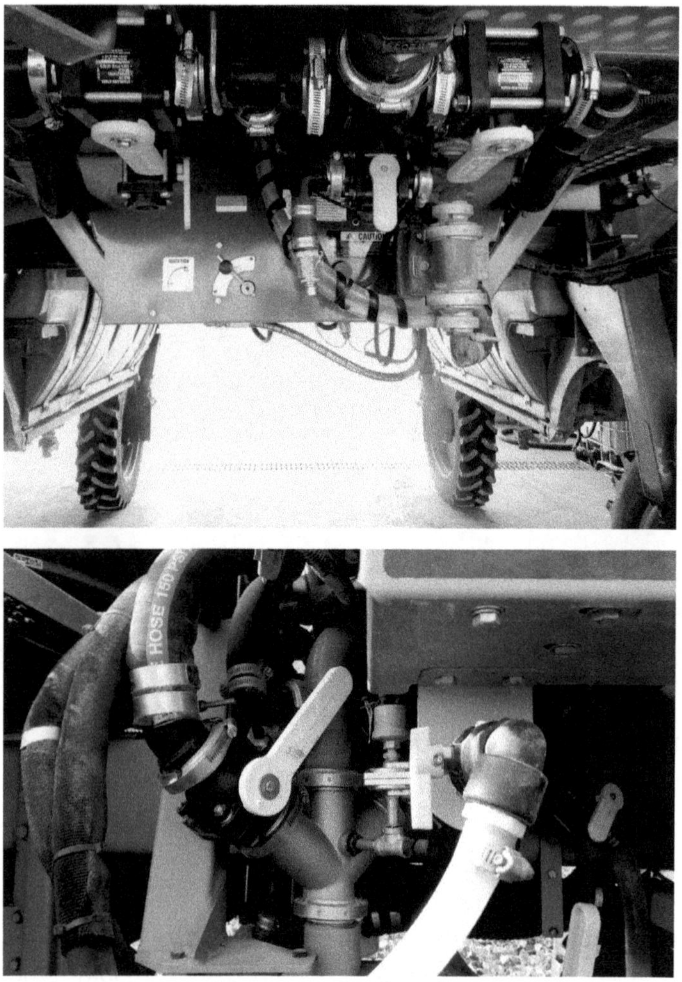

Figures 16.6–16.7. Understanding what the valves control is important to knowing if they hold back chemical residues.

ADVICE FROM AN APPLICATOR

Most of the dirty spray issues I have seen were caused by:

- Lack of dilution: a failure to completely drain the machine before and between flushes
- Dead ends on wet booms: a failure to remove end caps for thorough cleaning
- Screen contamination: a failure to completely clean screens (flushing doesn't remove putty)

Spray Tanks

A sprayer tank has a large surface area that can retain product on its interior surface. Sometimes herbicide residue can embed in the tank's lining (see chapters 2–4). If you don't clean the inside surface, residues can remain and accidently become part of the next spray mix.

A tank's design can also affect how much herbicide remains trapped inside it. Flat-bottom tanks have ledges and corners that can hold product. More important, many tanks do not drain completely. The suction or drain port leading out of the bottom of a tank is usually not flush with or below the tank's bottom.

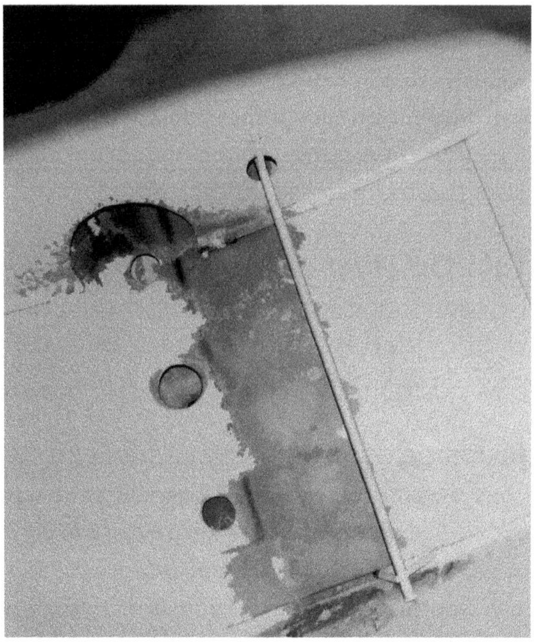

Figure 16.8. Pesticide products added to this sprayer adhered to the tank and did not fully suspend in the water. Dried-on herbicide residue on the bottom of a stainless steel tank required a pressure washer to remove.

Figure 16.9. Visual evidence of residue inside on the surface of a poly tank.

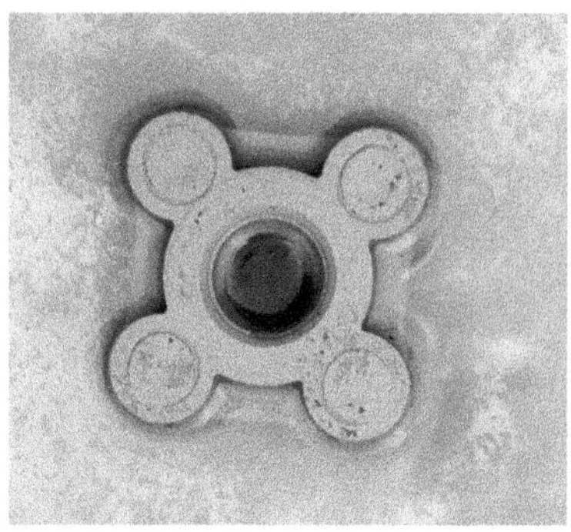

Figure 16.10. Typically, a tank's drainage opening is slightly above the bottom of the sump. The size of the sump and the area below the drain opening will determine how much residue may remain on the bottom of a tank.

Hoses

Hoses, particularly rubber ones, may hold embedded residues. While plastic hoses may also trap residue, plastic is much less porous than rubber. In addition, rubber hoses may develop small cracks and rough edges on the inside, which can also hold residue.

Also, many hoses do not make a straight run from one attachment point to the other. Some hoses sag between sections of the boom pump, which creates a trap for liquid or suspended products, similar to the trap under a sink. Liquid remains in the trap until more water pushes it forward. Many formulated herbicides can settle out of suspension in as little as 45 minutes. Avoid U-shaped hoses by using a tarp strap or zip tie so that they become self-draining.

Valves as Blind Alleyways

Depending on the system, you can turn valves on and off manually or electronically. Opening valves lets clean water and tank cleaners flow through all hoses, removing trapped products and residues.

Booms

Once the tank is empty, the pump will lose prime and cannot force all of the spray mix out of the booms. The only way to completely force the remaining herbicide mix out of the boom is to refill the tank with water and reprime the pump to build pressure. The clean water that runs through the system will push out any remaining herbicide from the pump to the hoses and booms, but any mixture that returns to the tank through the bypass could be contaminated with herbicide.

Many newer spray rigs are equipped with air systems that force air into the booms to push out any remaining spray mixture before you add clean water. Most of these systems clean one boom section at a time. These air systems do not necessarily help remove residues. Blowing out the booms is a first step, but you still need to flush the system.

Figure 16.11. You might think that after a pump loses its prime, hardly any product would remain in the system. But a 120-foot boom may hold 35 gallons of liquid, a 90-foot boom 25 gallons, and an 80-foot boom 18 gallons. Depending on the application rate, that's 1 to 3 acres of product left in the boom and hoses—a lot of potential damage waiting to happen!

Screens

Screens (also called *strainers* and *filters*) capture large pieces of herbicide products, undissolved bags, plastic shavings, sand, and grit before these large objects can reach and plug nozzle tips. Some spray systems also have suction screens on the fill lines going into the tanks. Suction screens pull liquid from the tank into and out of the pump, and then the liquid passes through another screen. Whether this screen is before or after the pump depends on the manufacturer's design.

CLEANING AND FLUSHING ARE DIFFERENT

Flushing water through a screen removes much of the herbicide residue but is not the same as taking off the screen and manually cleaning and removing the solid herbicide residue from it. If the residue didn't come out with all of the water already used, what makes you think that adding more water is going to do the job?

Screens are sized and classified by the number of squares per inch—the smaller the mesh number, the larger the mesh hole size. The mesh should be slightly smaller than the exit orifice of the nozzle. Fill screens often have relatively large holes designed to filter out larger objects. The screens before or after the pump are often 50-mesh, which means more holes that are smaller in size than the fill screen. As the fluid flows toward the nozzles, it will often pass through screens in the booms or at the nozzle tip.

In this example, any product smaller than the last screen before the spray tip will pass through the spray system, while the screens that preceded the nozzles will trap larger pieces. The trapped material will form a film on screens. Even if continuous agitation breaks down the dry product formulation into acceptable smaller particles, these fine particles can continue to build up on the screens because of the film or skin created by the larger particles. While this process protects the nozzle tips and does not affect the spray operation, herbicide residue lodged in the screens can result in coverage issues and lead to crop damage later.

Figure 16.12. An 80-mesh screen (left) and a 30-mesh screen (right).

Figure 16.13. The differences in screen mesh size. The mesh is the fine screen under the coarse structural part of the screen.

Figure 16.14. Screens may look different, but they all serve the same purpose: They prevent clogging of the nozzle tips by filtering out larger particles of herbicide and other foreign materials.

Figure 16.15. Screens immediately after the pump.

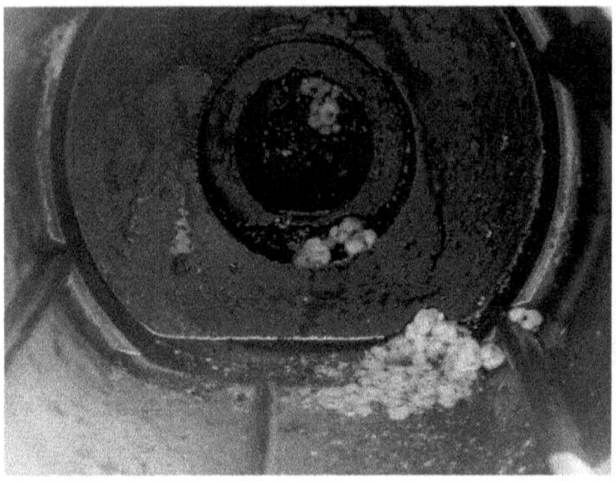

Figures 16.16–16.17. Materials accumulated in the filter between the tank and pump (left). The screen associated with this filter shows debris captured by the screen (right).

Figures 16.18–16.19. Herbicide residue was left in this drop nozzle from a previous use. There will be residue in the tip and the pressure shutoff.

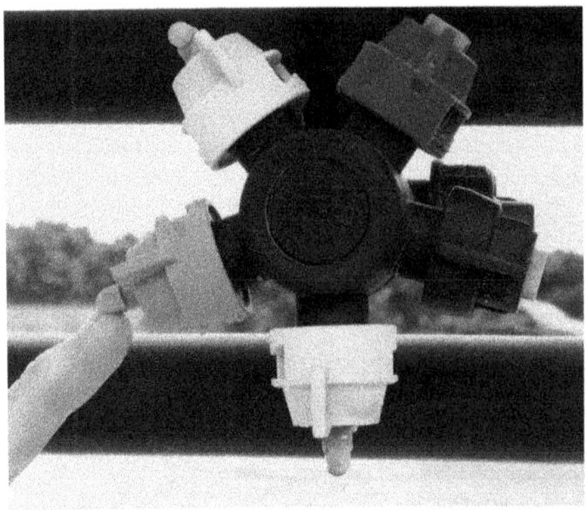

Figure 16.20. There are screens inside each nozzle of this turret. Nozzle turrets often hold herbicide residues even after you rinse the boom, so always rotate the turrets during the flushing process.

Figure 16.21. This photo shows a sprayer with 10-inch spacing between nozzles, but the producer used a 20-inch spacing (that is, only used every other nozzle). Field damage was traced back to unused nozzle clusters that had not been rinsed. After they started cleaning out the unused bodies, they reduced contamination.

End Caps or Cleanout Caps

Whether it's one boom or a series of sectional booms, each "pipe" has caps on the ends to contain the flow of water within the pipe so that the spray mix is forced through the nozzle tips. The last nozzle is often 2 to 4 inches from the end cap, which means that suspended herbicide can collect and remain undisturbed between the last nozzle and the end cap. These undisturbed particles can collect, gather, and bind together to form a milky, gummy, and sticky substance with a toothpaste-like consistency.

Sprayers normally have end caps on boom sections. End caps (also called *cleanout caps*) are designed to be taken off. You can clean the paste on the cap and inside the boom. If you do not remove the end cap and clean it, you can leave significant residue buildup trapped just inside the cap.

Figure 16.22. Considerable herbicide residue can build up on the inside of the end cap or in the boom. See page 532 for end caps that are easier to remove and clean.

Figures 16.23–16.24. (Left) This nozzle body is still a few inches from the end cap. (Right) This nozzle body goes out to the end of the boom section.

Figures 16.25–16.27. The accumulation of product at the end cap is a major problem (top). Using a dye and a transparent boom demonstrates this cleanout issue (middle and bottom). The operator filled the boom with red dye (representing an herbicide) and attached it to a sprayer boom via a nozzle body and quarter-turn attachment. The operator flushed the boom with water for 15 minutes while jostling and shaking the boom to emulate sprayer turns. The dye, compressed at the boom end, never moved but slowly mixed and was emitted from the final nozzle. Once we opened the small valve at the far end (representing any number of end-cap valves), it took 2 seconds to clear the dye.

INSURANCE CLAIM ON WINTER WHEAT: THE SHUTTLE

Two fields of winter wheat were treated for annual grasses. The applicator used shuttles for a water source, mixed with fertilizer. One of the shuttles had previously held glyphosate and had not been rinsed well. After the application, one field was killed, and the other field stunted. The insurance company paid nearly $14,000 in damages to settle the two claims.

ADDITIONAL SOURCES OF HERBICIDE CONTAMINATION

Shuttles and minibulks: Only use dedicated, labeled minibulks. Reusing a minibulk tank that has not been thoroughly cleaned can damage crops.

Inductor cones: Most of them have a valve that allows operators to rinse the tank with fresh water.

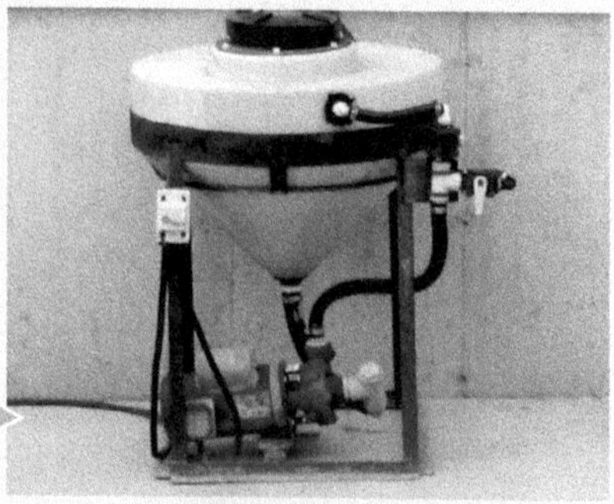

Hoses used to fill tanks with water: Typically, permanently mounted hoses move herbicide from bulk tanks and water. Most often, you can color-code these hoses: A yellow hose might indicate products used on corn, while a black hose might carry products used on soybeans. Color-coding hoses ensures that any residues remaining in the lines are applied only to a nonsensitive crop.

Tires on sprayers designed with booms in the front: Some applicators use tank cleaner and a brush to scrub the front tires and remove any residue.

Nurse tanks on trucks: Tender trucks are used to take premixed hot loads to the sprayer. If you use the same tender truck to haul multiple loads to an applicator spraying different crops, without cleaning the tank, it could lead to crop damage.

Figure 16.28. Sprayers have complicated interconnections of hoses, filters, and valves to control flow and pressure. A standard operating procedure and cleanout checklist will ensure that you omit no step.

TANK CLEANOUT: TURNING A LOT INTO A LITTLE

The procedure for removing herbicide residue from the inside of a sprayer tank is no different from the procedure for cleaning the inside of a 2.5-gallon plastic container or a 250-gallon minibulk. Each procedure relies on moving enough water over the interior surface to dislodge and dilute what's left inside the container.

Powered sprayers have additional components such as screens, hoses, pumps, and filters that may trap herbicide particles. In most instances, you may need to add a tank cleaner to help remove residue from the interiors of hoses and tank walls, especially for powered sprayers. This complete flushing, cleaning, and neutralizing process has proven to be the most effective method for removing herbicide residues from sprayers.

Cleanout is a somewhat misleading term because it implies we can remove all the residue. Actually, we can really only reduce the herbicide concentration to levels that don't injure crops or result in antagonism in the next mixture. The difficult question is: How clean is clean enough?

We know with certainty that it is impossible to completely purge a sprayer system of all herbicide residues. We also know from experience that proper cleaning can reduce any herbicide residue to such a low concentration that it is unlikely to injure the nontarget crop. So equipment is clean enough when the herbicide residues are diluted to concentrations low enough to prevent injury to sensitive crops.

Figure 16.29. These buckets contained antifreeze that had been used to winterize a sprayer in the fall. The antifreeze was drained the following spring to be recycled for the next year. After the operator drained the recycled antifreeze from the buckets, herbicide residue from the previous year was evident even though the sprayer had been thoroughly flushed. This demonstrates that completely removing all traces of herbicide is virtually impossible.

Removing residues is all about paying attention to the smallest of details. This is especially true with today's sophisticated sprayer systems where the herbicide mixture doesn't just flow from the tank, through the pump, to the booms. There are many pathways through which the tank mix moves to the booms. All of it moves through a series of hoses interconnected by opening and closing valves. Forgetting to flush these hidden areas may contaminate the next tank mix. Using a checklist will ensure that you flush your sprayer equipment systematically and completely.

Forgetting a single valve may leave enough herbicide in a section of hose to contaminate the next load. A few areas that are easy to forget are the front and side loaders, flowmeter hoses, and induction hoses. Knowing how your sprayer equipment is plumbed and what each valve controls is important. Paying attention to detail, flushing the sprayer with clean water, cleaning the screens, and adding a tank cleaner significantly lessen the possibility of injuring a crop. This level of detail is something everyone should expect and demand.

Always consult the herbicide and tank cleaner labels for specific information about how to use them.

Figure 16.30. An often-overlooked part of cleaning a sprayer is the system of hoses that move liquid through the flowmeters (the meters are marked with arrows).

Figure 16.31. Some sprayers are equipped with dual flowmeters for high and low flow rates. If you do not use the second flowmeter, the herbicide may get trapped on the outlet side of the valve. If you do not flush the inlet after use, it is possible that the herbicide will sit in that section until you open the valve.

Figure 16.32. Here the flow comes together at a Y juncture (arrows). Flow from one side of the Y can actually back up into the other. Unless you run clean water in each side, the remaining herbicide can become a contaminant.

THE STEPS FOR REMOVING HERBICIDE RESIDUES

Properly cleaning a self-propelled or pull-behind sprayer may take an hour or more. If you cut corners, you increase the probability that the next field you spray may exhibit an unwanted result.

THE POWER OF THREE

First Rinse.
This photo of the first rinse applied to soybeans shows stunted plants and cupped leaves.

Second Rinse.
This photo of the second rinse applied to soybeans shows no stunted plants, but cupped leaves are still evident.

Third Rinse.
This photo of the third rinse applied to soybeans shows no stunted plants or cupped leaves. This is why herbicide labels require you to triple-rinse your sprayers.

Step 1: Spray Out Booms Every Night

Applicators often comment that sprayers that sit overnight with herbicides in the hoses and booms for extended periods (more than eight hours) can result in crop injury issues even after they are properly cleaned. Mississippi State University research shows that when applicators leave herbicides overnight in rubber hoses, the herbicide can penetrate and reside in the hoses and ultimately becomes more difficult to remove.

GHOST DAMAGE

Sometimes crop damage can appear a few days after you rinse the sprayer equipment. The damage might come from herbicide residue that is released from hoses or fittings and works itself back into solution. That's why removing herbicides from equipment at the end of the day may be important to prevent damage.

Some herbicide products can settle out in 45 minutes if they are not agitated. Leaving sprayers idle overnight or longer allows more time for herbicides to penetrate the linings of the hoses or tanks. It also allows more time for herbicide residue to settle out and become more cemented to interior surfaces.

Even if you flush with more water and run tank cleaners for a few minutes, it may not be enough to remove this type of residue from the hose interior. First, completely empty the sprayer while in the field. It's OK to cover a previously sprayed area, because herbicides (as a general rule) are safe for crops at twice the labeled rate. Second, add 10 times the sump's remnant of clean water, circulate it, and spray it out in the field as well. Repeat. These two rinsing steps will take care of the majority of the cleaning and won't take very long.

Figures 16.33–16.35. This sprayer was traded in to the dealer. Residues had dried in the tank and were described as tacky like varnish (top). The tank was much cleaner after being soaked with a commercial tank cleaner followed by a rinse of 150 gallons of water (middle). Pressure washing then peeled off much of the crusted material from the side walls (bottom).

Step 2: Perform a First Rinse in the Field

The highest herbicide concentration left in the tank can be found when the pump loses its prime. It's important to remove as much of this concentrated spray mixture as possible before starting the actual cleaning process.

How much water should you flush through sprayers in the field? A general rule of thumb is to use 10 percent of the tank's capacity per rinse. Another general rule says to add at least two to three times the amount of gallons applied per acre. For example, if a spray application used 15 gallons per acre, 30 to 50 gallons should be sufficient to flush the sprayer. Add more water if the spray coming out of the nozzles during the flush process doesn't look clean.

Figure 16.36. This boom contains foam, which can hold herbicide residues.

Figure 16.37. The inside of a boom after the first flush with clean water.

Figure 16.38. Newer self-propelled sprayer models may be engineered with an air purge system to blow out the boom sections. Normally, each boom section is purged independently. Depending on the unit, you can retrofit some older sprayer models with this air purge equipment. The photo shows an example of a retrofitted air purge system designed by a farmer. It is best used when the pump loses its prime. Forcing products out of the hoses allows the first flush of clean water to do a better job of removing any leftover residue.

Figures 16.39–16.40. Side wall cleaners (left) and sparger/agitation bar (right).

Some applicators drain as much product as possible out of tanks before circulating clean water. They will open the main valves to drain the tanks and hoses. They will put in 3 to 5 gallons of clean water and recirculate to get the remaining material out of the tank by reopening the tank valve. This tactic will make the first rinse a cleaner rinse, because the water is circulated between pump and tank.

For systems equipped with inductor tanks and sparger tubes, it is important to open the valves to the sump, bypass, and inductor during the first flush. Activate the sparger tube to stir the bottom of the tank, and the tank rinser to spray water over the sides of tank, and allow enough time for the water to circulate through this closed loop of the system.

Make sure to open the front and side loading valves (including the plumbing) to the injectors a few times for a few seconds to allow water to circulate through the pumping system. This will push the remaining spray mixture through the booms and nozzles. Spray this mixture over the last crop until the pump loses its prime again—this allows the sprayer to discharge most of the spray mixture.

There's no question that you should spray the first rinse over the last crop you treated. Research from Colorado State University evaluated sequential spray tank rinses and found that the first rinse contained nearly the same herbicide concentration as the original solution. Cleanout efficiencies were determined by comparing the concentration of the active ingredient in the original spray solution with the concentration of the active ingredient in the rinse samples.

The two most common active ingredients in the spray solutions from this study were 2,4-D and dicamba. The first rinse contained an average of 92 percent and 91 percent of the original 2,4-D and dicamba solutions, respectively. Due to the high concentrations of active ingredients removed during the first rinse, you should apply this rinsate to a labeled site.

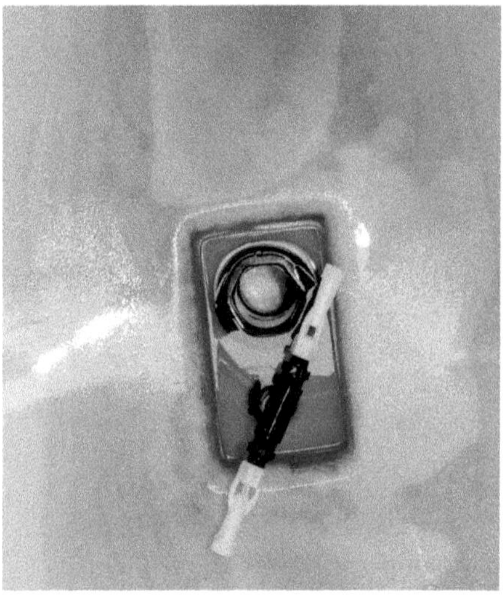

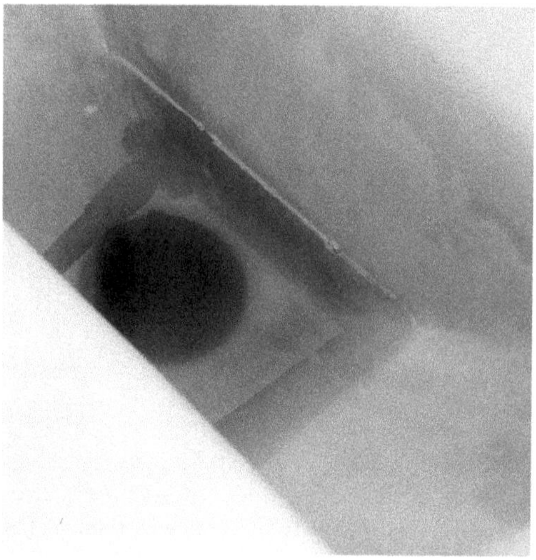

Figures 16.41–16.42. Some tank designs do not allow complete drainage. (Left) In this case, there is liquid contained below the drain opening that the pump will not be able to pull through the system. (Right) This tank design allows for liquids to completely drain out of the sump. The first 30 to 50 gallons of clean water will remove much of the tank mixture left in the sump area and in the hoses and booms.

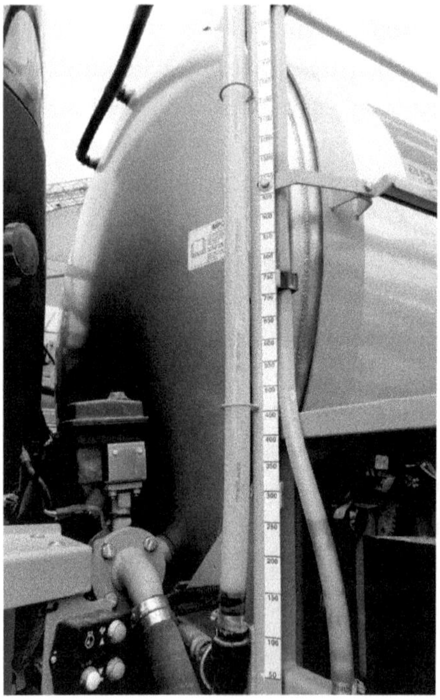

Figure 16.43. Replace sight gauges several times during the season. It's only a few bucks, but replacing these gauges can eliminate residues that can embed in the plastic. It's been suggested that unhooking the top and pulling the hose down will allow any trapped herbicides to exit.

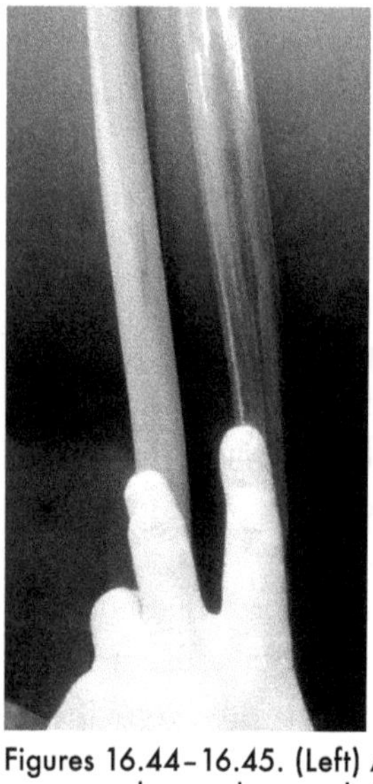

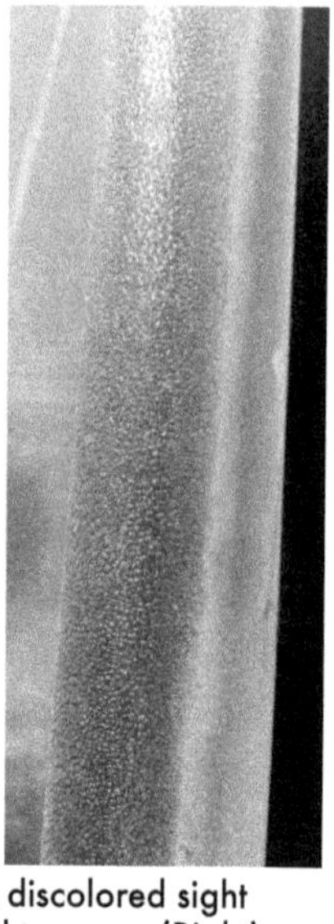

Figures 16.44–16.45. (Left) A discolored sight gauge and an undamaged sight gauge. (Right) A clear sight gauge that was damaged by chemicals and sunlight.

What to Do with Sprayer Rinse Water

For many retailers, one of the most pressing issues is where to dispose of sprayer rinse water. The question becomes even more complicated when you consider a statement included on many pesticide labels: "Do not contaminate water when cleaning equipment or disposing of equipment wash waters." The question of where to dispose of the first rinse is easy to answer: You should apply the rinsate to the outer rows of the field where you just made the last application. The question of where to clean booms, screens, and end caps and to dispose of the rinse water is more complicated and is up to the discretion of the applicator in most cases.

The best-case scenario is to perform the second and third rinses (steps 5 and 7) in the application field away from any ditches, streams, or other water sources. If you choose to perform the second and third rinses back at the retail center, take care to avoid contaminating any water sources. Ideally, you should perform such rinses over a contained area such as a mix/load pad. If a mix/load pad is not available, rotate the cleanout site to different areas of the facility so you do not concentrate herbicide residue in one particular spot.

Figure 16.46. A farmer commented, "Some farmers will not want retailers to dispose of rinse water in their fields. They think the retailers are making an unnecessary trip over the field, possibly knocking down more crop, possibly overapplying chemicals, and possibly causing crop injury. Retailers need to be able to address these issues."

Figure 16.47. This tank held rinsate from sprayer cleaning operations. Continually dumping even small amounts of herbicides (as might be found in second or third rinses) in the same spot can lead to off-site contamination. In this case, the residue flowed from the tank into a pond just north of the trees shown here. Water samples from the pond revealed low levels of many different herbicides. This represents a label violation because the wastewater contaminated a water source. It can be acceptable to apply second and third rinses to gravel areas as long as the herbicide residues do not get into waterways. Protecting water sources is key when cleaning out sprayers.

IMPORTANT MESSAGE

Check with your state department of agriculture to determine if the suggestions presented in this chapter are valid in your state.

Step 3: Remove and Clean All Screens

After performing the first rinse in the field, it's time to remove the concentrated herbicide from your sprayer equipment screens. Most of this is usually a pasty material that did not completely dissolve when it was added to the tank or that settled out of solution during the spray operation. This material will remain trapped in the screens until the next chemical is added or a solvent breaks down the herbicide residue.

Some sprayer units have only one screen, while others have multiple screens. Screens can be found before the pump, after the pump, in the boom sections, or at individual nozzles. Screens also capture rust, grit, debris, and even the occasional aluminum foil cap liner from a container. It's important to clean out all of these materials before the next application.

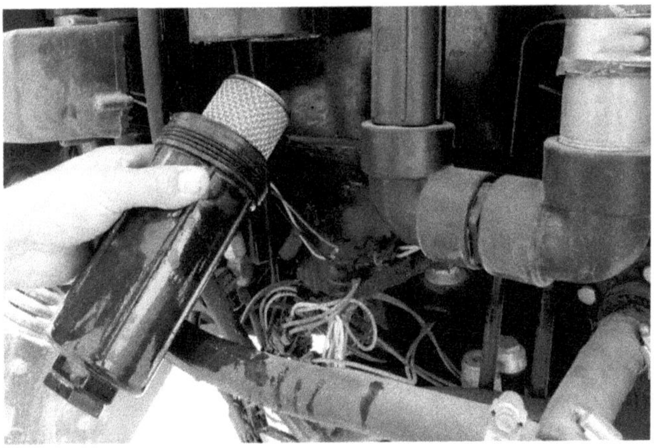

Figure 16.48. The screens collect herbicide particles, which is why they need to be thoroughly cleaned.

Figures 16.49–16.52. Some application equipment may have a ball valve at the end of the screen nearest the pump (top photos). The ball valve allows the screen to be easily flushed when you open the valve. While opening the valve can remove a significant amount of residue from inside the filter, you should never just open the valve and skip manually cleaning the screen (bottom photos). The only way to thoroughly clean a screen is to remove it and use a brush with tank cleaner and hot water.

WHY CLEAN SCREENS SO EARLY?

Currently, most applicators clean sprayer equipment screens just before or after adding a tank cleaner. But if you clean screens earlier in the sprayer cleaning process, it allows you to eliminate the last source of herbicide residue in the sprayer system. Cleaning the screens earlier allows you to use the rinse water in later steps and the tank cleaner to move and break down smaller particles of herbicides that may have escaped the cleaning and rinsing process.

A TIP FROM A COMMERCIAL APPLICATOR

This is one way a commercial applicator cleans screens in his sprayer equipment:

> Always have two complete sets of all screens for your sprayer equipment–from the largest ones to those on the nozzle. Keep one set in a bucket of tank cleaner. Put the set you're pulling from the sprayer back into the bucket containing the tank cleaner. This allows you to quickly replace the screens and helps speed up the cleanout process. If you keep the bucket with the extra screens on the outside of the sprayer, the sprayer movement will continually agitate the tank cleaner through the screen.

When changing to a different tank mix, it is important to actually remove and clean all screens—never just flush them. No amount of water flushed through the screens will remove large particles that have become lodged in the wire and coat the mesh. The best way to clean screens is to scrub the screen inside and out with a brush and soapy water to loosen particles. Then rinse the screens with clean water.

After cleaning the screens, replace them in your equipment. If you are still using screens (strainers) in the individual nozzle bodies, do not place them in your equipment until the very end of the cleaning process. This will allow you to push out any remaining particles instead of trapping them in the clean nozzle strainers.

There are other reasons to regularly remove and clean screens. Screens that become clogged with debris and algae can reduce the flow rate and alter the spray pattern and distribution.

Figure 16.53. This sprayer has two screens at the rear for each side of the boom. Note that the applicator is wearing personal protective equipment.

Figure 16.54. The buildup in a screen may not always contain active herbicide. This image shows a screen with clay particles from formulations and $CaSO_4$ from adding ammonium sulfate for hard water.

Figure 16.55. The number and location of screens in a sprayer vary by make and model. In this photo the gray cylinders are a bank of screens. Some argue that it takes hours to remove and clean all these screens. At $20 per hour, spending 3 hours cleaning screens will cost you $60. However, damaging a customer's field will cost you much more. When you do not clean each screen, you are rolling the dice.

Figures 16.56–16.57. Suction filters shown from two different viewpoints: (Top) Side fill. (Bottom) Front fill.

Figures 16.58–16.59. Hot soapy water and a brush can remove trapped herbicide.

Figures 16.60–16.61. Fewer applicators are using screens at the nozzle tips than they once did, because it takes too much time to clean that many screens. There are 96 nozzles on a 120-foot boom. Most manufacturers of new sprayers rely on the other screens to remove debris before it gets to the tips. However, for those other screens to be effective, they must be 50- or 100-mesh. Check the manufacturer's instructions—they suggest screen sizes.

Figure 16.62. A filter that has collected a lot of debris, which can influence the flow of water.

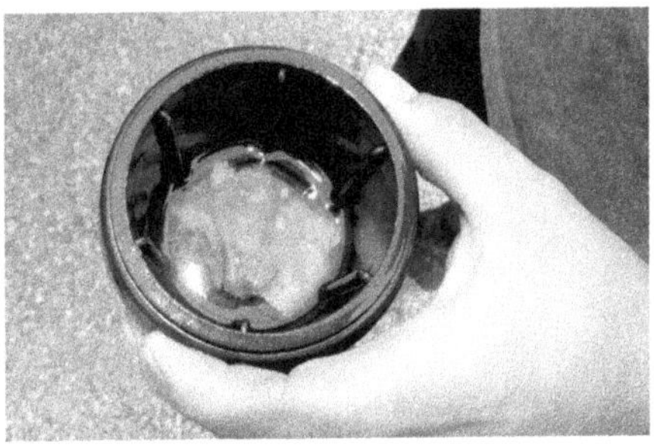

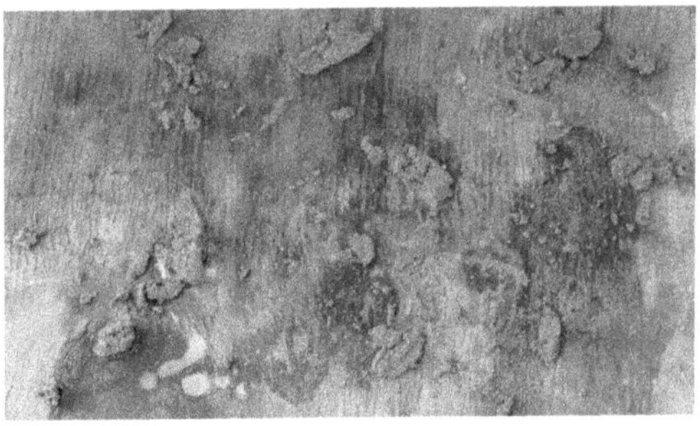

Figures 16.63–16.64. The screen in this water tank shows a buildup of algae (left) and rust (right).

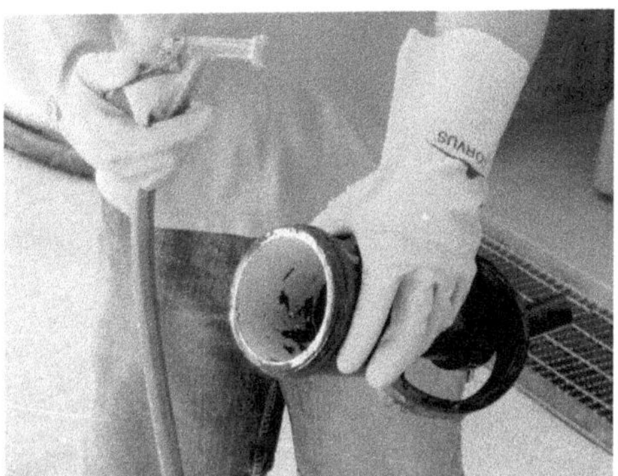

Figure 16.65. Be sure to clean the housing that holds the filter in place, because herbicide residues can also build up on the inside walls. Use a nozzle attachment on a hose (like the one shown here) to finish rinsing.

Figures 16.66–16.67. Examples of end caps. Note the accumulation of debris in the second one.

Step 4: Remove and Clean Boom End Caps

An end cap is the section of a boom from the last nozzle to the end of the boom. Just like screens, you need to remove and clean end caps before using your spray equipment again. To clean the boom section of your equipment, first remove the end caps and scrape away any debris and residue inside them with a brush, tank cleaner or detergent, and water. If you have seven boom sections, you will have 14 end caps.

Some companies clean out the boom sections of their equipment with steam. They will put the extended wand into the right end of the boom and extend it through the entire boom section. They then repeat the process by inserting the wand in the left end of the boom and again extending it through the entire boom section.

Figure 16.68. This end cap has been modified with a quick coupler to allow for easy removal and cleanout. Make sure to use a plastic tie on the lock rings to secure the coupler.

Figure 16.69. This boom section has been modified. The owner has removed the end cap and replaced it with a valve that can flush the end of the boom.

Figure 16.70. Attaching a quarter-turn valve at the end of each boom section allows for easy flushing.

Step 5: Perform a Second Rinse with Water

A second rinse of water will dislodge most, if not all, of the loose residue that remains on the tank walls, lines the hoses, or is trapped in low-lying places. Circulate the water through the hoses, pumps, tanks, fill lines, and flowmeter lines.

Turn on each boom section separately to flush out the water in each. Once you've flushed all the sections, replace the end caps and remove the caps on the opposite side of the boom section. Continue to run water through the boom, section by section. If you flush each section separately, that provides enough water volume and pressure to remove any remaining sludge built up in the system and booms. After you've rinsed the boom sections, replace and tighten all end caps. When you replace the end caps, it will probably be necessary to replace the rings because they are easily damaged. If your sprayer's end caps leak, the cause is often a damaged ring.

If your sprayer does not have end caps or quick couplers mounted on the boom or boom sections, leave the end nozzle out of each section, or consider installing a valve or quick coupler device to help with future cleanouts.

Figure 16.71. Nozzle bodies with multiple tips need to be rotated to ensure that you flush out any trapped residue in the body. Always wear gloves during the process.

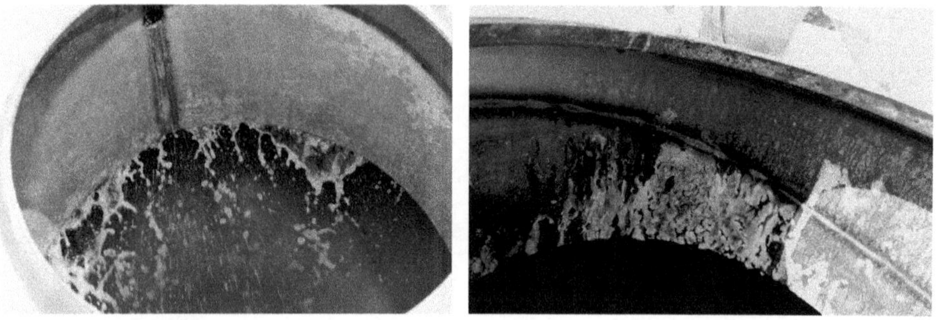

Figures 16.72–16.73. Some tank openings have locations that the side wall cleaner and agitation device cannot fully reach (left). In this example residues have built up on a piece of stainless steel that runs around the rim of the tank opening (right). The only way to remove this residue is with a hose or high-pressure sprayer.

Step 6: Add Tank Cleaner

Tank cleaners are sophisticated products that neutralize (break down) the herbicide molecule, increase the herbicide's solubility (so it can be flushed more easily), and remove any residues that may have penetrated the walls of hoses, tanks, or fittings. However, tank cleaners have limitations. Remember that tank cleaners work only on herbicides that are vulnerable to alkaline hydrolysis. And never forget that tank cleaners are no substitute for a thorough cleaning. View tank cleaners as the last step to break down or neutralize small amounts of residual herbicides that remain in the sprayer.

PUTTING TANK CLEANER IN THE RINSE TANK

Some retailers add tank cleaner directly to clean-water rinse tanks. They believe that having tank cleaner throughout the whole process from beginning to end helps break down residues. For these retailers, the extra cost of the cleaner is insignificant relative to the cost of crop damage claims.

There are various tank cleaner ingredients. One of the most widely used is sodium hydroxide, better known as caustic soda or lye. Most tank cleaners contain an ingredient designed to neutralize acids. Typically, these ingredients are alkaline-type constituents that have a pH value greater than 12 in concentrated form. Because many pesticides are neutral or acidic, alkaline-based tank cleaners help break down pesticides and make residues more soluble in water.

Typically, tank cleaners team caustic materials with detergents made from a variety of surfactant ingredients that are designed to keep residue suspended in the rinsate and to interact with oil substances to remove them from lines and tank walls. When you use any tank cleaner, pay attention to (1) how much to use and (2) how long to leave it in the system for optimal cleaning.

You normally add tank cleaners as pints or quarts of cleaner per 100 gallons of water. Add water to the spray tank, then add the appropriate rate of cleaner. Next, turn on the agitation and sparger tube and recirculate the solution between the pump and tank so the cleaner goes through all of the internal plumbing. Make sure the cleaner goes through all of the hoses, including the same pathways used to flush the system in the field.

Some operators use ammonia or bleach as tank cleaners (you should never use these products together, or they will create toxic fumes). However, ammonia and bleach are rather simple products and are poor substitutes for commercial tank cleaners. Properly formulated tank cleaners are far superior in their ability to clean sprayer equipment. Spray system cleaners should:

- Increase and hold rinsate pH levels above pH 12. This facilitates the chemical breakdown of herbicides that are vulnerable to alkaline hydrolysis and solubilizes residues for better rinsing.
- Contain detergents to penetrate and remove dried-on residues.
- Include surfactants to emulsify oil residues that can serve as anchoring sites for other oil-soluble herbicides.
- Protect equipment from corrosion by including corrosion inhibitors in the formulation.

THE LONGER YOU LEAVE THE TANK CLEANER IN, THE BETTER IT WORKS

One tank cleaner provides important insights on its label:

"Cleaning your tank is not always as easy as it may seem. With the tremendous amount of chemicals and adjuvant combinations, add to this the difference in water quality and fertilizer components, you have a large range of residues that can be formed. This creates a major challenge for the cleaner product."

On another part of the label it states:

"The solution should be allowed to remain on all surfaces to be cleaned so that the solution can have time to penetrate all existing residues. This process can take as little as a few minutes to as long as a day depending on the physical state of the deposit. Since it is impossible to know how long to wait, a person must use their best judgment. Since this is not 100% certain, you must determine whether the last rinsate is 'safe.' You can do this by applying a test solution to the next crop and determining by visual symptoms if it is safe to be used."

Figure 16.74. Oily residue removal test. Oil sludge can be difficult to remove, as is shown with using only ammonia. There are also differences in a tank cleaner's ability to remove oil residues.

Figure 16.75. Dried residue removal test. Ammonia has difficulty removing dried-on herbicide residues. Use high-quality tank cleaners that contain surfactants and other materials that can break down dried residues in sprayer systems.

Detergents can also help to clean any chemicals contained in oil adjuvants that might be stuck on tank walls or hose linings. While ammonia and bleach act as oxidizers and break down some residues, they may not be able to remove products that are embedded in hoses and tank walls.

How Long to Let Cleaners Work

There is an ongoing debate about how long tank cleaners should remain in sprayer systems. One thing is certain: You need to give cleaner products enough time in the tank and lines to effectively do their job. Too many operators immediately move the cleaning solution through sprayer systems after mixing. This limits its effectiveness. Thus, the key to using tank cleaners effectively is to leave products in tanks long enough to allow them to do the work they were designed to do.

Tank cleaner labels typically list a time for breaking down herbicides vulnerable to alkaline hydrolysis. However, allowing extra time may be valuable to help break down dried-on or oily residues. If you are cleaning out a system at the end of the day, we recommend that your tank cleaner work the second shift—that is, leave it in overnight and drain it the next morning.

After you have held the tank cleaner in the sprayer system for the specified time (or overnight), flush out the tank cleaner with another 30 or so gallons of water (or follow the cleaner's label recommendations). Not doing this could reduce the efficacy of the next application.

Figure 16.76. When circulating the tank cleaner through the sprayer system, open the valves to allow the cleaning solution to move into the inductor hoses. Some applicators add a small amount of tank cleaner directly to the inductor as an extra precaution.

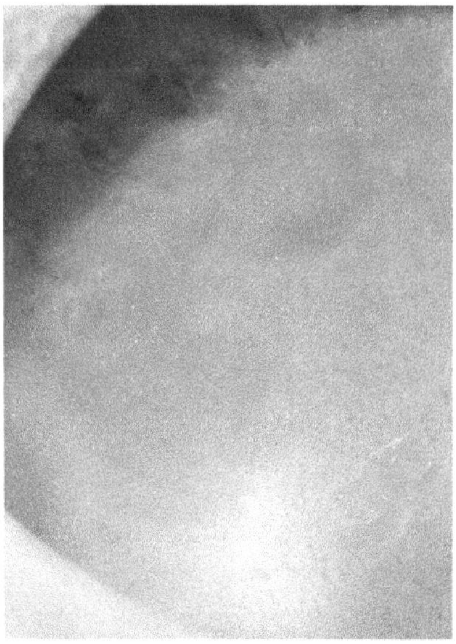

Figure 16.77. Agitating and swirling help tank cleaners work better.

Figures 16.78–16.79. (Left) You can readily see dried herbicide residue on the equipment. (Right) This operator uses a power washer to clean the outside of the sprayer.

Step 7: Perform a Final Rinse and Flush

Some retail operators perform one last flush of water to ensure the system is as clean as possible. Others even go an extra step by removing the screens again and using a pressure washer to clean residues from the sprayer surface. And still others use a tank-mix surfactant or fertilizer additive, which are good at removing herbicide residue that commercial cleaners can leave behind.

ADVICE FROM AN AG RETAIL MANAGER

"On rainy days, we clean out empty trucks and sprayers. We do this as a precaution. It also makes sure our equipment is ready to go wherever they are needed as soon as the fields are dry."

Figure 16.80. Is it worth the time and effort? Consider this 32-acre field of non-GMO popcorn that was damaged by glyphosate residue in an improperly cleaned tank. Properly cleaning a sprayer is time well spent. In the long run, you'll spend far less time and money cleaning a sprayer than you will dealing with this public relations and financial nightmare.

Figure 16.81. Take care to not introduce contaminants from the trucks that fill up the sprayers. There are a lot of separate lines to flush with this setup. Open and close each valve handle as part of the rinsing process.

ELIMINATING RESIDUES FROM TENDER OR NURSE TRUCKS

Transports (or as they are commonly called, *tender* or *nurse trucks*) are not nearly as sophisticated as herbicide sprayers. Nurse tanks have no filters, screens, booms, or end caps to clean. However, in some ways they are more difficult to decontaminate and clean. Many nurse trucks are not equipped with any type of internal mechanical agitation or side wall rinsers. Rinsing such tanks will take considerably more water to ensure you remove all the herbicide residues inside the tank. Therefore, many retailers dedicate tender trucks to specific herbicides throughout the season.

Do Not Add Herbicide to the Last Transport Load

When you use a transport or tender truck to supply tank mixes to the sprayer, don't make the last load that the transport carries of a particular tank mix a premixed (or hot mix) load. Sometimes that's easier said than done. As one farmer asked, "My transport holds 6,500 gallons of water. What am I to do with 3,250 gallons of water?" The answer is to use less water.

When you clean a tender truck, fill the tank only with water, activate the transfer pump, and then circulate the water through all the hoses and back into the opening at the top of the tank. Drive the vehicle around the field or on a road to get the water to slosh around the tank. Then use this water as fill water for the sprayer and add the chemical directly to the sprayer tank or inductor.

Add Tank Cleaner to the Transport

After bringing a final load of water to the sprayer (as described above), fill the nurse truck's tank half full of water. Add tank cleaner according to the label, and then circulate this solution well through the hoses and back into tank. Drive around again to circulate the tank cleaner solution around the inside walls of the transport tank. Release the water in the field away from streams, ponds, and other bodies of water.

Figure 16.82. A plastic sight gauge may help you determine if a tank has been cleaned. A cloudy sight gauge may indicate there is still herbicide residue in the tank.

A PROCEDURE CHECKLIST FOR CLEANING A SPRAYER

When you are performing any type of maintenance, it is better to follow a sequence of steps outlined on a written checklist or standard operating procedure (SOP). Whether the task is winterizing equipment, conducting an annual truck inspection, or checking multiple levels of fluids at the time of an oil change, following a checklist or SOP ensures nothing is missed.

Retailers should implement SOPs to keep their operators from having to memorize steps, processes, and other important items. But most farms and retailers do not have written SOPs for cleaning sprayers. You would think that cleaning the sprayer would be as high a priority as changing the oil in equipment. Following an SOP for cleaning a sprayer will ensure that you complete all tasks in a certain order and leave nothing to chance. By following the procedures step by step, you can clean the system each and every time so you can safely treat subsequent crops with an herbicide. Following an SOP also will make cleanings more time efficient.

DIFFERENT EQUIPMENT, DIFFERENT SOPs

Equipment makes and models differ significantly in what parts need to be rinsed. Retailers should ask those who are familiar with each sprayer rig to help develop an SOP checklist for their machine. Even if the exact machines are at different locations, it will be useful to compare the checklists and look for similarities and differences. Combine the best from each checklist into a single one that incorporates the best ideas from your experienced applicators.

CONCLUSION: WATER IS YOUR FRIEND

One thing remains constant from one year to the next: If something can go wrong, it probably will. Spring is about waiting for the soil to warm and dry enough to plant crops, apply fertilizers, and spray herbicides. While you can plan the work for a day or a week on paper, the weather inevitably will play a role. Weather is one thing you can't control—Mother Nature always wins.

In any given season, weather is usually the biggest obstacle. While a slow 2-inch rain may help a crop that's already planted, it wreaks havoc on those acres that still need to be tilled, planted, or sprayed. Rain will make you wait a day or two or longer to complete the work you feel should have already been done.

Farmers aren't the only ones feeling pressure at certain times of the year. Agricultural retailers feel that same pressure when fields get backlogged and put on the waiting list for applications due to weather delays. Each day that fields go unsprayed means that weeds will be tougher to control.

Field and environmental conditions like wet soils and winds prevent applications, which causes delays. So once the ground dries, retailers are compelled to move quickly. All customers want their fields sprayed first. It's understandable that growers want things done in a timely manner (and sometimes yesterday). Time is money, and delayed applications translate into lost dollars. So when customers call, it's obvious that their first question will be "When can you spray my fields?"

Figure 16.83. Talking to a grower who sees damage from a contaminated tank is never a pleasant experience to have to deal with as a commercial applicator.

"I am going to have my operators read through this, and together we are going to put together our own checklist per machine," said one applicator after reviewing this chapter.

Time is of the essence. What could have been plenty of time to get the thousands of acres sprayed becomes shorter with rain or excessive wind delays. Weather delays force changes to the workday. It means starting before the sun comes up and ending while the moon lights the way across the fields. The workweek expands to the weekend as applicators try to spray all the acres in a timely manner.

The last thing any commercial applicator needs or wants to do during these stressful and hectic days is to divert time from working those acres. Thoroughly cleaning sprayer equipment by the book and by the company's checklist means being down for an hour or two, but with more rain in the forecast, spending time in the fields may seem more important than thoroughly cleaning a sprayer. But that decision may cost you dearly.

The most common mistake applicators make is when they arbitrarily decide to skip a few steps in cleaning the sprayer tank, screens, hoses, and booms. For some, spending a few minutes doing a quick rinse to get the majority of the residues out of the tank seems like the right thing to do. How can it be wrong when it allows you to get back in the field and spray the acres assigned to you? Sometimes you get lucky. But when your decision damages crops, it is costly and embarrassing.

Proper cleanouts are time-consuming. There is nothing that can be done about that. It takes a lot of effort to ensure that sprayer equipment does not contain any herbicide residues that can injure the next crop. When crop injury does happen, it can result in a face-to-face confrontation with a farmer whose crops have been damaged by contaminated applications. Taking the time to clean out the sprayer properly not only keeps customers happy but keeps them coming back.

■ ■ ■

17

Winter Maintenance and Storage of Spray Equipment

With contributions from Corey Gerber, Dennis Nowaskie, Steve Hawkins, Pete Illingworth, Aaron Kreider, Jon Leuck, Kevin Shirk, and Kevin Leigh Smith

After a long season of applying pesticides to keep insects, weeds, and diseases at bay, it can take a lot of effort to make any needed repairs, complete fall maintenance, and winterize your sprayer. But with the ever-increasing cost of today's equipment, not servicing your sprayer for winter storage, or doing so improperly, can be a costly mistake.

Preparing your sprayer for winter and completing routine maintenance, to allow for a quick start-up in the spring, is simply part of the responsibility of owning and operating the equipment. The season is not really over until you properly clean your spray equipment, inspect it, complete any necessary maintenance, and winterize it. A few productive moments and timely repairs in the fall can save hours or days of delays in the spring.

This chapter is meant to help you plan for end-of-season maintenance on your spray equipment to keep it running at its best for the next season and many seasons after. Scheduling and performing fall maintenance has tangible benefits, including improved reliability, productivity, quality, and safety. Successful companies and farming operations consistently strive to attain these benefits. One of the most important advantages of performing maintenance in the fall is that you have more time to get the work done during the offseason. If you delay maintenance until spring, instead of making a timely, early-season application, you may have to wait to apply until repairs are made.

Here are some other advantages of performing sprayer maintenance in the fall:

- It is easier to remove pesticide residue from tanks, hoses, and screens. If you wait until spring, the residue will have dried and hardened, making it much more difficult to remove.

- Removing water from the system before temperatures drop below freezing ensures pumps, pipes, fittings, and flowmeters will not be damaged by frozen water.
- By changing fluids and filters at the manufacturer's specified intervals, you not only uphold equipment warranties but also keep the equipment running at maximum efficiency and extend the life of the sprayer.
- You reduce the need for emergency repair calls and unscheduled downtime in the spring and early summer.
- Properly maintained equipment is safer for employees to operate.
- Fully functional spray equipment ensures revenue potential and customer satisfaction by meeting scheduled applications.
- Repairing small problems now often prevents larger and more expensive repairs later.

Figures 17.1–17.3. Dirty oil and clogged filters lead to premature engine failures.

REVIEW THE WARRANTY REQUIREMENTS

At a minimum, you should begin any maintenance program by reviewing the owner's manual maintenance schedule. By following this schedule, you keep the equipment covered by the manufacturer's warranty. You should always follow the manufacturer's recommendations on how to properly maintain and store spray equipment. It may require you to access multiple manuals to review what you should do and when. Manuals can be either on paper or digital.

The manual recommendations provide information about fluid capacities, recommended fluid types, maintenance, and scheduled intervals for when to change fluids, filters, and screens and lubricate pivot points. Following these recommendations should prolong and preserve the life of the sprayer.

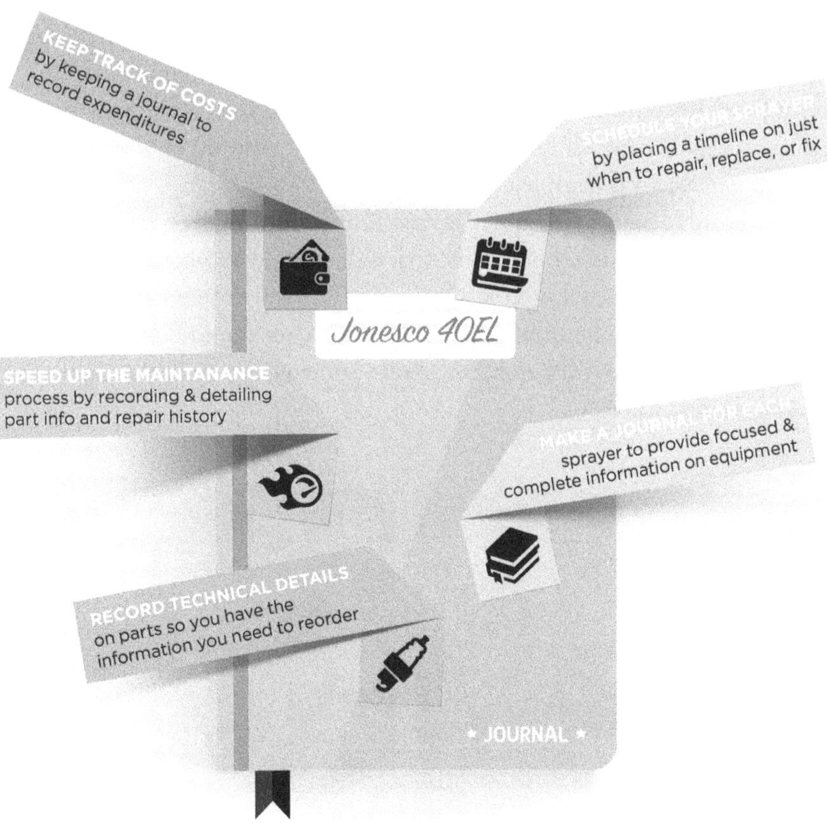

KEEP A MAINTENANCE JOURNAL

Keep a book on each sprayer to make it easy to reference important information and help keep machinery in top shape.

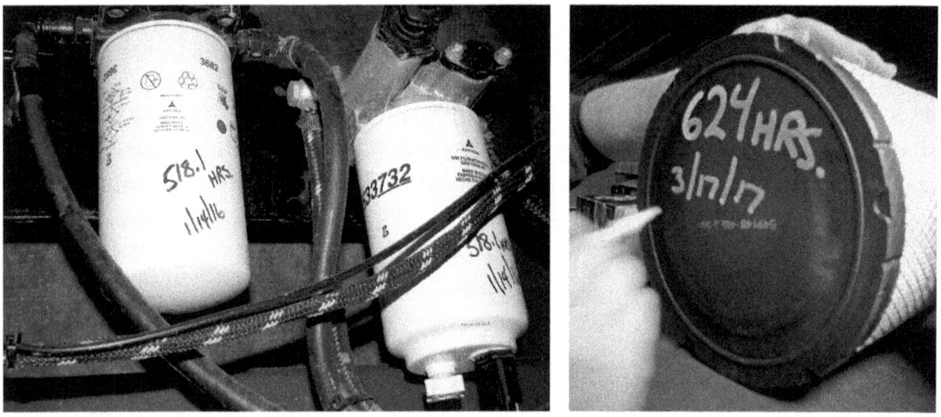

Figures 17.4–17.5. It's common to write the date on filters when you install the new one. It also helps to include the number of hours the sprayer has been in operation.

Don't just perform the maintenance; be certain to keep maintenance records. Create a paper or electronic folder for each sprayer to document the maintenance and repairs you completed. Keeping up-to-date maintenance records is critical if you want to keep track of costs. The records should quickly answer any questions about when and what type of maintenance was completed.

Service records also provide quick access to information. For instance, your own records should easily tell you how many quarts of a fluid you needed to add and what filter you used without needing to look it up in the owner's manual. In your records, be sure to write down part numbers, hours of service, serial numbers, model numbers, and VIN numbers. The VIN number allows someone to find the exact part that matches a specific sprayer.

Figures 17.6–17.7. Handle all waste fluids in an environmentally sound manner.

FALL INSPECTION AND MAINTENANCE

Fall is an excellent time to prepare your equipment for spring applications. In spring your equipment needs to be in the field, not sitting idle waiting for ordered parts to come in. Fall inspection and maintenance includes changing fluids, lubricating seals, replacing broken parts, adding antifreeze, and protecting electronic components from freezing temperatures. You should also thoroughly inspect the entire spray rig. This inspection should include the chassis and frame; the engine compartment; the electrical, hydraulic, and hose connections; the spray tank; and the mounting brackets.

Check Lights and Flashers

Warning lights and flashers alert other drivers that large, slow-moving spray equipment is on the road. Make sure all the lights on your equipment are working before you put the sprayer up for the winter. Turn on all of the lights (including the flashers) and walk around the sprayer. Visually inspect each light and flasher to ensure they are functioning properly.

Change Fluids and Filters

The intervals for changing fluids and filters depend on the make and model of the sprayer.

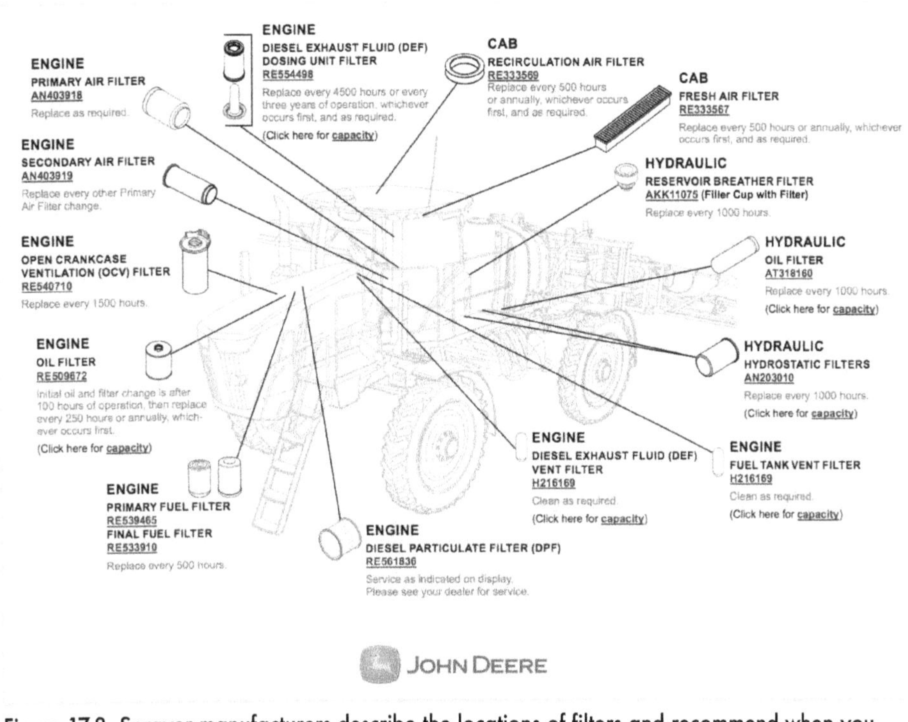

Figure 17.8. Sprayer manufacturers describe the locations of filters and recommend when you should replace them—like in this example from John Deere.

Figures 17.9–17.10. Routine maintenance (such as changing oil and filters like this one) will keep equipment running longer. Always double-check the oil filter surface that screws onto the filter base. Make sure you removed the old filter gasket from the equipment and didn't leave it behind. Leaving the old gasket will prevent a tight, secure seal. Apply new oil to the new gasket to improve the seal.

Figures 17.11–17.12. Many engines are equipped with a dual element engine air filter (left). Such designs keep the engine cleaner and help it run at peak efficiency. The outer filter removes larger particles, while the smaller (inner) filter removes smaller particles (right). Usually, the outer filter traps most of the particles found in "dirty" air. Follow your manufacturer's recommendations for replacing air filters.

It is acceptable to blow out the larger (outer) air filter, but be sure not to use too much pressure, because it can destroy the paper element. Many air compressors blow air at up to 150 psi, which can destroy a filter. Never blow out the smaller, inner filter. Replacing air filters is much cheaper than taking a chance of damaging your engine.

When you clean filters, always blow the air from the inside to the outside of the filter. If you blow from the outside in, it will only force the dirt particles into the filter material and permanently embed them into the element.

Figure 17.13. Change the cab air filter to help ensure that applicators breathe clean air during applications.

Lubricate Grease Fittings

A grease fitting means there is a pivot point of some kind that routinely needs to be greased. Bearings, bushings, hinge pins on folding booms, ball joints, and tie-rods with grease fittings continually lose grease during routine sprayer use. Replenishing grease minimizes the friction from metal-to-metal contact. Without grease, this friction will wear joints and lead to loose hinge points.

Figure 17.14. Keeping equipment greased is critically important in keeping equipment smoothly running.

Fill Fuel and Hydraulic Oil Tanks

Keep tanks full of fuel or hydraulic oil to help eliminate condensation. It can also help to add a fuel stabilizer to gasoline engines before you put them up for the season. Stabilizers keep the gas from getting old. After you add a stabilizer, run the engine long enough for the stabilizer-mixed fuel to work through the entire fuel system.

Inspect Engine Belts

Engine belts weaken and crack because of routine use, tension, heat, vibrations, chemical contamination, and age. Whether it is a V-belt that operates on separate pulleys or a serpentine belt that connects all of the engine components, it is important to check the condition of all belts.

A squeaking noise or grinding sound could indicate that the belt needs closer inspection. When inspecting belts, look for cracks, brittleness, and fraying. If belts are loose, tighten them and make sure they are in alignment. You might consider carrying spares of some of the smaller belts, especially the one for the alternator.

While you inspect belts, be sure to inspect the pulleys and idlers for rough surfaces. A rough pulley or idler can scrape against a belt and cause it to wear out prematurely. Replace rough or worn pulleys and idlers to get the most life from your belts.

Figure 17.15. A belt showing cracks is a clear indicator it needs to be replaced.

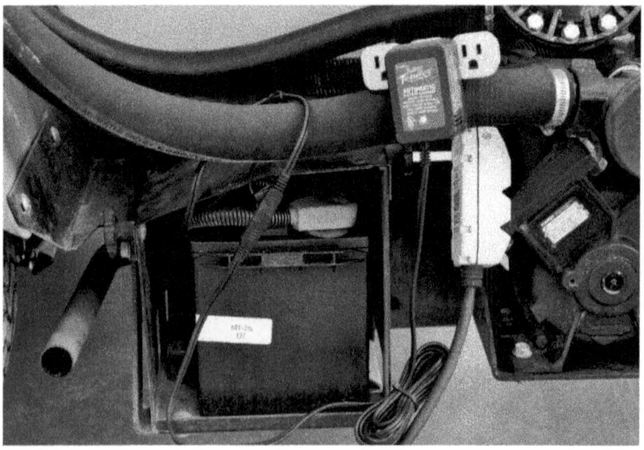

Figure 17.16. A trickle charger.

Remove, Clean, and Store Lead-Acid Batteries

Since batteries are very important (and rather expensive) parts, it is a good idea to take good care of them. Today's sprayers have advanced electronics, so it becomes more important to maintain the batteries. Computers can drain a battery even while the sprayer is in storage. For long-term storage, remove the battery from your equipment and store it in a heated location. At the same time, be sure to clean battery cables, connectors, and posts to remove any corrosion buildup.

Before storing a battery for the winter, make sure it is fully charged and check the water level. It is best to connect a trickle charger to the battery over the winter to keep it fully charged. If the batteries in your rig are difficult to access, you can purchase and attach leads that will make it easier to connect a trickle charger.

A partially charged battery will freeze at 32°F, but a fully charged battery can go as low as –76°F before freezing. A frozen battery can crack the housing, which can then release caustic acid into the surrounding area and damage brackets, hold-downs, and battery compartments.

Inspect Hoses for Wear and Deterioration

Hoses serve as the arteries of spray equipment. They move chemicals from the tank through the pump, out toward the boom, and eventually to the nozzles. When hoses fail, not only do you lose product, but it can also create an environmental hazard. In any case, a broken hose will definitely alter and delay your schedule.

Most accidental releases of pesticides or hydraulic oils originate from hoses that burst or that blow off fittings when clamps fail to secure them tightly. This is

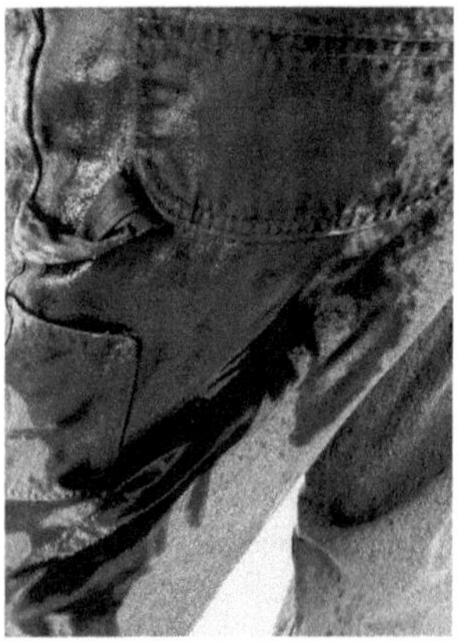

Figure 17.17. Broken hoses are more than just inconvenient. A blown hose can spray scalding hot fluids that can injure operators or bystanders.

why you must thoroughly inspect all pressure hoses and clamps as part of a routine equipment maintenance program.

At some point, all hoses will fail. They fail for a variety of reasons:

- A hose can rub against an object that abrades the outer cover, the underlying layers of wire, and the fiber that surrounds the inner core. The core (sometimes called the *intertube*) will not handle high pressures. Once the surrounding layers give way, the core will form a bubble and burst.
- Sharp objects can press on and into the hose over time.
- Sunlight or heat generated by an engine or pump can degrade the hose.
- Diesel fuel or oils can saturate the hose, which can then soften the rubber. Unlike gasoline, diesel fuel does not dissipate quickly.
- Any kind of constriction (such as a kink) in a hose will not only reduce the flow rate but also weaken the hose. That's because the reinforcing wires in the hose will not return to their original shape.

It is a good safety practice to annually replace the pressure hoses that carry spray solution. The cost of replacing a hose early outweighs the cost to the environment and the cost of cleanup when a hose fails. But if you do not replace hoses annually, closely inspect the outside of hoses to look for signs of abrasion, bulging, cuts, or slits. If you see a bulge in the outer rubber, that indicates that the core has been breached and that pressure is building between the layers as the intertube pushes outward.

You should also look for hoses that rest, rub, or vibrate on metal, wood, bolts, or any sharp objects. Closely inspect these points of contact to check if the hose is showing any signs of external abrasion and wear. You should always replace a hose that shows serious abrasion. When you do, reroute the replacement hose to bypass the point of contact or cause of failure. When rerouting is not possible, place something between the hose and the point of contact, such as a used hose split down the middle and wrapped around the new hose.

Figures 17.18–17.19. (Left) This cross section of a new hose shows that it has an inner core surrounded by rubber and reinforcement strands. (Right) Cutting away a layer of the hose reveals reinforced materials.

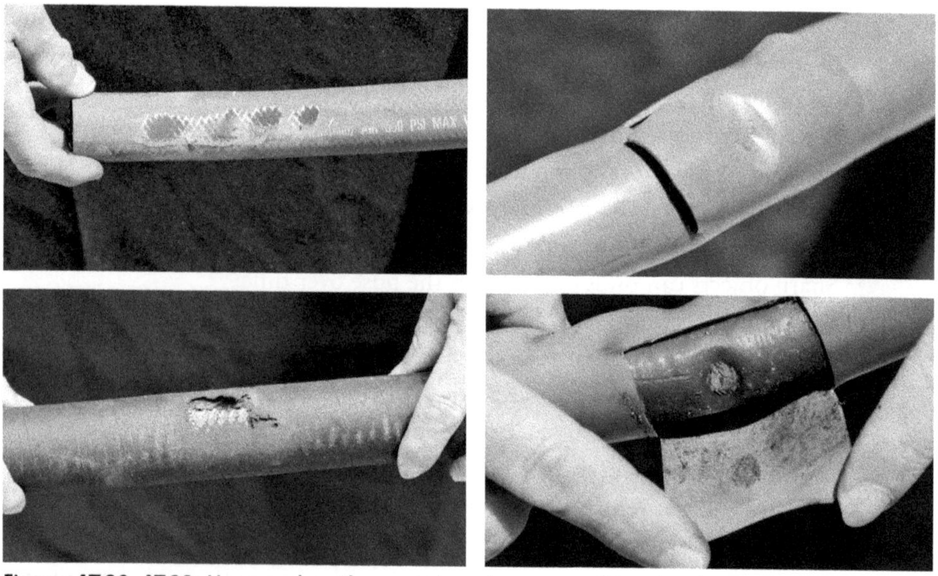

Figures 17.20–17.23. Hoses such as these are in such poor condition that the question is not if they will rupture but when.

Figure 17.24. A well-worn hose that needs replacing.

Figure 17.25. Replacing pressure hoses each year means you won't have to worry about problems occurring inside the hose.

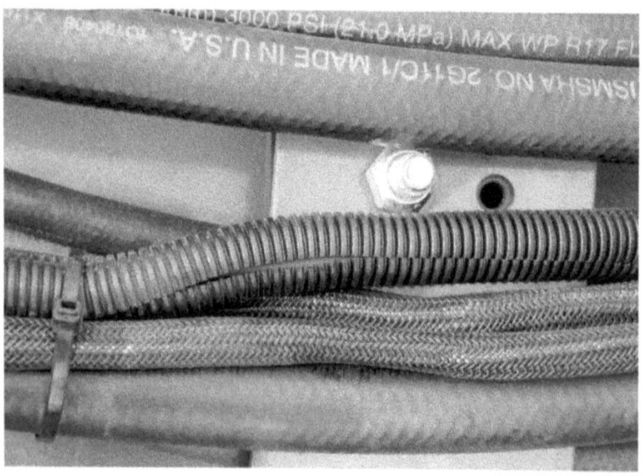

Figure 17.26. This hose is lying on a bolt, a point of contact that you need to visually inspect to see if the bolt is abrading the hose surface.

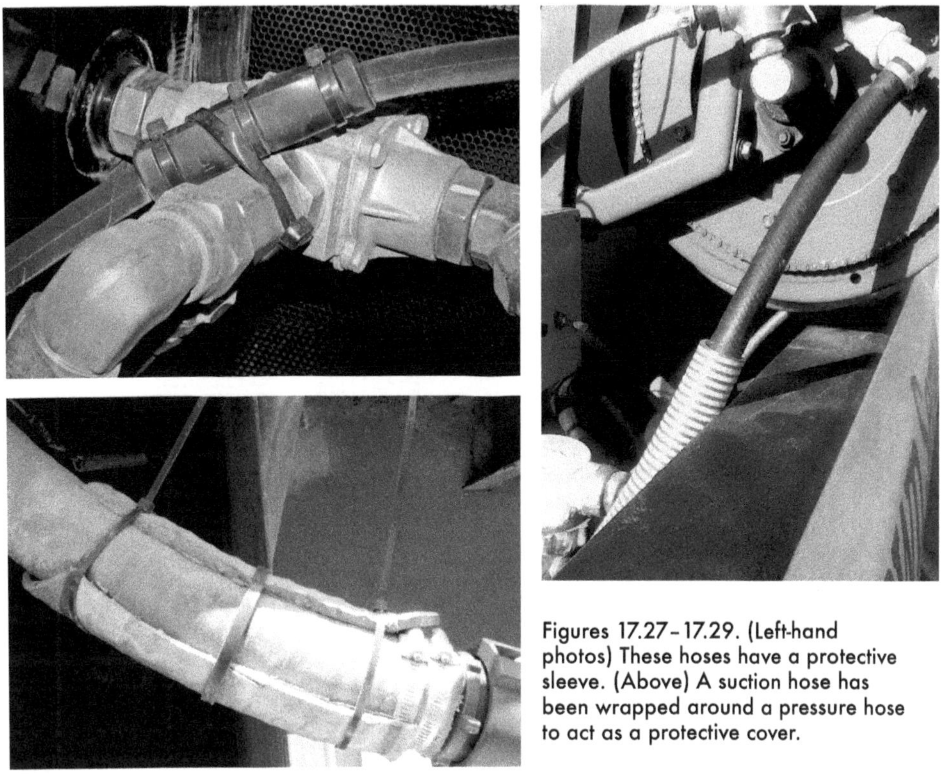

Figures 17.27–17.29. (Left-hand photos) These hoses have a protective sleeve. (Above) A suction hose has been wrapped around a pressure hose to act as a protective cover.

Upgrade to Better Hose Clamps

When you select hose clamps, consider what the environmental consequences would be if the clamp failed to hold the hose. Like hoses, clamps have different designs that serve specific purposes.

A standard worm gear hose clamp is popular due to its versatility, accessibility, and cost. It uses a screw to loosen or tighten the clamp around the hose. However, overtightening the screw can strip the slots in the band, rendering it ineffective. Worm gear clamps are made from different materials. Clamps made of mild steel can corrode from water, fertilizers, and pesticides. When you replace worm gear clamps for spray equipment, be sure they use stainless steel for the band, screw, and housing.

In many industries, T-bolt clamps have become the clamp of choice (currently the smallest T-bolt clamps are ⅜ inch). While more expensive, these clamps offer a more heavy-duty design and rigid frame to retain a tight fit around a hose. T-bolt clamps normally have wider bands that allow them to clamp down tighter than other clamps and to distribute the "squeeze" pressure more uniformly over a wider area, which produces a tighter seal.

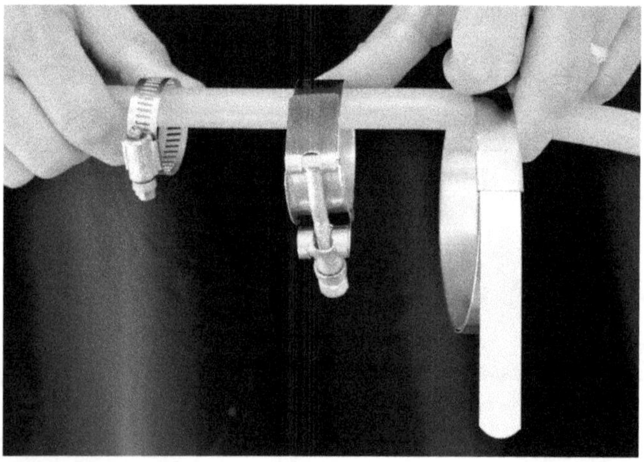

Figure 17.30. (From left) A worm gear clamp, a T-bolt clamp, and a Band-it clamp.

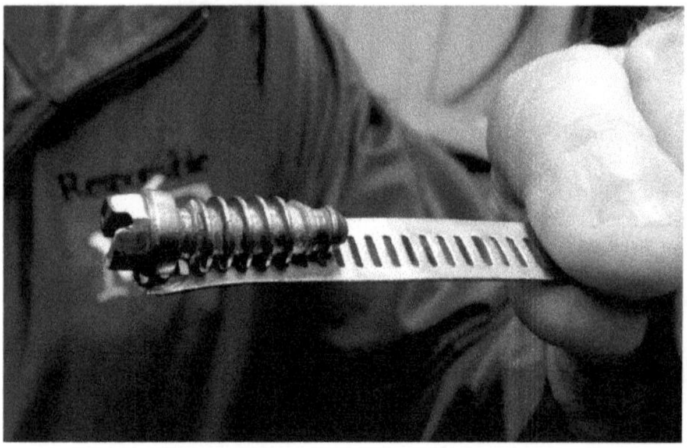

Figures 17.31 – 17.32. (Left) This disassembled worm gear clamp shows the interaction of the screw and the metal band. (Right) This rusted clamp needs to be replaced immediately.

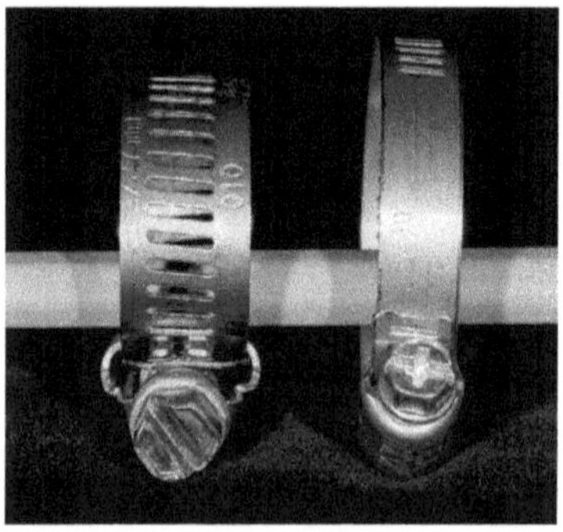

Figure 17.33. The wider the band on a worm gear clamp, the stronger the hold.

Figure 17.34. A T-bolt clamp is a very strong and reliable clamp. Unlike other clamps, the bolt is tightened around a solid frame.

In many instances, a second clamp is used to hold the hose to the fitting as an added safeguard. Using a second clamp requires a longer hose barb. When double clamping, offset the clamps by having the tension portion of the clamps opposite each other (that is, rotated 180 degrees from each other).

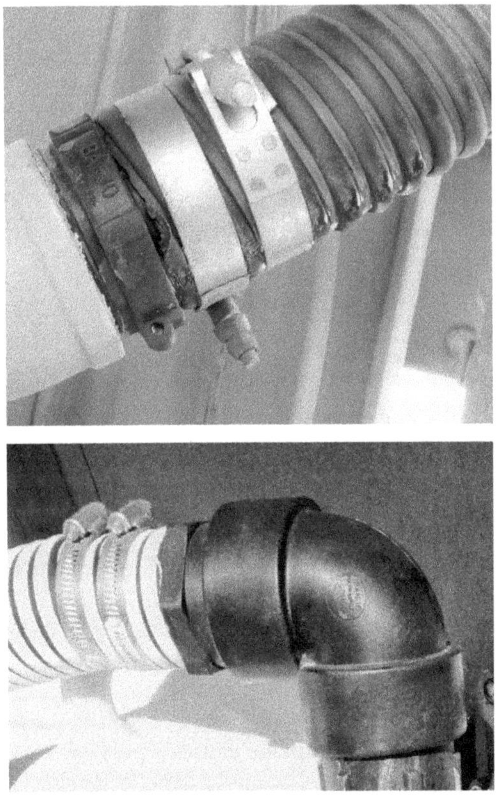

Figures 17.35–17.36. (Top) A good example of double clamping. Note that the screws are on opposite sides of the hose. (Bottom) In this photo the two clamps were put on with the same orientation. This negates some of the benefits of double clamping.

Figure 17.37. Keeping extra replacement clamps on your rig can save you time if a clamp fails in the field.

Figure 17.38. Make sure to read the suggestions provided by the manufacturer of the metal straps or hoops on how to tighten them when attaching a tank to a wagon or trailer.

Tighten Straps, Hoops, and Bolts on Tanks

It is not unusual for the bolts and straps that hold the tank to the frame to loosen during the spray season. At least once a year, retighten the bolts to make sure they hold the spray tank tightly to the frame. You should also inspect the tank baffles or cushions. Check for wear, chafing, and evidence of shifting. Before you tighten the straps or pipes, make sure the tank is full. This ensures the clamps function properly when the tank is at its most vulnerable.

Look for Signs of Sun Degradation

The number-one enemy of poly tanks is sunlight. Ultraviolet (UV) radiation degrades the tanks, which is difficult to see. UV damage causes a poly tank to lose its elasticity and become hard and brittle. Once this happens, the liquid moving back and forth in the tank or twisting the tank can cause it to break apart.

Just looking at a poly tank won't always indicate an imminent failure. Check for damage by rubbing it with a water-soluble marker or by hitting it with a baseball bat (see chapter 2 for details on how to check poly tanks for invisible damage). Always replace a tank that shows evidence of cracking. Or, at a minimum, don't use such a tank for fertilizers or chemicals.

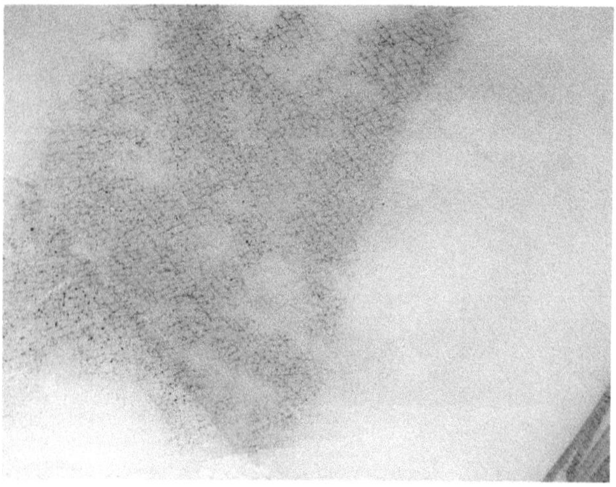

Figure 17.39. The walls of this spray tank show initial signs of ultraviolet radiation damage.

Pay Attention to Other General Items

Don't forget other general tasks during your fall inspection:

- Wash off grease and grime. Use a high-pressure washer to clean the outside of the sprayer. If you wait to wash the grime, dirt, grease, and chemicals off the sprayer in spring after they have dried all winter, it will take a lot more work and time. Furthermore, leaving residues on equipment for extended periods can accelerate corrosion and rust. Avoid putting high-pressure water directly on bearings and seals.
- Replace lost and damaged decals. Slow-moving-vehicle decals are vital for communicating potential hazards to the public. Decals wear and fade over time, so take the time to replace old decals.
- Touch up areas where paint has been scratched during the season. Touching up these areas with spray paint will minimize corrosion and rust.
- Protect electronic equipment. Upload all the data you recorded this year for safekeeping. If possible, remove electronic components from the rig (which can include rate controllers and monitors), and store them in a dry and warm location for the offseason.
- Check tire pressure and tighten lug nuts.
- Check sprayer airbags. Once you store the sprayer, lower the pressure in the airbags.

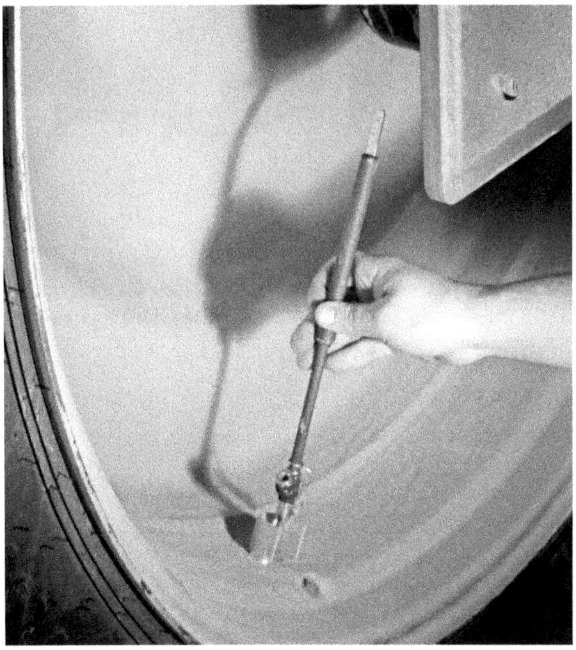

Figure 17.40. There are so many reasons why air pressure should be checked often.

Figures 17.41–17.46. Don't forget to check hitches. Also inspect hitch components on trailers that pull sprayers.

(A) This bolt is no longer attached.

(B) This hitch has a bent receiver.

(C) This receiver is rusted.

(D) This chain repair is illegal and unsafe.

(E) This chain is worn and weak.

(F) Another hitch with a bent receiver.

CONSIDER RODENT TRAPS AND BAIT STATIONS

Farm equipment can provide multiple resources to rodents. Grain and plant material can accumulate in the machinery's nooks and crannies. And rodents often chew on electrical wires, mistaking the wires for plant stems.

Air filters and engine components are also excellent nesting sites for small rodents. Cabs and the foam insulation that surround them are also places rodents might take up residence. To prevent unwanted tenants and protect your equipment, you need to consider using some deterrent.

When you use rodenticides to protect equipment, your first consideration should be to minimize exposure to nontarget animals. Farm dogs and cats may eat rodenticides placed in and around farm machinery. Protect those animals by placing the packets or bars in areas only rodents can access.

If you store your machinery in an enclosed building:

- Control clutter so rodents have fewer places to hide or nest.
- Maintain well-stocked, tamper-resistant bait stations around the exterior and interior perimeters of the storage building.

Whether indoors or out, you can secure rodenticides in inaccessible areas inside machinery, such as trucks and combines. Secure the boxes in place with wire. Do not place loose baits in the open cab. Here are a few places you can secure baits out of reach of nontarget animals:

- In engine compartments but away from objects that get hot enough to ignite a wax block or soft bait pouch
- Inside control panels and other compartments
- In latched glove compartments
- In tamper-resistant bait stations in truck cabs and next to the wheels of machinery stored outdoors for long periods

Figure 17.47. A bait station filled with rodent bait is an excellent example of protecting the wiring on equipment.

WINTERIZING MEANS REMOVING WATER

You should have two goals when you winterize spray equipment: (1) to remove as much water as possible from the system (including the tank, hoses, and booms); and (2) to add antifreeze to prevent the water that couldn't be drained from freezing and damaging sprayer parts. When water freezes, it expands (in contrast to other liquids that contract when they freeze). That expansion can apply enough pressure to crack steel and plastic parts, including valves, pipes, nozzles, pumps, flowmeters, and filters. If you add antifreeze, you lower water's freezing point.

Choose the Right Type of Antifreeze

A number of different antifreeze products are available. Check your sprayer's operator manual or manufacturer to determine the best winterizing solution for your needs. The three main ingredients in most antifreeze products are:

1. Ethylene glycol (called automotive antifreeze)
2. Propylene or ethanol glycol (called RV antifreeze)
3. Diethylene glycol (called deicers)

Ethylene glycol, which contains corrosion inhibitors, is typically used in truck and car radiators, where the coolant is expected to remain for many years and thousands of miles. In most cases, manufacturers do not recommend ethylene glycol antifreeze as a winterizing agent for sprayers, because the chemicals that prevent corrosion can deteriorate plastic parts and pump seals.

Another concern is that ethylene glycol has been a source of accidental poisonings. Ethylene glycol is a colorless and odorless substance, but it tastes sweet, so it is attractive and deadly to pets and wildlife that ingest the material. This is why it is important to properly dispose of ethylene glycol.

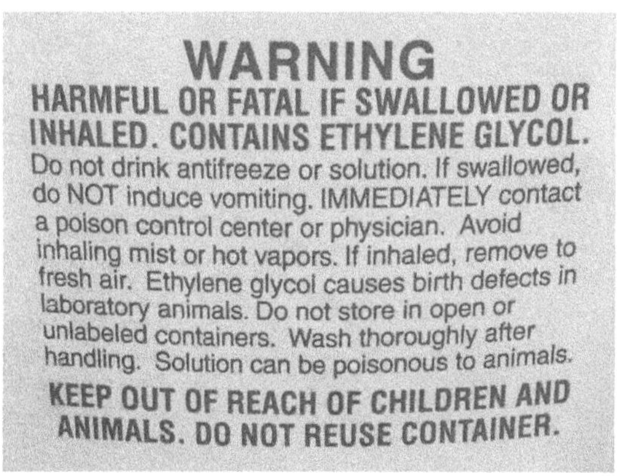

Figure 17.48. It is always important to read the warnings on antifreeze labels as one way of protecting people, wildlife, and the environment.

RV antifreeze usually contains propylene glycol. This form of antifreeze lacks the corrosion inhibitors found in engine coolants; therefore, the product should remain in the system only for short periods before being flushed out. RV antifreeze made from propylene glycol has many positive benefits:

- It works well.
- It is relatively inexpensive.
- It is easy to use.
- It is readily available.
- It should not damage crops.
- It does not damage plastic or metal parts.
- It is safe to people, pets, and the environment.

However, there are RV antifreezes that contain ethanol glycol instead. While ethanol glycol is effective at lowering water's freezing point, there are two concerns. First, it is flammable. Second, in certain situations it can dry out rubber seals, which can cause leaks next season.

People sometimes use windshield washer fluid as a winterizing agent although it is technically not an antifreeze. Windshield washer fluids contain ethyl glycol or methanol, dyes, and soaps. Thus, they may not be the best product to use in your situation. In any case, always check to see if you are buying a winter formulation that will protect equipment at colder temperatures.

Figure 17.49. RV antifreeze is generally use to winterize equipment.

Figure 17.50. Always check that the temperature indicated on windshield wiper fluid is low enough to protect equipment during the coldest months.

Check Freeze and Burst Pressures

Two cold-temperature values associated with a winterizing agent (such as RV antifreeze or windshield washer fluid) are its freeze point and its burst point. The freeze point is the temperature at which antifreeze begins to form ice crystals and turns slushy. The burst point is more critical. It is the temperature at which a copper pipe filled with antifreeze will burst. A product's label will prominently display its burst point. Always check that the burst point temperature on the product's label will provide the protection you need to prevent water from freezing in the sprayer you are winterizing.

The Three Steps of Winterizing

As we have mentioned, the key point of winterizing equipment is to remove water and prevent any remaining water from freezing. We recommend three steps for winterizing:

Step 1: Clean Chemical Residues

After you complete the last pesticide application of the year, focus on removing any remaining pesticide residue from the spray system or sprayer itself (see chapter 16). As we've already discussed, allowing residues to remain in or on the sprayer throughout the winter will only make it harder to remove them in the spring.

Most pesticides will sink to the bottom of tanks, hoses, and screens after you stop moving or agitating the solution. If you leave any pesticides in the sprayer, they can form a layer of residue on the tank walls, screens, elbows, and corners. These residues can also dry in rubber hoses if they remain in the sprayer until spring, which makes them extremely difficult to remove.

Step 2: Remove Water

Water can get trapped in many places in a sprayer, including the tank, hoses, screen bodies, tips, and many other nooks and crannies. In other words, there is always the possibility that there could be water somewhere in the sprayer's plumbing—from the tank to the nozzle tips.

Figure 17.51. Removing as much water as possible allows the antifreeze to work best.

Figure 17.52. Opening all valves allows more water to be removed.

Figure 17.53. Remove nozzle bodies from the sprayer to protect O-rings. Always use gloves.

Figure 17.54. Remove the strainer and empty any water inside.

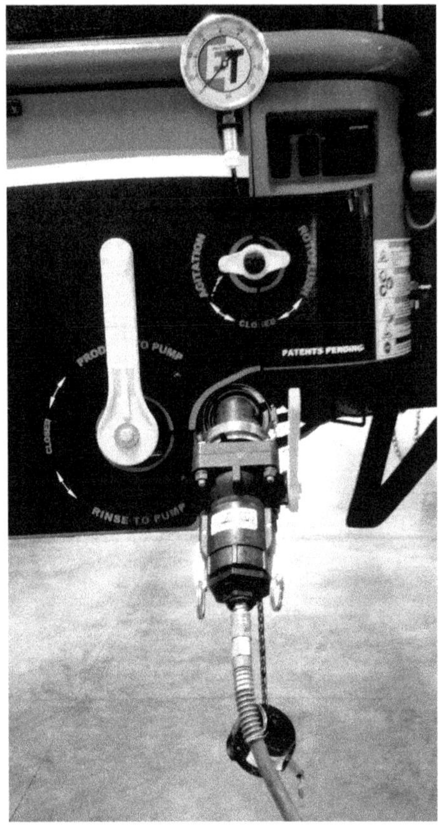

Figure 17.55. The operator is using an air hose and attachment to remove water from this sprayer.

When removing water from your equipment, follow these five steps:

1. Let the pump push water out of the tank until it loses its prime. Once you clean and flush the spray tank, spray as much of the water out of the system as you can, or spray until the pump no longer has enough water to push out of the tips. Even after you have pumped all the water from the tank, water may still remain in the hoses, valves, and screens from the pump to the nozzle tips. Close down or shut some sprayer sections to help force out more water.
2. Open valves to remove water. Open all valves slowly to allow any remaining water to drain. Make sure to remove and empty all in-line strainers.
3. Use air to push water in the lines out toward nozzle tips. Be aware that many air compressors can deliver more than 150 pounds per square inch (psi) of pressure. Exceeding the pressure capacity of the sprayer may damage sensitive components. Check with the manufacturer or the owner's manual to see if they provide any warnings about excessive pressures.

4. Drain the pump. Your pump may have a plug that you can remove to drain any remaining water within the pump. (You can also use this plug to prime a pump if needed.)
5. Tip the booms. If possible, tip spray booms at a 45-degree angle to push water from the end of the booms toward the center, or from the center to the boom ends. Take off individual pressure caps to relieve the pressure and allow the water to flow unrestricted.

Figure 17.56. When possible, drain the water out of the pump.

Step 3: Add Antifreeze

Despite your best efforts to remove all the water, some water will inevitably remain somewhere in the system. It could be in the bottom of hoses that droop between sections of the boom, in elbows or valves, or, depending on the design, in the bottom sump of the tank. Keep this trapped water from freezing by mixing it with antifreeze.

Be sure to read the antifreeze label to determine whether you need to dilute it or whether it is a ready-to-use product (which requires no dilution). Generally, RV antifreeze and windshield washer deicers are ready-to-use products.

The question is, How much antifreeze should I add to the sprayer? A general rule of thumb is to add 1.5 to 2 times the capacity of the sprayer's plumbing. Thus, the amount of antifreeze you add depends on the make and model of the sprayer and the length of the spray boom and hoses. You want enough antifreeze to push the water out of the system and to make sure that any trapped water is thoroughly mixed with the antifreeze.

Here are six steps to follow when adding antifreeze:

1. Pour antifreeze into the spray tank. Sparge or circulate the antifreeze from the tank to the pump numerous times. Then turn on the booms so that the antifreeze moves through the hoses and exits the nozzles.

   Be sure to also winterize the rinse tank if your sprayer is equipped with one. Start by pouring antifreeze solution into the rinse tank. Then turn on the system to pump antifreeze into the larger spray tank. This will protect the rinse lines from freezing. If your sprayer is equipped with a rinse tank, you may want to start with this step before pumping the antifreeze through the entire system.

Figure 17.57. Adding RV antifreeze to the tank.

2. Open and close valves while the sprayer is operating. This allows the antifreeze to go into inaccessible spots where water might remain trapped. Open bypass valves, product valves, inductor valves, and agitator valves to get the antifreeze into those areas.
3. Observe the color of the antifreeze as it comes out. At first, the antifreeze coming out of the nozzles will be lighter in color than what you poured into the tank, because it will have become diluted with the water remaining in the system. After a few minutes, the color of the liquid coming out of the nozzles should be more similar to the color when you poured it into the tank. Only then will you know that the antifreeze has flowed through the entire underbelly of the sprayer, through the pump, screens, flowmeter, booms, and nozzle tips.

   Remember, you need enough antifreeze to keep the pump running or primed and to force any remaining water out of the system.
4. Don't forget the fencerow nozzle circuit. If the machine is equipped with a fencerow nozzle at the end of the boom, open that nozzle when running the antifreeze solution through the booms to allow those lines to flush as well.
5. Record how much antifreeze it took to winterize the sprayer. You can refer to these records the next time you winterize the sprayer.
6. Remove unused solution from the foam marker with a wet-dry vac. Some foam marker systems draw water from the rinse tank. If your sprayer uses this setup, you will need to purge the water from that water line. If there is any foam left in the tank, use a wet-dry vac to suck it out and save it for reuse. Then pour about half a gallon of antifreeze into the foam tank and run it through the lines and out the marker ends.

BE READY IN THE SPRING

Preparing the sprayer for spring operations is much easier than winterizing the sprayer. Before operating the sprayer in the spring:

- Review the owner's manual.
- Reconnect batteries.
- Reinspect belts.
- Check oil and coolant levels.
- Remove rodent bait stations and inspect equipment for rodent damage.
- Inspect the steel on the booms, looking for defects or cracks that might need to be welded.
- Check tire pressure.
- Place electronic equipment back in the sprayer.
- Run clean water through the lines to remove antifreeze residues.

After you check everything, fill the sprayer with enough water to run the pump for 15 minutes. As the sprayer flushes out the antifreeze, check for leaks in the plumbing. Look closely for leaks in these locations:

- Nozzle bodies
- Plastic fittings
- Hoses
- Hose clamps
- Tanks

Recycling Antifreeze

There are three schools of thought on what to do with leftover antifreeze:

1. Discard it. Since RV antifreeze is nontoxic, some will simply rinse it out of the tank. While this is an acceptable practice, perform this operation a safe distance from any water sources. The rinse water may contain pesticide residues that you do not want to contaminate any water sources.
2. Collect and dispose of it. A second option is to collect as much of the antifreeze as possible, pour it into a collection drum, and then have a legally licensed firm collect and dispose of it at a later date.
3. Collect and reuse it. A third option is to collect as much of the antifreeze as possible, store it in an appropriately labeled container, and reuse it the following winter. One concern is that over time the antifreeze will become diluted and get progressively weaker. At lower concentrations, the antifreeze will be less effective and freeze at a higher temperature. If you do reuse antifreeze, you need to test it first to ensure it will be effective.

Figures 17.58–17.60. You can (top) collect and dispose of antifreeze, (middle) collect and reuse it, or (bottom) simply rinse it out of the tank away from water sources.

CONCLUSION

You've probably heard the old saying that the job is not over until the paperwork is complete. In the same way, the spray season not over until the spray equipment has been cleaned and winterized. Being proactive with your maintenance during the winter will allow you to bring out the equipment in the spring ready to go. Although you may not like making repairs, performing such tasks will add to the life of your sprayer. It only makes sense to complete these repairs during the offseason, since time is in such short supply during spring and summer.

GET OUT THERE!

Conditions are fickle – when the time is right you need to be in the field, not in the garage.

If you do not maintain equipment properly, you will experience more problems and will need to replace your sprayer sooner. If you are in the middle of spraying and on a tight schedule to complete applications, the last thing you need is a breakdown. There is nothing worse than waiting on a part to repair your sprayer when daylight is burning or rain is approaching. Daylight will quickly become dark, and you'll miss the opportunity to finish the application if your machine is sitting idle while you wait on parts to repair it. Unless you are fortunate enough to buy or use a new machine every year, the more preventative maintenance you can do in the offseason, the smoother and less stressful the busy spraying season will be.

INDEX OF SUBJECTS

INDEX OF CONTRIBUTORS

www.ingramcontent.com/pod-product-compliance
Ingram Content Group UK Ltd.
Pitfield, Milton Keynes, MK11 3LW, UK
UKHW060351180326
469106UK00006B/3011